Manufacturing Reliability Growth for Automotive Engineering

Volume III

Manufacturing Reliability Growth for Automotive Engineering

Volume III

YOUNG J. CHIANG

SAE INTERNATIONAL®

Warrendale, Pennsylvania, USA

SAE INTERNATIONAL

400 Commonwealth Drive
Warrendale, PA 15096-0001 USA
E-mail: CustomerService@sae.org
Phone: 877-606-7323 (inside USA and Canada)
                724-776-4970 (outside USA)
Fax: 724-776-0790

Library of Congress Catalog Number 2025942183
http://dx.doi.org/10.4271/9781468608052

ISBN-Print 978-1-4686-0804-5
ISBN-PDF   978-1-4686-0805-2
ISBN-epub 978-1-4686-0806-9

To purchase bulk quantities, please contact: SAE Customer Service

E-mail: CustomerService@sae.org
Phone: 877-606-7323 (inside USA and Canada)
                724-776-4970 (outside USA)
Fax: 724-776-0790

Visit the SAE International Bookstore at books.sae.org

Publisher
Sherry Dickinson Nigam

Product Manager
Amanda Zeidan

**Production and
Manufacturing Associate**
Michelle Silberman

From the Publisher

I am thrilled to introduce our groundbreaking series, *Design of Experiments for Product Reliability Growth with Automotive Applications,* which is divided into four distinct volumes. *Manufacturing Reliability Growth for Automotive Engineering* marks the third installment of our exploration of this critical subject.

We are publishing each volume separately to ensure in-depth coverage of each topic, allowing readers to focus on specific aspects of the subject matter for a deeper understanding. After careful consideration of how our readers will be using the content, we opted to continue with the page numbering system throughout the volumes for continuity.

Volume I   Fundamentals of Design of Experiments for Automotive Engineering
Volume II   Product Design and Testing for Automotive Engineering
Volume III   Manufacturing Reliability Growth for Automotive Engineering
Volume IV   Business Reliability Growth for Automotive Engineering

Thank you for joining us on this journey. We look forward to your continued support and engagement with the upcoming volumes in the series.

Sincerely,
Sherry Nigam
Publisher, SAE Books

Contents

CHAPTER 12

Structural Joints   1

CHAPTER 15

3D Printing and Additive Manufacturing   241

Preface

An enterprise resorts to driving profitability and capital discipline through reliability growth and innovation, finding ways to spend less for the same results with an unwavering focus on the customer. Reliability and profitability of a product are directly impacted by its manufacturing process. There are three principal methodologies of fabricating parts, i.e., subtractive manufacturing such as controlled removal of material by turning, formative manufacturing such as reshaping of plastic materials by molding, and additive manufacturing such as adding material to form the finished product. Design of experiments (DOE) is employed to identify control factors affecting process performance and propose alternative solutions. Thus, the subsequent statistical inferences and related computer-aided engineering (e.g., finite element methods) are used to validate the product and process improvements through a limited number of experimental data. Applications of DOE to structural joints, mechanical fabrication, electronic fabrication, and three-dimensional (3D) printing are addressed in this volume.

Chapter 12 (Structural Joints): Assembly of an automobile from its constituent parts involves either bonded or mechanically fastened joints, or both. DOE for structural joints based on interference fits, bolts, screws, rivets, adhesives, spot welding, high-strength welding, ball joints, and seals is narrated using real-world examples.

Chapter 13 (Mechanical Fabrication): DOE for mechanical fabrication including injection molding, transfer molding, metal forming, clinching, turning, drilling, boring, milling, grinding, heat treatment, diffusion, and coating/printing are narrated using real-world examples.

Chapter 14 (Electronic Fabrication): DOE for electronic fabrication including through-thickness interconnection, ball grid arrays, solder reflow, stencil printing, integrated circuit (IC) interconnects, printed circuit boards, coating of electronic wafers, chemical vapor deposition, epitaxial deposition, insulation of windings, and electric connectors are narrated using real-world examples.

Chapter 15 (3D Printing and Additive Manufacturing): Additive manufacturing technology, such as 3D printing, enables a manufacturer to produce geometrically complex parts including shapes and textures, allowing the production to use less material than subtractive manufacturing methods. Rheological behaviors of materials used for 3D printing have a great impact on the printed parts. Dimensional integrity including surface roughness and shrinkage of 3D-printed parts is explored. Inherent orthotropic material properties of 3D-printed parts are theorized. Applications such as printed acoustic noise attenuators, porous structures and triply periodic minimal surfaces, printed electrochemical storage devices, and printed sensors including strain gauges and antennae are exploited. DOE is deployed throughout all these in this chapter.

Commonly used data types found in the manufactured products are time-to-event data, degradation data, and recurrent event data. It is advantageous to utilize reliability engineering techniques based on DOE to improve manufacturing quality with decisions based on the integration of historical data and current dynamic events. A production line can be digitized, and the digitalization services in combination with Industry 4.0 technologies, such as Internet of things, simulations, and cloud computing. Digital shadow, as applied to production processes, may be developed to collect, analyze, and identify potential issues. The combined technology can be used to digitize the knowledge of ongoing manipulations based on statistical facts.

Acronyms

2_R**k-p** - 2-level fractional factorial design having k variables with resolution R

3_R**k-p** - 3-level fractional factorial design having k variables with resolution R

3 Spheres - Quality Management, Assurance, and Control

3T's - Task, Treatment, and Tangibles in Service Design

360° Evaluation - Evaluating Performance with inputs from supervisors, Peers, and Other Employees

4P - People, Place, Procedure, and Policies

4S - Surroundings, Suppliers, Systems, and Skills

5M - Manpower, Machine, Method, Material, and Measurement

5R - Responsiveness, Reliability, Rhythm, Responsibility, and Relevance

5S - Select/Sort, Straighten/Set in Place, Shine, Standardize, and Sustain

5w2h - Who, What, When, Where, Why, How, and How Much

6M - Manpower, Machine, Method, Material, Measurement, and Mother Nature

6S - Select, Set in Place, Shine, Standardize, Sustain, and Safe

7P - Proper Prior Planning Prevents Pitifully Poor Performance

7 Factors - Team, Provider, Patience, Task, Work environment, Equipment and Technology, and Organization and Management

7 Wastes - Over-production, Transport, Waiting, Inventory, Defects, Over-Processing, and Unnecessary Movement

8D - 8 Disciplines for Structured Step-by-Step Problem-Solving

80/20 Rule - 80% of the Problems Resulting from 20% of the Causes

A2LA - American Association of Laboratory Accreditation

AAI - Appearance Approval Inspection

AALA - American Association for Laboratory Accreditation

AAR - Appearance Approval Report

ADV - Analysis/Development/Validation (GM Specific)

AEB - Automatic Emergency Braking

AI - Artificial Intelligence

AIAG - Automotive Industry Action Group

AIQ - Average Incoming Quality; Average quality level before inspection

ALT - Accelerated Life Testing

ANCOVA - Analysis of Covariance

ANN - Artificial Neural Network

ANOM - Analysis of Means

ANOVA - Analysis of Variance

AOQ - Average Outgoing Quality (Average quality level leaving the inspection point after rejection and acceptance of a number of lots)

AOQL - Average Outgoing Quality Limit (Maximum value of the AOQ)

AOZ - Accept on Zero

AP - Advanced Product

APQP - Advanced Product Quality Planning

AQC - Attribute Quality Characteristic

AQL - Acceptable Quality Level

AQP - Advanced Quality Planning

ARIMA - Autoregressive Integrated Moving Average

ARL - Averaged Run Length

ARMA - Autoregressive Moving Average

ASIL - Automotive Safety Integrity Level

ASME - American Society of Mechanical Engineers

ASQ - American Society of Quality

ASTM - American Society for Testing and Materials

AVL - Approved Vendor List

B7 (Q7) - Basic Seven QC Tools, i.e., Fishbone Diagram, Histogram, Pareto Analysis, Flowchart, Scatter Plot, Run Chart, and Control Chart

BCWP - Budgeted Cost of Work Performed

BCWS - Budgeted Cost of Work Scheduled

BDD - Binary Decision Diagram

BEV - Battery Electric Vehicle

BIST - Built-in Self-Test

BIT - Built-in Test

BJT - Bipolar Junction Transistors

BLDC - Brushless Direct Current

BN - Bayesian Network

BOM - Bill of Materials

BPI - Business Performance Improvement

CA - Corrective Actions

CAAM - China Association of Automobile Manufacturers

CAD - Computer-aided Design

CAE - Computer-aided Engineering

CAM - Computer-aided Manufacturing

CAN - Controller Area Network

CAPA - Corrective and Preventive Action

CBP - Customer Benefits Package

CC - Critical Characteristic

CCD - Central Composite Design

c-Chart - Count Chart (Attribute Quality Control Chart for Number of Defects)

CCPM - Critical Chain Project Management

CDF - Cumulative Distribution Function

CF - Carbon Fiber

CFRP - Carbon-Fiber-Reinforced Polymer or Carbon-Fiber-Reinforced Plastics

CFT - Cross-Functional Team

CI - Confidence Interval

CIP - Continuous Improvement Process

CLT - Central Limit Theorem

CMM - Capability Maturity Model

CMM - Coordinate Measuring Machine

CONC - Cost of Non-conformance

COP - Customer-Oriented Process

COQ - Cost of Quality

COR - Cost of Reliability

CPM - Critical Path Method, either PERT or Gantt Chart

CPP - Critical Process Parameter

CPPD - Collaborative Product Process Design

CPV - Cost per Vehicle

CQI - Continuous Quality Improvement

CQM - ASQ Certified Quality Manager

CRD - Completely Random Design

CRM - Customer Relationship Management

CRS - Constant Returns to Scale

CS1 - Control Shipping 1

CS2 - Control Shipping 2

CTQ - Critical to Quality

CUSUM - Cumulative Sum Charting

CV - Coefficient of Variation

DCO - Document Change Order

DCP - Dimensional Control Plan

DCP - Dynamic Control Plan (Ford)

DDDM - Degradation Data-Driven Method

DEA - Data Envelopment Analysis

DETMAX - Determinant Maximizing Algorithm (for D-Optimality)

DF or DOF - Degree of Freedom

DFA - Design for Assembly

DFM - Design for Manufacture

DFMA - Design for Manufacture and Assembly

DFMEA - Design Failure Mode and Effects Analysis

DLM - Direct Linearization Method

DMADV - Define, Measure, Analyze, Design, Verification

DMAIC - Define, Measure, Analyze, Improve, and Control (Kaizen- five steps)

DMU - Design Making Unit

DOC - Depth of Charge

DOD - Depth of Discharge

DOE - Design of Experiments

DOE-E - Design of Experiments Based on Exponential Distribution

DOE-F - Design of Experiments Based on F-Distribution

DOE-F - F-Distribution-Based Design of Experiments

DOE-ln - Design of Experiments Based on Lognormal Distribution

DOE-t - Design of Experiments Based on t-Distribution

DOE-W - Design of Experiments Based on Weibull Distribution

DOE+ML - Design of Experiments and Machine Learning, Joint Application

DRACAS - Data Reporting Analysis and Corrective Action System

DRF - Datum Reference Frame

DSC - Dynamic Stability Control

DSD - Definitive Screening Designs

DVP&R - Design Verification Plan and Report

ECFA - Event and Causal Factor Analysis

ECN - Engineering Change Notice

ECR - Engineering Change Request

EMI - Electromagnetic Interference

EMOO - Evolutionary Multi-Objective Optimization

EMS - Expected Means Squares

ERP - Enterprise Resource Planning

ES - Earned Schedule

ES - Engineering Specification

ESC - Electronic Stability Control

ESD - Electrochemical Storage Devices

ESS - Environmental Stress Screening

EV - Electric Vehicle

EVM - Earned Value Management

EVOP - Evolutionary Operation Process

EWMA - Exponentially Weighted Moving Average

EWMV - Exponentially Weighted Moving Variance

EWS - Early Warning System

Exp, exp, e - Exponential Function

FAI - First Article Inspection or Final Acceptance Inspection

FAST - Function analysis System Technique Diagram

FDI - Fault Detection and Isolation

FDM - Fused Deposition Modeling

FEA - Finite Element Analysis

FEM - Finite Element Methods

FFA - Functional Failure Analysis

FFF - Fused Filament Fabrication

FHA - Functional Hazard Analysis

FIFO - First In First Out

FIM - Fisher Information Matrix

FIR - Fast Initial Response

FMEA - Failure Mode and Effects Analysis

FMECA - Failure Modes, Effects, and Criticality Analysis

FMEDA - Failure Modes, Effects, and Diagnostic Analysis

FMVT - Failure Mode Verification Test

FOZ - Feature of Size

FPGA - Field Programmable Gate Array

FQPR - Field Quality Problem Report

FRACAS - Failure Reporting, Analysis, and Corrective Action System

FRB - Failure Review Board

FSLT - Full System Life Test

FSR - Functional Safety Requirement

FTA - Fault Tree Analysis

FTC - First Time Capability

GA - Genetic Algorithm

Gage R&R - Gage Repeatability and Reproducibility

GaN - Gallium Nitride

GD&T - Geometric Dimensioning and Tolerancing

GD$_3$ - Good Design, Good Discussion, Good Dissection

GERT - Graphic Evaluation and Review Technique

GF - Glass Fiber

GFRP - Glass-Fiber-Reinforced Polymer or Glass-Fiber-Reinforced Plastics

GK - General Knowledge

GNSS - Global Navigation Satellite System (e.g., GPS + IMU)

GPa - Giga Pascal

GPC - Gage Performance Curve

GPS - Global Positioning System

GR - Gage Repeatability

GR&R (G_{RR}) - Gage Repeatability and Reproducibility

GTD - Getting Things Done

GTO - Gate Turn-off Thyristor

GVDP - Global Vehicle Development Process (GM Practice)

H_1 - Alternative Hypothesis

HACCP - Hazard Analysis and Critical Control Points

HALT - Highly Accelerated Life Test

HARA - Hazard Analysis and Risk Assessment

HASS - Highly Accelerated Stress Screening

HAZOP - Hazard and Operability

HCF - High Cycle Fatigue

HEV - Hybrid Electric Vehicle

HFCV - Hydrogen Fuel Cell Vehicle

H_o - Null Hypothesis

IATF - International Automotive Task Force

IC - Integrated Circuit

IDOV - Identify, Design, Optimize, Validate

IEEE - Institute of Electrical and Electronics Engineers

IGBT - Insulated Gate Bipolar Transistor

IGCT - Integrated Gate Commutated Thyristor

iid - Independent and Identically Distributed

IM - Induction Motor

IMDS - International Material Data System

IMP - Integrated Master Program

IMU - Inertial Measurement Unit

IoT - Internet of Things

IPMA - International Project Management Association

ISIR - Initial Sample Inspection Report

ISO - International Organization for Standardization

IsoPlot - 3D Iso-surface Plot

ITPV - Incidents per Thousand Vehicles

JIT - Just in Time

KAIZEN - Change for the Better

KBF - Key Business Factors

KCC - Key Control Characteristic

KISS - Keep it Simple, Stupid

KLT - Key Life Tests

KPC - Key Product Characteristic

KPI - Key Product (Process) Indicator or Key Performance Index

LCF - Low Cycle Fatigue

LCL - Lower Control Limit

LFP - LiFePO$_4$ Battery

Li-Air - Lithium-Air (Battery)

Lidars - Laser Light Detection and Ranging Systems

Li-ion - Lithium-Ion (Battery)

LKA - Lane Keeping Assistance

LMC - Least Material Condition

LN, Ln, or ln - Lognormal

LRP - Logistics Requirement Plan

LRU - Line Replaceable Unit

LSL - Lower Specified Limit

LSM - Linear Scheduling Method

LTBF - Longest Time between Failures

LTPD - Lot Tolerance Percentage Defective Acceptable in Production Lots

M&TE - Measurement and Test Equipment

MBD - Management by Decree

MCO - Manufacturing Change Order

MEOST - Multiple Environment Over Stress Tests

MILTFP-41 - Make it Like the Finished Print- For Once

ML - Machine Learning

MLE - Maximum-Likelihood Estimator

MMC - Maximum Material Condition

MME - Method of Moments Estimator

MOPSO - Multi-Objective Particle Swarm Optimization

MORT - Management Oversight and Risk Tree

MOSFET - Metal-Oxide-Semiconductor Field-Effect Transistor

MPa - Mega Pascal

MRR - Median Rank Regression

MRS - Marginal Rate of Substitution

MSE - Mean-Squared Error

MTBF - Mean Time between Failures

MTBOF - Mean Time between Operational Failures

MTS - Make to Stock

MTTF - Mean Time to Failure

MTTR - Mean Time to Repair

MVBs - Manufacturing Validation Build Saleable (GM specific)

N_2 - An Interface Matrix, Representing Interfaces between System Elements

N7 - New Seven QC Tools (Affinity Diagram/Brainstorming, Relation Diagram, Tree Diagram, Matrix Diagram/QFD, Arrow Diagram/PERT, Process Decision Chart, and Matrix Data Analysis/Principal Component Analysis)

NCP - Nonconforming Products

NCR - Nonconformance Report

NDT - Non-Destructive Test

NIST - National Institute of Standards and Technology, Dept. of Commerce, United States

np-chart - Attribute Quality Control Chart for the Number of Defective Products

NPI - New Product Introduction

NTF - No Trouble Found

NTSA - National Transportation Safety Administration, Dept. of Transportation, United States

O&SHA - Operational and Support Hazard Analysis

OACD - Orthogonal-Array Composite Designs

OC Curve - Operating Characteristic Curve

OCV - Open Circuit Voltage

OEE - Overall Equipment Effectiveness

OEM - Original Equipment Manufacturer

OVAT - One Variable at a Time

PAA - Product Application Agreement

PAT - Process Analytic Technology

PBS - Product Breakdown Structure

PCA - Permanent Corrective Actions

PCB - Printed Circuit Board

PCBA - Printed Circuit Board Assembly

P-chart - Attribute Quality Control Chart for Proportion of Nonconforming Units

PCI - Process Capability Index

PCP - Production Control Plan

PDAS - Plan–Do–Study–Action

PDCA - Plan–Do–Check–Act Cycle

PDF - Probability Density Function

PDF, pdf - Probability Density Function

P-Diagram - Parameter Diagram

PDM - Product Data Management

PDPC - Process Decision Program Chart

PERT - Program Evaluation and Review Technique

PEV - Plug-in Electric Vehicle

PFD - Process Flow Diagram or Probability of Failure on Demand

PFMEA - Process Failure Mode and Effects Analysis

PHEV - Plug-in-Hybrid Electric Vehicle

PHR - Part Handling Review

PL - Product Line

PLM - Product Life Management

PMA - Program Management Administration

PMBoK - Project Management Body of Knowledge

PMHF - Probabilistic Metric for Hardware Failure

PMI - Project Management Institute

PMO - Preventive (or Planned) Maintenance Optimization

PMSM - Permanent Magnet Synchronized Motor

Poka Yoke - Mistake Proofing; Fool Proofing

POSEC - Prioritize by Organizing, Streamlining, Economizing, and Contributing

PP100 - Problems Reported per 100 Vehicles

PPAC - Product Performance Agreement Center

PPAP - Production Part Approval Process

PPC - Production Planning and Control

PPHR - Parts per Hundred Rubber Parts

PPM - Parts per Million

PPV - Product and Process Validation

PQP - Product Quality Planning

PQRR - Program Quality Readiness Review (GM)

PRR - Problem Reporting and Resolution

PSO - Particle Swarm Optimization

PSS - Passive Safety Systems

PSW - Part Submission Warrant

PV - Part-to-Part Variation

PWM - Pulse Width Modulation

Q1 - Ford Q1, considered worldwide as an indication of exceptional quality

QA - Quality Assurance

QALT - Quantitative Accelerated Life Test

QbD - Quality by Design

QC - Quality Circles

QC - Quality Control

QCC - Quality Control Circle

QCCAR - Quality Control Corrective Action Request

QFD - Quality Function Deployment

QIP - Quality Improvement Program

QIT - Quality Improvement Team

QL - Quality Lever

QLS - Quality Loss Function or Quadratic Loss Function

QMD - Quality Measurement Data

QMS - Quality Management System

QS 9000 - Quality System Requirements 9000

QSA - Quality System Assessment

QSB - Quality System Basics

QSR - Quality System Requirements or Quality Service Reviews

QVR - Quality Verification Report

R&C - Reliability and Confidence Level (e.g., a combination of R ≥ 95% and C = 75% means that someone is 75% confident that reliability ≥ 95%)

R, R_{adj}, R_{pred} - Regression, Adjusted, and Predictive Correlations

R@R - Run at Rate (GM specific)

RBD - Reliability Block Diagram

RCA - Root Cause Analysis

RCDQ - Reactive Customer-Driven Quality

R-chart - Chart Used to Monitor the Dispersion of a Process

RCL - Robustness Checklist

RCM - Reliability-Centered Maintenance

RDM - Reliability Demonstration Matrix

RDOE - Reliability Design of Experiments

RDT - Reliability Demonstration Test

Red X - Primary Cause

REML - Residual Maximum Likelihood

RFD - Reliability Function Deployment

RFQ - Request for Quotation

RFS - Regardless of Feature Size

RFT - Right First Time

RG - Reliability Growth

RGCA - Reliability Growth Cause Analysis

RMA - Return Material Analysis

RMS (rms) - Root-Mean-Squared (i.e., Root-Sum Squares Method)

ROCOF - Rate of Occurrence of Failures

RoHS - Restriction of Hazardous Substances

ROI - Return on Investment

RPN - Risk Priority Number

RQL - Rejectable Quality Level

RQT - Reliability Qualification Tests

RSM - Response Surface Method

RSS - Root-Sum Squares Method (i.e., Root- Mean-Squared Method)

S Chart - Control Chart for Standard Deviations

SAE - Society of Automotive Engineers

SASIG - Strategic Automotive product data Standards Integration Group

SBU - Strategic Business Unit

SCM - Supply Chain Management

SEV - Smallest Extreme Value

SFDC - Shop Floor Data Collection

SFF (S_{FF}) - Safe Failure Fraction

SFMEA - System Failure Mode and Effects Analysis

SiC - Silicon Carbide

SIF - Stress Intensity Factor

SLA - Supplier Launch Audit (GM specific)

SLAM - Simultaneous Localization and Mapping

SLM - Ship Logistic Management

SLS - Selective Laser Sintering

SMED - Single Minute Exchange of Die for flexible manufacturing

SOC or SoC - State of Charge

SOH - State of Health

SOP - Start of Production

SOP - Standard Operating Procedure

SORP - Start of Regular Production

SOTIF - Safety of the Intended Functionality

SPC - Statistical Process Control

SPOF - Single Point of Failure

SQA - Supplier Quality Assurance

SQE - Supplier Quality Engineer or Supplier Quality Engineering

SQIP - Supplier Quality Improvement Program

SS (6-Sigma) - Six-Sigma

STBF - Shortest Time between Failures

Supplier CR - Supplier Customer Readiness

SVHC - Substances of Very High Concern

SWIP$_3$E - Standard, Workpiece, Instrument, Person/Procedure/Policy, and Environment (six essential elements of a generalized measuring system)

SWOT - Strengths, Weaknesses, Opportunities, and Threats

TAAF - Test, Analyze, and Fix

TAAT - Test, Analyze, and Test

TBD - To Be Decided or To Be Determined

TCO - Total Cost of Ownership

TE - Tooling and Equipment Certification

TGR - Things Gone Right

TGW - Things Gone Wrong

TNI - Trouble Not Identified

TOP - Time Optimized Process

TOPS - Team- Oriented Problem-Solving (Ford Motor Company specific)

TPM - Total Productive Maintenance

TPMS - Triply Periodic Minimal Surfaces

TQM - Total Quality Management

TRIZ - Russian, i.e., Theory of Inventive Problem-Solving (TIPS)

TV - Total Variation

TWI - Training within Industry

UAV - Unmanned Aerial Vehicle

u-chart - Attribute Quality Control Chart for Frequency of Defects

UCL - Upper Control Limit

UCS - Ultimate Compressive Strength

USL - Upper Specified Limit

UTS - Ultimate Tensile Strength

VDA6.3 - Verband der Automobilindustrie (German); Automotive Quality Management System Process Audit

VEVA - Value Engineering and Value Analysis

VIF - Variance Inflation Factor

VMEA - Variation Mode and Effect Analysis

VOC - Voice of Customer

VRS - Variable Returns to Scale

VSM - Value Stream Mapping

WBS - Work Breakdown Structure

Nomenclature

a, b, c, …	Coded variables
A, B, C, …	Design factors
C	Confidence level
[C]	Conference matrix
C_b	Battery capacity
C_p	Process capability index, usually for symmetric distribution; two-sided
C_{pk}	Process capability index, usually for skewed distribution; one-sided
C_{Pm}	Process capability index, a departure of mean μ away from process target
[D]	Design matrix
D_c	Depth of charge
E (GPa)	Modulus of elasticity or Young's modulus
E[ ]	Expected value
E_{11}, E_{22}, E_{33} (GPa)	Moduli of elasticity, namely Young's moduli, for orthotropic materials
E_T (GPa)	Tensile modulus of elasticity
F	F-distribution
F (N)	Force (Newton)
f( )	Function of ( )
G (GPa)	Shear modulus of elasticity
G_{23}, G_{31}, G_{12} (GPa)	Shear moduli of elasticity for orthotropic materials
H, h, hr	Hour
H_V (kgf/mm², GPa)	Vickers hardness: rectangular pyramidal indenter
[I]	Information matrix or Fisher information matrix
I_{zod} (kJ/m² or J/m²)	Izod notched impact strength at 23°C (ISO 180/1A)
L( )	Likelihood
L_{ln}( )	Likelihood of natural-log-transformed data
Ln( ) or ln( )	Natural logarithmic transformation of ( )
Log( ) or log( )	Logarithmic transformation of ( )
LR	Likelihood ratio
P_p	Process performance
P_{pk}	Process performance index for a stable process
R	Reliability

R	Resistance, electric
R	Model correlation, e.g., regression model
R_a (μm)	Surface roughness average
R_{adj}	Adjusted model correlation
R_{pred}	Predictive correlation, based on predicted values
R_{rms} (μm)	Surface roughness root-mean square (rms)
R_z (μm)	Surface roughness maximum
S	Sample standard deviation or sample error
S^2	Sample variance
S_c	State of charge
S_h	State of health
SS	Sum of squares
t	t-distribution (Student's t-distribution)
t (sec)	Time
T (°C), Temp (°C)	Temperature in °C
T (ton)	Mass for millimeter (mm)–second (sec)–ton system
T_k (°K)	Temperature in °K
u, v, w (m; mm)	Displacements in x-, y-, and z-directions, respectively
V	Volts
(X, Y, Z), (x, y, z)	Cartesian coordinate system
Y	Response or objective function
Y_p	Predicted value for Y
α	p-Value
α (μm/m/°C)	Coefficient of linear thermal expansion; $\perp$ and // to mold or casting flow
$\alpha_1, \alpha_2, \alpha_3$ (μm/m/°C)	Coefficients of linear thermal expansion in 1, 2, and 3 directions
$\alpha_x, \alpha_y, \alpha_z$ (μm/m/°C)	Coefficients of linear thermal expansion in x-, y-, and z-directions
β	Coefficients of linear moisture expansion
$\beta_1, \beta_2, \beta_3$	Coefficients of linear moisture expansion in 1, 2, and 3 directions
$\beta_x, \beta_y, \beta_z$	Coefficients of linear moisture expansion in x-, y-, and z-directions
ε	Residual
ε	Strain
$\varepsilon_1, \varepsilon_2, \varepsilon_3$	Principal strains
$\varepsilon_{11}, \varepsilon_{22}, \varepsilon_{33}$	Normal strains defined in the (1, 2, 3) coordinate system
$\varepsilon_{11c}, \varepsilon_{22c}, \varepsilon_{33c}$	Ultimate compressive strains along the primary orthotropic material axes
$\varepsilon_{11t}, \varepsilon_{22t}, \varepsilon_{33t}$	Ultimate tensile strains along the primary orthotropic material axes
$\varepsilon_{23}, \varepsilon_{31}, \varepsilon_{12}$	Shear strains (tensor) defined in the (1, 2, 3) coordinate system
$\varepsilon_{23u}, \varepsilon_{31u}, \varepsilon_{12u}$	Ultimate shear strains in primary orthotropic material coordinates (1, 2, 3)
ε_{Creep}	Creep rupture strain
ε^e and ε_e	Elastic strain

ε_{eq}	Equivalent strain
$\varepsilon_{eq}{}^{p}$	Equivalent plastic strain
ε^{p} and ε_{p}	Plastic strain
ε_{ucs}	Ultimate compressive strain
ε_{uts}	Ultimate tensile strain
$\varepsilon_{xx}, \varepsilon_{yy}, \varepsilon_{zz}$	Normal strains
$\varepsilon_{yz}, \varepsilon_{zx}, \varepsilon_{xy}$	Shear strains
μ	Mean
ν	Degrees of freedom
ρ, ρ_{f}, ρ_{m} (g/cm^3)	Density (overall), density of fiber, and density of matrix, respectively
σ	Standard deviation
σ (MPa)	Stress
σ^2	Variance
$\sigma_1, \sigma_2, \sigma_3$ (MPa)	Principal stresses
σ_{eq} (MPa)	Equivalent stress, e.g., von Mises stress
σ_{f} (MPa)	Fatigue limit, also called endurance limit
$\sigma_{f}{}'$ (MPa)	Fatigue strength coefficient
σ_{ucs} (MPa)	Ultimate compressive strength
σ_{uts} (MPa)	Ultimate tensile strength
$\sigma_{xx}, \sigma_{yy}, \sigma_{zz}$ (MPa)	Normal stresses
$\sigma_{xy}, \sigma_{yz}, \sigma_{zx}$ (MPa)	Shear stresses
$\sigma_{y}, \sigma0.2\%$ (MPa)	Yield strength, i.e., stress at 0.2% (tensile) or -0.2% (compressive) strain
σ_{YC} (MPa)	Yield strengths in compression
σ_{YT} (MPa)	Yield strengths in tension
$\tau, \tau_{yz}, \tau_{zx}, \tau_{xy}$ (MPa)	Shear stresses
χ^2	Chi-square distribution

Structural Joints

S tructural joints with bolts/screws, crimping, form-locking (e.g., snap fits), rivets, clips, welding, adhesives, and interference fittings are widely used in primary structures like vehicles, with their high load-carrying capability and excellent availability, repairability, replaceability, traceability, and cost efficiency. The exploration of the properties of structural joints, by means of closed-form solutions, finite element modeling, or physical tests, is itself an area of research, but some product reliability of structural joints has resorted to DOE.

12.1. Mechanical Joints

Joining means to assemble parts of a tangible mechanism or machine via connections that can be detachable such as assembling with bolts/screws, crimping, and form-locking or permanent such as assembling with rivets, clips, welding, adhesives, and interference fittings. Design of mechanical joints inevitably has to deal with the interdigitated features such as notch, friction, inhomogeneity, contact/impact, anisotropy, creep, corrosion, surface roughness, and compatibility of dissimilar materials, in addition to the work environment.

Mechanical joints play an important role in a mechanism or machine such as aircrafts and on-ground vehicles. The resultant motion on operating a vehicle is determined by the mechanical joints that connect part members. Mechanical joints allow motion in some directions and constrain it in others. These types of motions allowed and constrained are related to the degrees of freedom of joints. Mechanical joints are meant to resist tensile loads, shear loads, and bending moments by delivering the following engineering virtues:

stiffness, vibration resistance, damping effect, dynamic stability, and finally the load capacity. Specifically, attention must be paid to the following common causes of potential failures of mechanical joints:

1. Tribological behavior
2. Stress concentration
3. Differential thermal expansion and phase transformation
4. Galvanic corrosion

12.1.1. Reduction in Stress Concentration

One fundamental problem to deal with in mechanical joints is stress concentration. Here is an example of how to reduce the stress concentration via stress relief grooves based on DOE. Stress relief grooves are often utilized to reduce the stress level at the snap ring cut of an axle shaft in a vehicle as exhibited in Figure 12.1.1 and thus to maximize the fatigue life of the drive axle without failing the stress relief grooves first [Shin et al. 2014]. A snap ring is usually a C-shaped thin ring that works as a retaining fastener for assembling parts onto a shaft and retaining the parts inside the circular bore. Nominal dimensions of the width and depth of the snap ring bore of an axle shaft are 1.85 mm and 1.25 mm, respectively. The shaft is made of steel SCM440H, of which Young's modulus = 205 GPa, Poisson's ratio = 0.29, yield strength = 911 MPa, ultimate tensile strength = 1025 MPa, and ultimate tensile strain = 21%.

The geometric dimensions of each of the two stress relief grooves are identified using locating parameter D and feature radius R, which designate the distance measured from the arc center of the snap ring bore and the radius of each individual stress relief groove, respectively. It is of great interest to understand how locating parameter D and feature radius R of these two stress relief grooves influence the maximum equivalent stresses (e.g., von Mises stress) observed at the snap ring bore and stress relief grooves themself, designated as variables C and G, respectively. The torque is applied to the spline of the axle shaft with a torque of 2499 Nm at a rotating speed of 900 revolutions per minute (rpm). As listed in Table 12.1.1, the maximum equivalent stresses C (for sap ring cut) and G (stress relief groove) are obtained from nonlinear finite element analyses (FEAs) with the following data.

FIGURE 12.1.1   Axle shaft with snap ring bore and two stress relief grooves [Shin et al. 2014, Fig. 4].

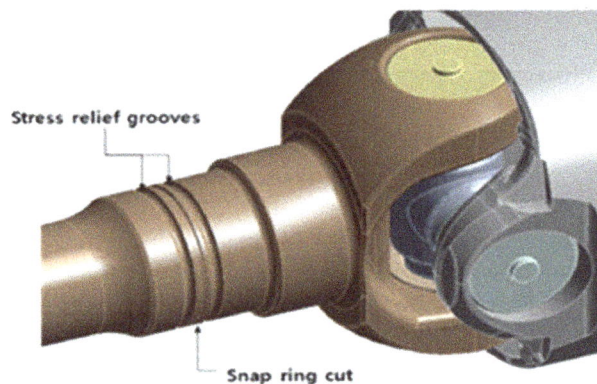

TABLE 12.1.1 Design matrix for balancing stress concentrations at the snap ring bore and stress relief grooves of an axle shaft.

Run	R	D	B	H	X	Y
1	0.25	3.7	1.85	1.25	931.5	632.9
2	0.5	3.7	1.85	1.25	900.7	646.1
3	0.75	3.7	1.85	1.25	895.2	675.5
4	1	3.7	1.85	1.25	872.9	677.2
5	1.25	3.7	1.85	1.25	822.4	695.1
6	1.5	3.7	1.85	1.25	755.7	706.7
7	1.75	3.7	1.85	1.25	616.4	726.6
8	2	3.7	1.85	1.25	533.7	745.2
9	2.25	3.7	1.85	1.25	410.8	765.2
10	0.25	5.55	1.85	1.25	940	688.5
11	0.5	5.55	1.85	1.25	922.5	690.5
12	0.75	5.55	1.85	1.25	913.5	705.4
13	1	5.55	1.85	1.25	899	711.7
14	1.25	5.55	1.85	1.25	879.7	727.2
15	1.5	5.55	1.85	1.25	868.7	738.5
16	1.75	5.55	1.85	1.25	807.8	751.2
17	2	5.55	1.85	1.25	775.1	766.4
18	2.25	5.55	1.85	1.25	703	784.1
19	0.25	7.4	1.85	1.25	952.5	709.6
20	0.5	7.4	1.85	1.25	947.1	712.3
21	0.75	7.4	1.85	1.25	934.1	729.8
22	1	7.4	1.85	1.25	925.3	731.9
23	1.25	7.4	1.85	1.25	912.2	753.6
24	1.5	7.4	1.85	1.25	888.9	756.6
25	1.75	7.4	1.85	1.25	883.4	768.5
26	2	7.4	1.85	1.25	873.8	785.3
27	2.25	7.4	1.85	1.25	858.3	801.2

where
R (mm): Radius of each stress relief groove
D (mm): Distance from the center of the snap ring bore to each stress relief groove
B = 1.85 mm: Width of the snap ring bore, kept constant
H = 1.25 mm: Depth of the snap ring bore, kept constant
X (MPa): Highest equivalent stress at the snap ring bore, as loaded
Y (MPa): Highest equivalent stress at stress relief grooves, as loaded

The data analysis is carried out using Minitab (Stat → Regression → Regression → Fit Regression Model). The "most insignificant" factor or interaction that comes with the largest p-value is weeded out one by one until all p-values are less than 10%, which is here defined as the parting significance level based on the F-distribution. In light of the analysis of variance (ANOVA) presented in Table 12.1.2, the final predictive equations for the highest equivalent stress at the snap ring bore and stress relief groove are, respectively,

$$X_p = 1018.6 - 282.5\ R - 71.0\ R^2 - 2.694\ D^2 + 57.98\ R\ D$$

$$(R = 97.4\%,\ R_{adj} = 97.0\%,\ \text{and}\ R_{pred} = 95.2\%), \tag{12.1.1}$$

and $$Y_p = 504.2 + 61.87\ R + 39.87\ D + 7.18\ R^2 - 1.748\ D^2 - 4.877\ R\ D$$

$$(R = 99.5\%,\ R_{adj} = 99.4\%,\ \text{and}\ R_{pred} = 99.1\%). \tag{12.1.2}$$

These two predictive equations are effective and adequate with high correlations, including unadjusted model correlation, adjusted model correlation, and predictive correlation.

Three-dimensional surface plots to show how the highest equivalent stresses at the snap ring bore and stress relief grooves vary with respect to the radius of the stress relief grooves and the distance from the center of the snap ring bore to each individual stress relief groove are exhibited in Figure 12.1.2. According to Equations (12.1.1) and (12.1.2), it can be seen that

(a) The stress relief at the snap ring bore decreases quickly as the radius of the stress relief grooves and the distance from the center of the snap ring bore to each individual stress relief groove increase. Their interaction also helps reduce the stress level of stress concentration at the snap ring bore.

(b) The stress level at the stress relief grooves increases with respect to increasing radius of the stress relief grooves.

(c) The stress level at the stress relief grooves decreases slightly with respect to increasing distance from the center of the snap ring bore to each individual stress relief groove.

The design of an axle shaft (with a snap ring bore) that goes from the power source (e.g., electric motor) to the power destiny (e.g., tire wheel) is a balance between the highest equivalent stresses at the snap ring bore and stress relief grooves.

TABLE 12.1.2 ANOVA for balancing stress concentrations at the snap ring bore and stress relief grooves of an axle shaft.

X:	Source	SS_{adj}	DOF	MS_{adj}	$F_{u,v}$	p-Value
	Regression	431,929	4	107,982	103.7	0.0%
	R	26,783	1	26,783	25.71	0.0%
	R^2	18,212	1	18,212	17.48	0.0%
	D^2	12,110	1	12,110	11.63	0.3%
	RD	89,260	1	89,260	85.69	0.0%
	Error	22,917	22	1042		
	Subtotal	454,846	26			
	Grand average		1			
	Total	—	27			
	Regression	45,036.9	5	9007.37	426.6	0.0%
	R	1267.4	1	1267.44	60.03	0.0%
	D	868.4	1	868.35	41.13	0.0%
	RR	186.2	1	186.21	8.82	0.7%
	DD	214.8	1	214.80	10.17	0.4%
	RD	610.7	1	610.65	28.92	0.0%
	Error	443.4	21	21.11		
	Subtotal	45,480.2	26			
	Grand average		1			
	Total	—	27			
Y:	Source	SS_{adj}	DOF	MS_{adj}	$F_{u,v}$	p-Value
	Regression	45,036.9	5	9007.37	426.6	0.0%
	R	1267.4	1	1267.44	60.03	0.0%
	D	868.4	1	868.35	41.13	0.0%
	R^2	186.2	1	186.21	8.82	0.7%
	D^2	214.8	1	214.80	10.17	0.4%
	RD	610.7	1	610.65	28.92	0.0%
	Error	443.4	21	21.11		
	Subtotal	45,480.2	26			
	Grand average		1			
	Total	—	27			

FIGURE 12.1.2 3D surface plots of highest equivalent stresses at the snap ring bore and stress relief grooves.

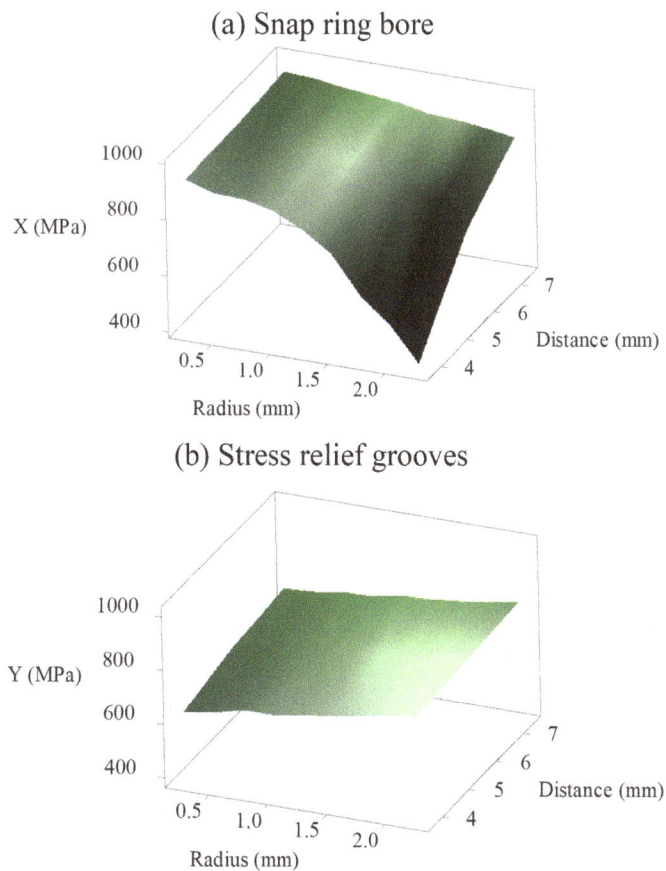

(a) Snap ring bore

(b) Stress relief grooves

12.1.2. Tribological Behavior

One fundamental problem is that it is extremely hard, if not impossible, to measure or predict what the absolute value of friction or its temporal and spatial context is in a mechanical joint. Given the nature of current finite element modeling, it requires the development of constitutive equations for mechanical joints that both adequately describe their distinct tribological behaviors and lend themselves to an effective computational process [Segalman 2001].

12.1.3. Differential Thermal Expansion and Phase Transformation

The variation in thermal expansion shall affect the amount of press fit at an elevated working temperature. The situation can be further complicated if constituent materials undergo phase transformation with an associated volume change on heating and cooling.

12.1.4. Galvanic Corrosion

A galvanic cell is formed between two dissimilar metals in the press-fit joint. In the presence of aggressive electrolytes, galvanic cells induce dissolution of the less noble metal in the joint. This phenomenon is called galvanic corrosion.

12.2. Interference Fit

Interference fit, also called press fit or friction fit, is suitable to assemble mechanical components that need a high load-carrying capacity using the frictional contact between mating parts. Either heating up the outer part or cooling down the inner part will make a potential interference fit. On cooling, the temperature can be attained using liquid nitrogen (−196°C), liquid oxygen (−184°C), and dry ice (−80°C). One typical example is to install high-performance rolling element bearings. When a separate inner race that requires an interference fit on a shaft is used, thermal fitting is usually the preferred assembly method.

Interference-fit interconnects are also adapted to support a wide range of electrical applications such as printed circuit board (PCB)-to-PCB stacking interconnects (e.g., 0.4×0.5-mm pins), as a matter of electronification and miniaturization of automotive electronics in electrical vehicles (e.g., PCB thickness ≤ 1.0 mm). Press-fit assembly is more reliable than most other joining methods such as termi-point clips, solder (by machine), wire bond, crimp, insulation displacement, screw, and clamp, while only next to wire wrap. Furthermore, it can be automated easily.

Consider a general case of inserting a hollow shaft through a hub of dissimilar materials. The nominal contact pressure between the shaft and hub can be [Loc and Phong 2000] written as

$$P = \frac{\delta}{2\, r_c \left\{\left[\dfrac{r_c^2 + r_i^2}{E_s\,(r_c^2 - r_i^2)} - \dfrac{\nu_s}{E_s}\right] + \left[\dfrac{r_c^2 + r_i^2}{E_h\,(r_o^2 - r_c^2)} - \dfrac{\nu_h}{E_h}\right]\right\}}, \qquad (12.2.1)$$

where
   P is the nominal contact pressure
   δ is the radial interference
   r_c is the nominal radius of the contact surface
   r_i is the inner radius of the shaft
   r_o is the outer radius of the hub
   E_s and ν_s are Young's modulus and Poisson's ratio of the shaft material, respectively
   E_h and ν_h are Young's modulus and Poisson's ratio of the hub material, respectively

The above equation based on linear elastic deformations reveals the important structural parameters that must be considered for the insertion of a shaft into the hub as the first approximation. The contact pressure distribution of a press-fit joint can be measured using ultrasonic devices [Marshall et al. 2011]. However, the following factors must be examined before finalizing the reliability of a shaft–hub joint:

   (a) Boundary effect at contact edges

   (b) Softening of materials due to plastic deformation

   (c) Changes in material roughness at contact surfaces

   (d) Galvanic corrosion

   (e) Differential thermal expansion and phase transformation

The load capacity of a press-fit is often decided upon the torsional durability in terms of its fatigue life cycle [Biron et al. 2012], regarded as a viable index of joint reliability. Automotive application examples include rotating turbine components, transmission shafts, and railway axle and wheel assemblies. Stress concentrations at notches and stress intensities in contact surfaces have a profound yet complex effect on press-fit joints.

12.2.1. Torsional Durability of Press-Fit Joints

Torsional durability of press-fit joints subjected to a dynamic torsion load was investigated using DOE by Rajakumar [2012]. The purpose of the study is to figure out which design parameters significantly affect the durability of press-fit joints in terms of torsional fatigue cycles. The design matrix and test results are given in Table 12.2.1.

TABLE 12.2.1 Experimental design for studying torsional durability of press-fit joints [Rajakumar 2012].

Run	A	B	C	Y	Y_p
1	10	17	180	396	399
2	10	20	250	450	449
3	10	23	420	583	583
4	15	17	250	426	416
5	15	20	420	522	531
6	15	23	180	468	462
7	20	17	420	600	594
8	20	20	180	438	453
9	20	23	250	474	467

where
A (mm): Contact length
B (μm): Interference
C (Hv): Vickers hardness
Y: Number of torsional cycles to break the joint
Y_p: Number of torsional cycles to break the joint, as predicted using the DOE model

The data analysis is carried out using Minitab (Stat → Regression → Regression → Fit Regression Model). The "most insignificant" factor or interaction that comes with the largest p-value is weeded out one by one until all p-values are less than 10%, which is here defined as the parting significance level based on the F-distribution. In light of the ANOVA presented in Table 12.2.2, the final predictive equation for the life cycle of each treatment (run) based on the factorial design is

$$Y_p = 319.4 + 1.044\,A^2 + 1.017\,B^2 + 0.002177\,C^2 - 1.517\,A\,B - 0.0428\,B\,C$$

(12.2.2)

$$(R = 99.3\%,\ R_{adj} = 98.1\%,\ \text{and}\ R_{pred} = 85.0\%).$$

The predictive equation given above is effective and adequate as evidenced by high unadjusted model and adjusted model correlations, as well as the predictive correlation.

According to Equation (12.2.2), the torsional life cycle of a specific press-fit joint increases quadratically with respect to each individual factor, i.e., contact length, interference, and material hardness. This is coincident with Figure 4 presented in Rajakumar [2012]. Nevertheless, it decreases with the interaction between interference and contact length and the interaction between interference and material hardness.

TABLE 12.2.2 ANOVA for studying torsional durability of press-fit joints.

Source of variance	SS_{adj}	DOF	MS_{adj}	$F_{u,v}$	p-Value
Regression	38,883.5	5	7776.7	42.16	0.6%
A^2	2511.1	1	2511.1	13.61	3.5%
B^2	4720.9	1	4720.9	25.59	1.5%
C^2	2639.5	1	2639.5	14.31	3.2%
AB	2388.9	1	2388.9	12.95	3.7%
BC	1101.2	1	1101.2	5.97	9.2%
Error	553.4	3	184.5		
Subtotal	39,436.9	8			
Grand average	—	1			
Total	—	9			

12.2.2. Boundary Effects

A significant change in the contact pressure distribution occurs at the edges of the fit where the contact pressure first arises to a local peak before abruptly falling away, and so does the shear stress on the contact zone. This is called the free boundary effect. The stress concentration due to stress singularities appears on (near) the edges in the contact zone because the free-edge effect at a press-fit joint shall be taken into consideration as a design parameter [Parsons and Wilson 1970], especially for estimating the fretting fatigue life of a press-fit joint.

Specifically speaking, the press fit between an axle and a wheel can introduce compression stresses beneath the press-fit and axial tensile stresses in the fillets, which may cause fretting corrosion fatigue due to the microslip between the press-fitted parts in the long run. It is feasible to alleviate free-edge stress intensity by removing material near notches, creating a groove at the hub that is called a relief notch. A stress relief notch cut (mostly elliptical/circular) along the free edges of the hub can reshape the contact stresses and overwhelmingly reduce the stress intensity [Bijak-Zochowski et al. 1991] as demonstrated schematically in Figure 12.2.2 but may not be able to eliminate it. Let subscripts shaft and hub refer to inside and outside components, respectively, in a polar coordinate system. The stress singularity is mostly pronounced in the hub (external cylinder), and the stress intensity increase in radial stress $\sigma_{rr,hub}$ is more severe than tangential (circumferential) stress $\sigma_{\theta\theta,hub}$, radial stress $\sigma_{rr,shaft}$, and tangential stress $\sigma_{\theta\theta,shaft}$ according to Pérez Cerdá et al. [2014]. The use of several smaller relief notches instead of a single long one may facilitate smoother stress flow. Furthermore, a gradual transition in the cross-sectional variation can be also helpful. Finite element methods turn out to be an effective analysis tool because it is hard to set up proper boundary conditions for a closed-form solution. It is worthwhile to apply the DOE to the shape optimization of a specific hub–shaft joint to reduce the stress intensities in both $\sigma_{rr,hub}$ and $\sigma_{r\theta,hub}$ simultaneously.

Variations of stress intensities are more complex if the hub and shaft in a press-fit joint are made of two dissimilar materials or composites [Chiang and Rowlands 1991].

FIGURE 12.2.2   Radial stresses σ$_{rr,hub}$ (MPa) at the hub of a press-fit joint.

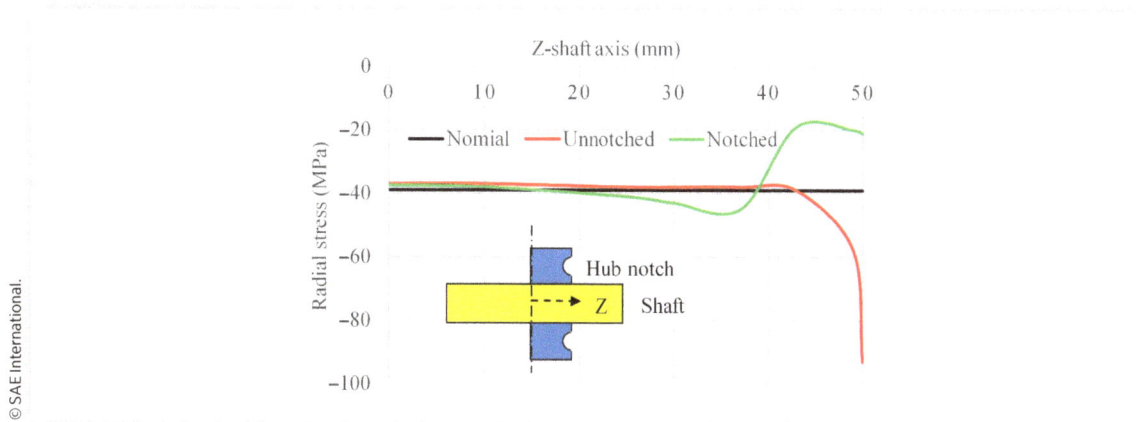

12.2.3. Reduction in Stress Intensity due to Boundary Effect at Contact Edges

A stress relief groove can be also introduced into the shaft for the reduction in stress intensity in a hub–axle joint per EN 13104 specification, as shown in Figure 12.2.3. Influences of a stress relief groove on the fretting wear and fatigue crack initiation of a 35CrMo steel used in a railway axle–hub joint axle are investigated using finite element methods [Zeng et al. 2018]. The depth and width of the wear scar in response to the applied groove radius and depth are listed in Table 12.2.3. The stress pattern reshapes as the fretting wear develops. The simulation results show that the stress concentration at the contact edge is gradually relieved with an increase in fretting fatigue cycles due to the fretting wear as a new stress concentration pattern appears at the edge of the fretting wear scar, which gradually extends toward the inner side of the contact area. The stress concentration in the inner side eventually leads to the fatigue crack initiation.

FIGURE 12.2.3   Introducing a stress relief groove into a hub–axle joint [Zeng et al. 2018].

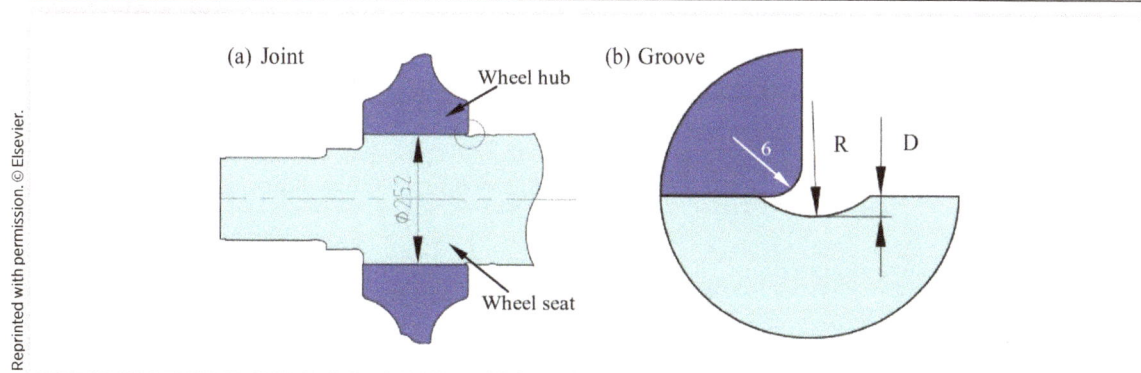

TABLE 12.2.3 Influences of stress relief groove on fretting wear of a steel axle [Zeng et al. 2018].

Run	R	D	X	Y	X_p	Y_p
1	16	1	3.175	12.8	3.011	12.81
2	16	2	2.313	12.4	2.623	12.40
3	16	4	2.091	10.8	2.115	10.76
4	16	6	1.899	8	1.963	8.01
5	24	4	2.132	11.2	2.115	11.26
6	32	4	2.173	11.6	2.115	11.55
7	40	4	2.264	11.6	2.115	11.62

Notes:
R (mm): Radius of stress relief groove (Figure 12.2.5)
D (mm): Depth of stress relief groove (Figure 12.2.5)
X (µm): Depth of wear scar from FEA
Y (µm): Width of wear scar from FEA
X_p (µm): Depth of wear scar by Equation (12.2.3)
Y_p (µm): Width of wear scar by Equation (12.2.4)

The data analysis is carried out using Minitab (Stat ➜ Regression ➜ Regression ➜ Fit Regression Model). The "most insignificant" factor or interaction that comes with the largest p-value is weeded out one by one until all p-values are less than 20%, which is here defined as the parting significance level based on the F-distribution. In light of the ANOVA presented in Table 12.2.4, the regression models for the width and depth of wear scar are given, respectively, as follows:

$$X_p = 3.487 - 0.521\ D + 0.0445\ D^2$$

$$(R = 92.12\% \text{ and } R_{adj} = 87.92\%),$$

(12.2.3)

and $$Y_p = 11.310 + 0.1292\ R - 0.001664\ R^2 - 0.13721\ D^2$$

$$(R = 99.97\%, R_{adj} = 99.94\%, \text{ and } R_{pred} = 99.42\%).$$

(12.2.4)

The predicted value of both depth and width of wear is listed in Table 12.2.3. Because the design matrix in Table 12.2.3 does not meet the requirement of a rotatable and balanced design, a loose p-value (=20%) is set up to accommodate the arbitrary randomness.

The depth of wear varies quadratically with respect to groove depth as exhibited in Figure 12.2.4, given that the statistical significance level is set to 20%. The width of wear varies quadratically with respect to groove radius and groove depth, respectively, as exhibited in Figure 12.2.5. As the wear extends into the inner side of the joint (i.e., toward the center of the joint), the stress intensity eventually causes the fatigue crack initiation.

TABLE 12.2.4 Influences of stress relief groove on fretting wear of a steel axle [Zeng et al. 2018].

X (Width of wear)	Source of variance	SS$_{adj}$	DOF	MS$_{adj}$	F$_{u,v}$	p-Value
	Regression	0.86185	2	0.431	11.22	2.3%
	B	0.26344	1	0.263	6.86	5.9%
	B^2	0.09309	1	0.093	2.42	19.5%
	Error	0.15367	4	0.038		
	Lack of fit	0.13724	1	0.137	25.06	1.5%
	Pure error	0.01643	3	0.0055		
	Subtotal	1.01552	6			
	Grand average	—	1			
	Total	—	7			
Y (depth of wear)	Source of variance	SS$_{adj}$	DOF	MS$_{adj}$	F$_{u,v}$	p-Value
	Regression	14.7111	3	4.904	1644	0.0%
	A	0.1162	1	0.116	38.94	0.8%
	A^2	0.0563	1	0.056	18.87	2.3%
	B^2	14.2498	1	14.25	4777	0.0%
	Error	0.0089	3	0.003		
	Subtotal	14.7200	6			
	Grand average	—	1			
	Total	—	7			

FIGURE 12.2.4   Influences of stress relief groove on the depth of fretting wear of a steel axle.

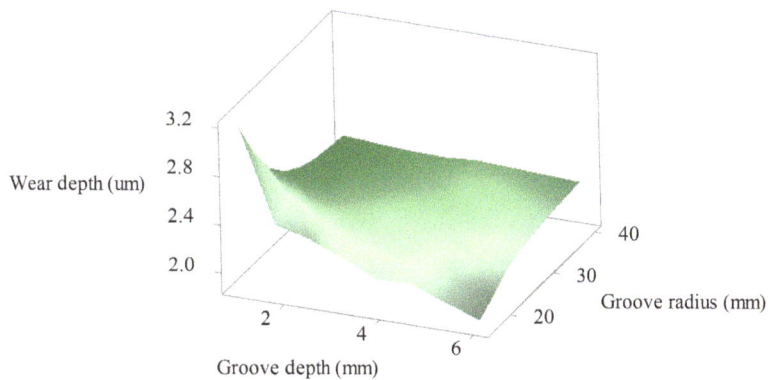

FIGURE 12.2.5 Influences of stress relief groove on the width of fretting wear of a steel axle.

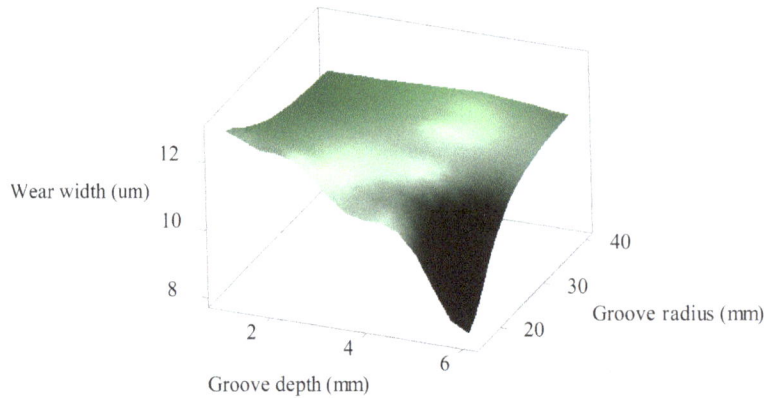

12.2.4. Changes in Material Roughness at Contact Surfaces

Surface roughness has a dramatic impact on the insertion and retention force. An experiment was conducted to examine the impact of repeated assembling and disassembling of the press-fit joint of steel shaft (C45) and brass hub (C2680). The shaft is extracted from the hub after the first insertion. After being inserted and extracted once, the surface roughness of the shaft is literally unaffected, while the surface roughness of the brass hub changes dramatically [Loc and Phong 2000] as shown in Table 12.2.5. The change in the inner surface of the brass hub after the first cycle of insertion/extraction operation can be found out regressively as

$$\Delta R_1\text{-}B = 0.1891 - 0.005512\ \delta - 0.75461\ R_0\text{-}B$$

$$(R = 99.995\%, R_{adj} = 99.995\%, \text{ and } R_{pred} = 98.25\%), \tag{12.2.5}$$

where
    δ (μm) is the radial interference
    R_0-B is the surface roughness (average) of brass hub as fabricated

It is shown by the above regression model that the material roughness of the inner surface of the brass hub reduces after the first cycle of insertion/extraction operation. Nevertheless, the material roughness does not reduce significantly after the second cycle of insertion/extraction operation, except that excessive asperities are further crushed (run 2), as exhibited in column ΔR_2-B of Table 12.2.5. This explains why the press-fit pin insertion process is a fast, cheap, and reliable manufacturing process that allows a repair of the press-fit pin up to two times.

TABLE 12.2.5 Design matrix for changes of inner surface roughness of brass hub after insertion/extraction operations [Loc and Phong 2000].

Run	δ (μm)	R_0-S	R_1-S	R_2-S	R_0-B	R_1-B	R_2-B	ΔR_1-B	ΔR_2-B
1	33	1.24	1.07	1.34	0.53	0.14	0.17	−0.39	0.03
2	33	2.98	3.02	2.86	3.35	0.83	0.19	−2.52	−0.64
3	42	2.68	2.92	2.95	0.83	0.15	0.17	−0.68	0.02
4	45	2.35	2.43	2.50	0.67	0.11	0.12	−0.56	0.01
5	47	1.01	0.95	1.13	0.76	0.12	0.15	−0.64	0.03

Notes:
1. Nominal diameter of the contact surface: 29.9 mm
2. Nominal insertion length: 14.5 mm
3. δ (μm): Radial interference
4. R_0-S, R_1-S, and R_2-S: Material roughness (average) of the contact surface of the steel shaft as fabricated, after one insertion/extraction, and after two insertion/extraction operations, respectively
5. R_0-B, R_1-B, and R_2-B: Material roughness (average) of the contact surface of the brass hub as fabricated, after one insertion/extraction, and after two insertion/extraction operations, respectively
6. ΔR_1-B: Change in surface roughness at the contact surface of the brass hub after one cycle of insertion/extraction operation
7. ΔR_2-B: Change in surface roughness at the contact surface of the brass hub after two cycles of insertion/extraction operations

12.2.5. Softening of Materials Due to Plastic Deformation

Shaft–hub joints are a widely accepted press-fit joints. One example is to attach a gear to a shaft. FEAs and physical tests of a specific shaft–hub joint (8 mm in diameter at the joint interface surface) are conducted by Madej and Sliwka [2021]. The shaft is inserted into the hub with a joint length of 15 mm. Both parts are made of S235JR steel with a Young's modulus of 200 GPa, the yield point of 225 MPa, and the ultimate tensile strength of 360 MPa. Four types of joints with different interferences are considered in the study, as shown in Table 12.2.6. The accuracy of finite element models is validated using a physical test of joint J5 as depicted in Figure 12.2.6.

TABLE 12.2.6 Assembly and retention forces of shaft–hub (steel S235JR) joints as a function of interference δ, with a nominal fit diameter of 8 mm [Madej and Sliwka 2021].

Joint	Fit (type of tit)	δ (μm)	$F_{assembly}$ (kN)	$F_{retaining}$ (kN)
J1	H7/k6 (locational transition)	4	2.293	2.092
J2		6	3.438	3.137
J3	H7/p6 (locational interference)	8	4.516	4.22
J4		10	5.549	5.125
J5		12	6.437	5.801
J6		14	7.174	6.006
J7		16	7.784	6.004
J8	H7/s6 (medium drive fit)	18	8.285	6.01
J9		20	8.682	6.01
J10		22	8.986	6.002
J11		24	9.209	5.999
J12		26	9.373	6.001
J13	H7/u6 (force fit)	28	9.474	5.999
J14		30	9.560	5.960
J15		32	9.632	5.974

FIGURE 12.2.6   Retention force of press-fit joint J5 by physical tests [Madej and Sliwka 2021].

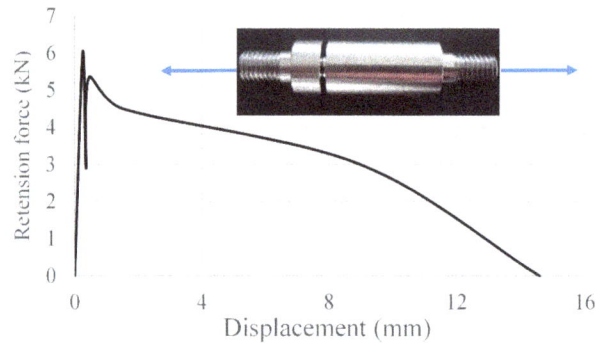

An increasing interference may require a larger insertion force to complete the assembly but does not necessarily increase the value of the retention force (i.e., breaking force) as pointed out by Hüyük et al. [2014]. As shown in Figure 12.2.7, the peak retention force (breaking force) is 6.01 kN that occurs around an interference between 18 and 20 μm. Note that the retention force is an important performance index of shaft–hub joints.

A sequential approximate multi-objective optimization approach to designing interference-fit assemblies subjected to fatigue loads, including finite element modeling, DOE, and multi-objective optimization techniques, was taken by Biron et al. [2012]. It sums up to be a valuable way of product optimization in the stochastic domain. Nevertheless, a more realistic model is needed for the fatigue life prediction and round-off radius at edges has to be accounted for in the finite element modeling with material and geometric nonlinearities involved in press-fit assemblies [Chiang 2019]. Safety factor is not a magic number for indexing product life [Ashely 1993]. For example, the product life prediction based on a safety factor calculated using strains, stresses, strain energy, strain intensities, creep, corrosion, diffusion, or their combined effect would not lead to a unique number.

FIGURE 12.2.7   Breaking force of a press-fit joint by finite element methods [Madej and Sliwka 2021].

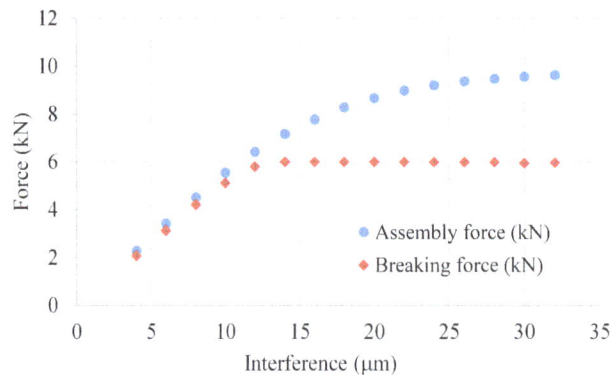

12.3. **Bolted Joints**

Bolted fasteners are commonly used to join or secure components and parts in the automotive industry. One advantage of using bolted fasteners is easy to assemble and disassemble components and parts at a low cost. High-strength bolts are well established as economical and efficient devices for connecting structural parts of on-ground vehicles. The principle of slip-critical connections relies upon the clamp force applied in each bolt in a connection. Shear loads are expected to be transferred by frictional resistance in the axial contact surfaces rather than by in-plane bearing load exerted by the hole faces onto bolt shanks.

12.3.1. **Tightening by Torquing**

Bolted fasteners can be tightened using simple tools. However, in order to maintain the required clamping force in the joint, the proper torquing procedure must be applied. Proper torquing, including torque size and torquing sequence, is crucial to secure the fastening force and contact pressure distribution around a multi-bolted joint. It is expected to always have a direct correlation between bolt elongation and bolt tension until the bolt yields.

The torque, as a function of time, required to tighten a bolted fastener varies, depending on the following conditions: bolt size (e.g., diameter), bolt material, thread type, thread fit, nut type, nut size, nut material, washer type, washer size, washer material, lubrication, fastened-member materials, torquing speed, torquing sequence, bearing surfaces, and working temperature. Torque charts are only permitted to be used for the installation or inspection of bolted joints tightened with the same conditions mentioned above. Tightening steps are suggested as follows:

1. Theoretical Prediction: Theoretical calculations and some tryout tests will tell the torque in need at the given turning speed in terms of rpm, at which the transition phase ends and the pseudo-linear portion of the trace begins to grow.

2. Torque Control Period (Snug Tight Torque): This is to determine the dynamic torque at the end of torque control period. It provides the accommodation of part-to-part variations. Snug tight is necessary after all the members in a connection have been assembled into firm contact by the bolts in the joint, and the bolts have been tightened sufficiently to prevent removal of the nuts without a tool as suggested by the American Institute of Steel Construction.

3. Angle Control Period: It starts from the snug position with a specified angle of rotation that shall be performed with sufficient accuracy. This is done using tightening tools that are equipped with angle encoders in light of the respective empirical joint stiffness, i.e., clamp force–angle curve specified for this type of bolted joint. As bolt length-to-diameter ratios get smaller, more tension is generated per degree of rotation, so greater accuracy is required. It is required to check if the joint stiffness deviates from the pseudo-linear relationship.

4. Residual Torque after Assembly: The residual torque is to be checked right after tightening operation in a slow manner so that the dynamic effect is mitigated. The amount of fastening tension can vary by as much as 20% after the bolted joint is left standing for several days. Residual torque can be higher than dynamic torque. At the end of the production line, the measurement of residual torque is a crucial step for completing the quality control system. Another factor that has an influence on the residual torque is "mutual interaction" in a multibolted assembly, which occurs when the torque applied to one bolt affects another. Changing the installation sequence of fasteners may resolve the issue.

5. Torque Auditing: Joints may take a long time to settle down. Torque auditing detects loosened fasteners that no longer provide the desired fastening force. It also detects joint relaxation. In many cases, torque auditing is the final judge of the harmony between the fastening tools and product design and materials in an assembly process.

Even if testing is required to establish angle control or torque control schemes on all but the simplest joints, joint stiffness tests and verified friction factors are needed beforehand [Archer 2009]. On different joint designs that even have the same bolts and nuts, the torque-tightening process including the magnitude of torque and torquing sequence for each joint can be different.

Finite element dynamic simulations can be used to simulate the installation process of the bolt by gradually applying a torque until the bolt fails. Based on simulations including material nonlinearity and geometric nonlinearity, and contact nonlinearity for capturing potential joint failure modes, the torque–angle curves under various DOE treatments are generated and used to decide the installation torque of the joint [Hwang 2013].

12.3.2. Potential Unreliability Inherent to Tightening

Consider the case of tightening a single bolted joint, of which the fastened members are by default well-aligned flat plates. The torque, without taking relaxation into consideration, will drive the bolt into the following four working ranges: rundown stage, alignment stage, elastic prevailing stage, and torque-turn-to-yield stage. Unexpected unreliability can occur at each stage.

1. Rundown Stage: The prevailing torque is initiated before the bolt head and nut contact their bearing surfaces.

2. Alignment Stage: The snugging zone wherein the bearing surfaces struggle to fit in a snug position. Alignment is so important, because the bolt preload acting on the thread developed in this stage is usually slanted and it creates a torque in the circumferential direction that may result in self-loosening of the nut. This could be a major source of unreliability.

3. Elastic Prevailing Range: The desired clamping force develops, and the slope of the torque–angle curve is essentially constant. The higher the prevailing torque, the higher the resistance to self-loosening, but the threads may get yielded. Although the bolt stretches in the elastic range, the plastic deformation of some real contact "points" because of surface roughness accumulates. This is a latent source of unreliability.

4. Torque-Turn-to-Yield Stage: It begins with an inflection point at the end of the elastic range because of apparent yielding of fastened members (including gaskets), such as yielding of nut threads, yielding of bolt threads, or even the bolt head–stem fillet. When a bolted joint is tightened beyond the yield point, it exhibits strain hardening and plastic deformation grows progressively at higher torque each time as they are reused. This makes the reuse of such a bolt undesirable. This could be another source of unreliability.

The peak contact pressure occurred at points away from the bolt hole, as attributed to the effect of the edge of the bolt head on the pressure distribution [Ito et al. 1979]. One more concern about unreliability is the unevenly distributed pressure because of flange effect or tilting slope that can be induced by the contact surface profile (e.g., uneven flatness), uneven contact surface roughness, or uneven material texture. Fortunately, this can be overcome in theory by a sufficient amount of preload, so that loosening movement will not occur; otherwise, a locking device (e.g., locking nut) may be employed to guard against the nut self-loosening.

Another concern of unreliability of a bolted joint is the usage environment after the bolt is installed. It has been known for decades that a relative axial or transverse movement between bolted joint members (parts, bolts, nuts, etc.) may partially, or sometimes completely, loosen the joint. It has been identified that in the automobile industry, 23% of all service problems were because of loosened bolted joints [Grabon et al. 2018]. The report further indicates that loosened fasteners were found in 12% of all new cars surveyed. The report evaluates the efficacy of various locking screw fasteners, including nylock nut, aerotight nut, chemical lock, cleveloc nut, flat washer, nylon washer, serrated washer and spring washer with bolts of different materials, sizes, and types with different initial clamping forces under the accelerated vibrating conditions obtained in an indigenously made testing rig [Bhattacharya et al. 2010]. It was found out that chemical locking exhibits the best anti-loosening characteristics among all, followed by nylock and aerotight nut.

12.3.3. Loosening by Axial Vibration

An investigation of the loosening of bolted joints subjected to axial loading and vibration was conducted by Goodier and Sweeney [1945]. Loosening of bolted joints was attributed to radial contraction of the bolt because of tensile loads and simultaneous radial dilation of the nut wall when dynamically loaded in the axial direction with respect to the bolt. Nonlinear normal contact micro-vibrations may be excited by harmonic forces [Grudzinski and Kostek 2007].

The axial motions of either bolt or nut are linearly related to the rotation, as long as they have the same constant pitch or lead angle. If one pitch, either bolt or nut, is designed to follow the rotation around the bolt axis nonlinearly but still operates in the elastic-contact range, it becomes a self-locking fastener due to the vertical interference between the bolt and nut threads. A cubic-functional variation along the bolt axis in one-pitch advancement, i.e., per 360° rotation, was proposed by Ranjan et al. [2013].

12.3.4. Bolted Joints under Transverse Movement

Besides fatigue, self-loosening induced by dynamic transverse vibration is another major cause of failure in bolted joint subjected to dynamic loads. Once an external force in one direction (e.g., transverse) overcomes the force of friction between the two engaged threads, a force smaller than the original static frictional force (e.g., radial repulsive force between threads in contact) can cause movement to occur in other directions.

Once relative motion occurs between the threaded surfaces and/or other contact surfaces of the clamped parts, the bolted connection would get loosened and it seems almost completely free of friction in the contact surface circumferentially. When cyclic transverse slip occurs in a bolted joint, the self-loosening rotation occurs and the bolt preload (axial force) gets weaker and weaker [Eccles 2009], as depicted in Figure 12.3.4 schematically. In the meanwhile, an initial preload will help retard the loosening in life cycles as shown in Figure 12.3.5 until the fastening force is reduced to be less than a critical load [Bickford 1995]. Fasteners may be placed at antinodes of an assembly when subjected to vibrational loadings so tight that the influence of the local exciting shear force is mitigated [Pai and Hess 2003].

FIGURE 12.3.4   Loss of bolt preload when subjected to a relative transverse movement [Eccles 2009].

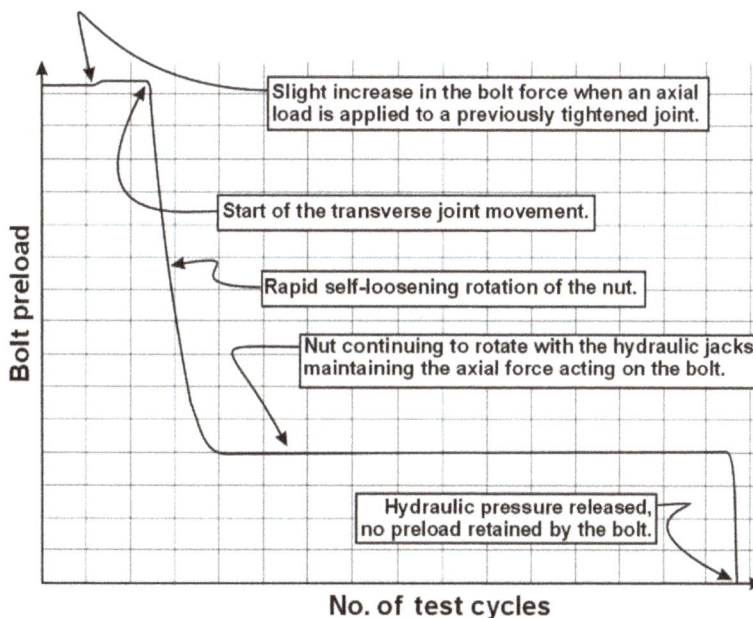

Slight increase in the bolt force when an axial load is applied to a previously tightened joint.

Start of the transverse joint movement.

Rapid self-loosening rotation of the nut.

Nut continuing to rotate with the hydraulic jacks maintaining the axial force acting on the bolt.

Hydraulic pressure released, no preload retained by the bolt.

Bolt preload

No. of test cycles

FIGURE 12.3.5 Influence of preload on the loosening of bolts in life cycles [Jiang et al. 2003].

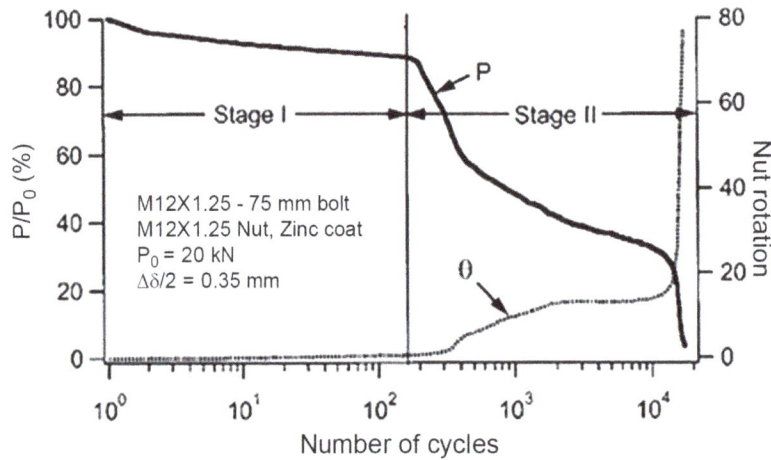

12.3.5. Bending Effect on Bolt Loosening

Repetitive bending of a bolted joint changes the normal contact stresses and results in repeated microslip at the interface and eventually causes fretting wear between contact surfaces. This wear will decrease the axial fastening force predetermined by the preload and reach the critical slip, which is defined as the relative movement between two bearing surfaces in contact. Loosening then occurs because of a significant amount of transverse force. It is recommended to use chemical thread lock, such as liquid epoxy-acrylate chemical adhesive, between the threads in contact to prevent thread movement and the resulting consequent loosening [Haviland 1981]. In the meanwhile, differential thermal expansion arising from temperature differences in dissimilar materials of joints can also cause the joint to slip.

12.3.6. Material Degradation in Bolted Joint

Preload relaxation is also a prevailing failure mode of bolted joints, especially in fastened plastic composites subject to cyclic loadings. The preload degradation mechanisms of bolted joints can be attributed mainly to the following potential failure modes [William 2010]:

(a) Compressive creep of fastened materials

(b) Varied interfacial contact condition because of surface roughness and texture

(c) Fretting fatigue manifested as localized microslip

A complete pretension loss could occur under even relatively small dynamic transverse displacements when thick paint films are used on the clamped surfaces of the bolted joint components. This happens because of the creep-embedded effect that leads to another kind of self-loosening [Satoh et al. 1997].

12.3.7. Bolts at Elevated Temperature

Most fastener materials are temperature-sensitive; that is to say, their properties are influenced by temperature variations. When a significant amount of fastening force is placed on a bolt and it is then exposed to a high temperature, the bolt begins to relieve itself of some of the stress and ultimately reduces the clamping force (i.e., preload). Since the stress and the preload are related, this implies that the clamping force with which the bolt holds the joint together will be significantly reduced.

Differential thermal effect is one of the most problematic issues of bolted joints resulting from a significant temperature rise. Different coefficients of thermal expansion between the fastener and fastened members may fail the joint by the induced thermal stress. Even if the bolt and fastened members are of the same material but heated up at different rates, the differential thermal expansion will also cause a problem to the bolted joint.

Bolted joints get degraded as the material strengths of both bolts and fastened members decrease when the working temperature increases. Coatings may break down, and material corrosion (oxidation) and creep (stress relaxation) may occur at a significantly high temperature. Each kind of bolt material has a temperature limit, above which it is unsafe to use. For example, grade 8.2 bolt (boron steel) is expected to work under 340°C.

12.3.8. Loosening Bolts

Loosening involves more quasi-static resistance, while tightening involves more dynamic resistance. Consequently, the residual torque is often lower than the dynamic torque. The difference is keeping something moving versus getting something to move.

12.3.9. Measuring Clamp Force for Design for Reliability

Torque is not a reliable measure of bolt tension, while the clamp load is the primary health monitoring parameter in a bolted joint. The following methods directly quantify the clamp load developed in a bolted joint: (a) direct measurement of elongation, (b) tension indicator washer, (c) Belleville washer, (d) supersonics, and (d) fiber optics. Some of the internally gaged clamp force sensors (working like load cells) meet the needs for research, testing, measurement, and control applications that require durability, reliability, and accuracy in space, on land, and even underwater.

In-service monitoring of bolt tension can be useful to ensure safe operation of structures subjected to dynamic loading such as remote in-service bolt tension monitoring that highlights the application in wind turbines.

12.3.9.1. Direct Measurement of Elongation

As a general rule, a bolted joint is designed with sufficient fastening to apply the required clamp load at around 70% of the fastener proof load stress, i.e., well below the fastener's yield point. Bolt elongation is directly proportional to axial load when the applied stress still falls within such an elastic range of the material. If both ends of a bolted joint are accessible, a micrometer measurement of bolt length made before and after tightening will ensure the required axial force is applied.

Bolt elongation can be consciously monitored using strain gages bonded inside the bolt as shown in Figure 12.3.6 [Mekid et al. 2019]. For example, the strain gage is installed into a hole of 1.5 mm in diameter, over a length of approximately 20 mm or more, for an M12 bolt of length 50 mm. Further instrumentation, as schematically depicted in Figure 12.3.6, makes the entire tension measurement system available for remote monitoring. In the device presented by Mekid et al. [2019], the required electric energy is stored in a capacitor that can be remotely charged using radio frequency (RF) harvesting board with a transmitter operating at a frequency between 850 and 905 MHz.

With the proposed technology, each critical bolt is equipped with the self-contained, miniature wireless transmitter, which periodically reads and reports bolt tension without the involvement of manpower. Local wireless transceiver is used to collect bolt tension readings from a great number of bolts and transmits the acquired data via Internet or another network means to a secure server. The system allows continuous monitoring of bolt tension and automatic report generation. Furthermore, it can also deploy an alarm notifying stakeholders via email, text message, or sound in case the bolt tension suddenly falls below a predefined functional level.

Temperature compensation with quarter-bridge strain gauge circuit is required for the operation of smart bolts. As schematically shown in Figure 12.3.7, it can be easily seen that the strain output is quite nonlinear, especially when operating at low voltage. It is imperative to maintain the operating voltage in the quasi-linear range, i.e., between 0.75 and 1.5 V.

FIGURE 12.3.6   Remote monitoring using wireless smart bolts [Mekid et al. 2019].

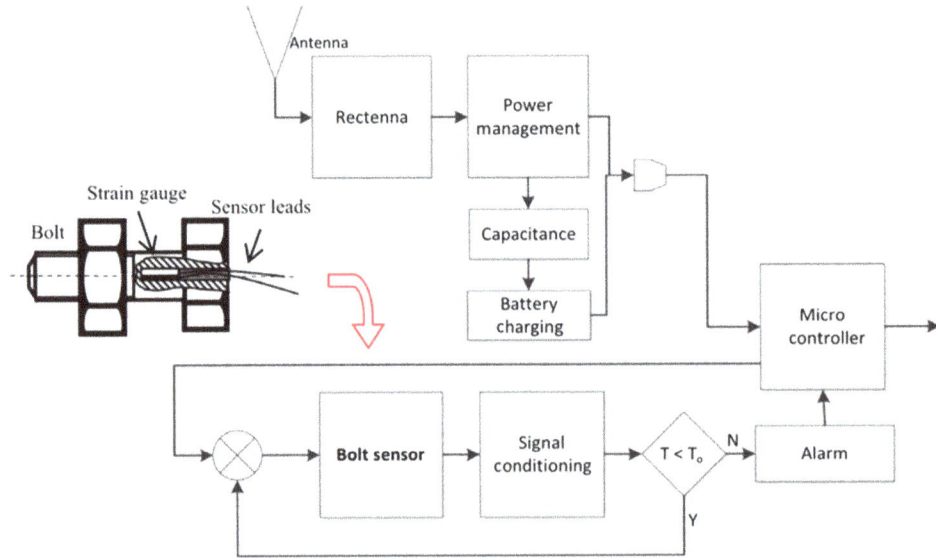

FIGURE 12.3.7   Temperature compensation with quarter-bridge strain gage circuit [Mekid et al. 2019].

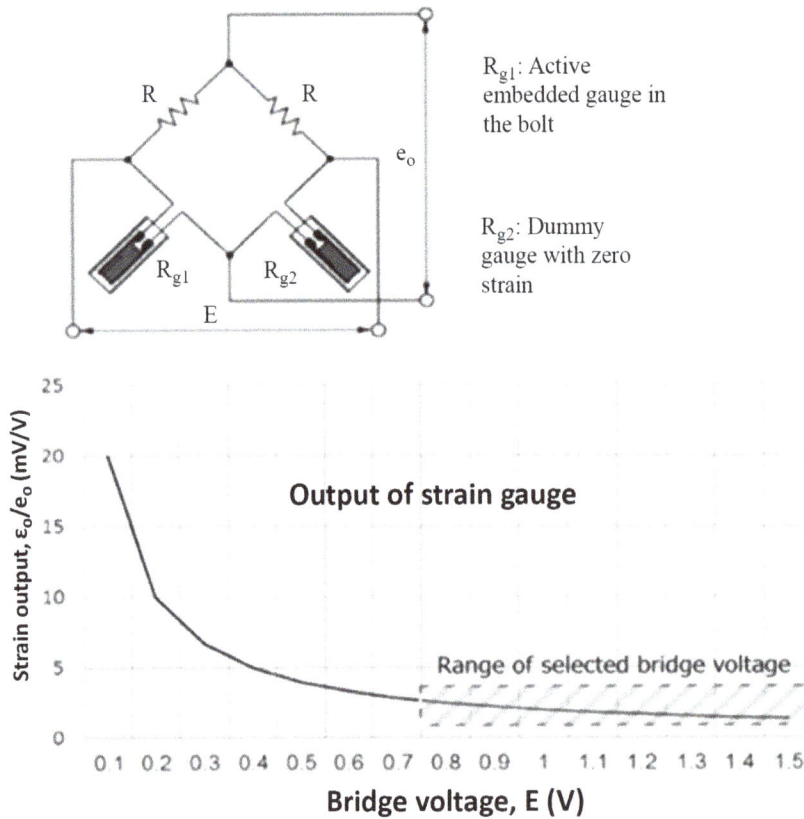

R_{g1}: Active embedded gauge in the bolt

R_{g2}: Dummy gauge with zero strain

12.3.9.2. **Direct Tension Indicators (DTIs)**

A DTI measures the fastening force in a bolt in spite of the amount of torque applied. As shown in Figure 12.3.8, one version of DTIs is a hardened, washer-shaped device with protrusions, i.e., bumps pressed out on one face of the indicator, manufactured according to the provision of ASTM F2437/F2437M-17. It functions like a load cell or transducer while looking like a washer. So far, it is the least expensive and simplest detecting device to measure the bolt tension.

When tightened by turning the nut, the indicator is placed between a fastened member and the bolt (screw) head with its bumps being placed against the bolt (or screw) head. As torquing proceeds, the gap between the bolt and fastened member reduces (mainly due to the elastoplastic deformation of bumps) but is still discernible. The appropriate sensing gage must be refused in a given number of gaps between the bumps to assure a proper tightening [TurnaSure 2004]. Some other ways of using DTIs, such as torquing the bolt instead of nut, can be found in TurnaSure [2004] and ASTM F2437/F2437M-17 [2022].

An innovative way to directly tell the clamp force is to make it visible on the bolt head, as shown in Figure 12.3.8(c), which can be observed comfortably at a certain distance.

FIGURE 12.3.8   DTIs [MaxBolt n.d., TurnaSure 2004].

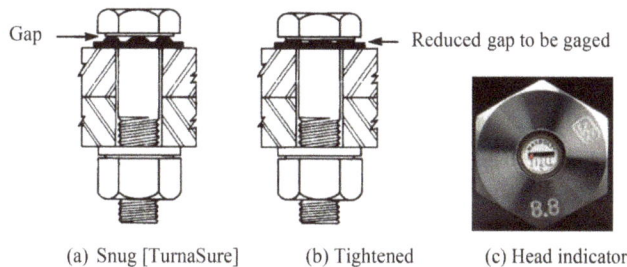

Gap → | ← Reduced gap to be gaged

(a) Snug [TurnaSure]   (b) Tightened   (c) Head indicator

Technical background provided by Wrought Washer Mfg., Inc., featuring TurnaSure® Direct Tension Indicators.

12.3.9.3. **Belleville Washers**

A simple way to measure the tension developed at a bolted joint is to place a stack of Belleville washers between the nut (or head) and its supposedly to-be-contacted fastened member (Figure 12.3.9). Bellevilles are not only used in gasketed joints but also for electrical bus conductors that carry high voltage and current. Using springs to compensate for unloading due to differential thermal expansion in electrical bus conductors is a common application. When the fastener is tightened, the overall height of the clamped Belleville washers decreases in compliance. Belleville washers are designed to deform elastically, so the compressive force can be figured out according to their axial movement and the overall "spring constant" of the washer stackup. The washers can be used singularly or stacked as a means to reach the desired overall spring constant. A Belleville stackup has a higher spring constant when aligned in the same direction as putting individual springs in parallel and vice versa when aligned in an alternating direction as being equivalent to adding springs in series.

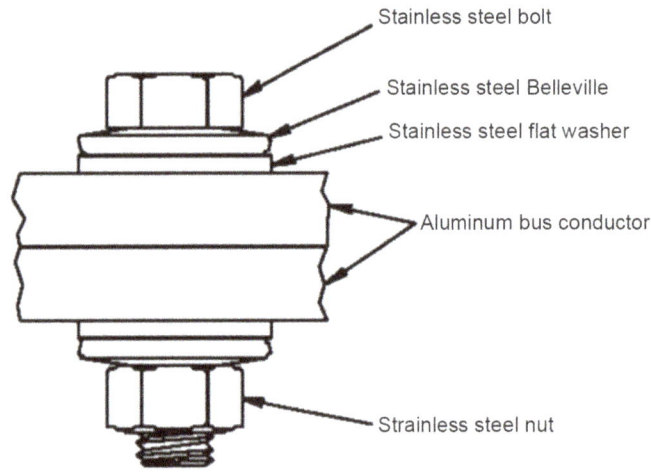

Courtesy of Solon Manufacturing Company.

12.3.9.4. **Ultrasonics**

Ultrasonic testers are the most reliable way to accurately measure bolt tension, including preload and in-service load. They are especially useful for difficult-to-reach bolts. An ultrasonic instrument, either portable or lab-sized, sends a brief burst of ultrasound through a bolt and measures the time required for the sound to echo off the far end and return to the transducer. As a bolt elongates in tension, the time duration required for the ultrasonic signal to complete its travel circuit increases. The phase shift is computed by comparing the phase of the returned signal with that of the original tone burst at the moment when the bolt is tightened. With ongoing research, the following four ways of checking the reliability of bolted joints using ultrasonic instruments are applied:

(a) Multiple Echo: With transducers on both ends of the bolt, having "multiple echo," the measurement can be recalibrated according to the physical properties of the bolt. Pattern matching, keeping the reference echo in memory for subjective comparison with the final echo, is used to improve measurement accuracy. For lab-sized equipment, computer analysis and synthesis, grip length, bolt material, and thread run-out lengths are used to elaborate the change in transmittal time and relate it to bolt tension.

(b) Reusable Sensor: A low-cost, reusable "glue-on" ultrasonic tension sensor called a UTensor (Micro Control, Inc., Bloomfield, MI, US) has been developed and patented that can be used in conjunction with the measuring device to produce reasonable accuracy and repeatability profoundly needed in the automotive industry. This device has been in use at Ford and General Motors (GM).

(c) For Short Bolts: Short bolts are so-called when the ratio of bolt length to bolt diameter is less than 4. Uncontrollable factors such as surface roughness, thread length, and shank length may have significant influences on the measurement uncertainty, as being dominated by the stochastic nature of grip and thread lengths that constitute the effective length of the fastener.

(d) Smart Bolts: Sensors and actuators are embedded in the joint to monitor its health, such as measuring minute dimensional displacements and indicating when bolts are loosened or tightened.

Fiber Optics

Light is fed into the fiber, and the fiber Bragg grating reflects the light with one given wavelength back to an optical interrogator. Once the tension in the optical fiber changes, the reflected wavelength of light correspondingly changes. The change is then detected by the optical interrogator and converted into a tension measurement as calibrated.

Brainy bolts, developed and patented by Cleveland Electric Laboratories (Cleveland, OH, US) as an innovative sensing solution, utilize fiber-optic technology to measure tension in fasteners. A brainy bolt insert may be integrated into virtually any bolt, rivet, stud, or screw that has a bolt diameter of 12.7 mm (0.5″) or more, making it a smart fastener and allowing measurement of tension in any location including those that are remote or difficult to access. It also provides individual temperature compensation capability using the same internal fiber and optical interrogation methodology. The optical interrogator may be located at distances up to 25 km from sensors.

12.3.10. Multivariable Effects on Tightening Bolted Joints

Torque turn to yield means to utilize the bolt to the most by torquing up to the inflection point at the end of the elastic range in accordance with the load–deflection curve. It may put the fastener in jeopardy because of potential stress concentrations in a bolt head fillets, the thread run-out point (where the threads meet the body), and the first thread that engages the nut. However, there has never been defined with regard to the exact location of yielding, whether the torque curve (load–deflection) begins with an inflection point at bolt thread roots, bolt thread flanks, or the bolt head–stem fillet. Classical closed-form solutions, experimental mechanics, analysis charts, and finite element methods have been developed for selecting and designing bolts and nuts for various applications.

Multivariable effects of screw-torquing processes were experimentally examined by Malek et al. [1993]. Applications of bolted joints to plastic composites were investigated by Chiang and Rowlands [1991]. Further applications to wood were completed by Rahman et al. [1991]. Fatigue life predictions due to stress concentrations (e.g., thread roots) were attempted by Lee et al. [1995]. Fracture mechanic approaches to bolted joints were published by Chiang and Rowlands [1991]. Investigation into multibolted joints was conducted by Rowlands et al. [1982]. Durability of self-tapping screws subjected to thermal stresses for plastics was studied by Chiang and Barber [1997].

The objective of the study addressed hereon is to identify the individual main effects and multifactorial interactions on the initiation of turn to yield at the thread root due to such inherent stress concentration. Herein, nonlinear FEAs were employed to determine the load transfer path and magnitude along the bolt, nut, and fastened members. Eight design factors that were thought to have an impact on the stress concentration are prescribed in Table 12.3.1. Modeling is detailed in Chiang et al. [2006]. Parametric finite element meshes for the 3D bolted joint were automatically generated using proprietary software written in C/C++. Models of the bolt, nut, and fastened members were modeled using 8-node isoparametric solid element. Factorial analyses of stress concentration at bolt thread root based on 3D finite element models are then carried out.

TABLE 12.3.1 List of factors and design levels for FEA of bolted joint [Chiang et al. 2006].

	Variables	Level (−)	Level (+)
A	Major diameter (mm)	8 (M8)	10 (M10)
B	Pitch (mm)	1.0 (fine)	1.5 (coarse)
C	Thread angle (deg)	60 (standard)	75
D	Bolt head fillet radius (mm)	0.5	1.0
E	Bolt head thickness (mm)	3.0	3.5
F	Bolt head diameter (mm)	16	20
G	Nut outside diameter	18	20
H	Lateral clearance between threads (mm)	0.075	0.1

Notes:
Bolt material: SAE Class 4.6 (σ_y = 240 MPa, σ_{uts} = 400 MPa, ε_{uts} = 22%)
Bolt material: SAE Class 4.6 (σ_y = 240 MPa, σ_{uts} = 400 MPa, ε_{uts} = 22%)
Fastened members: SAE 1010 HR (σ_y = 180 MPa, σ_{uts} = 320 MPa, ε_{uts} = 25%)

© SAE International.

FIGURE 12.3.10 Validation of FEA results with empirical data [Chiang et al. 2006].

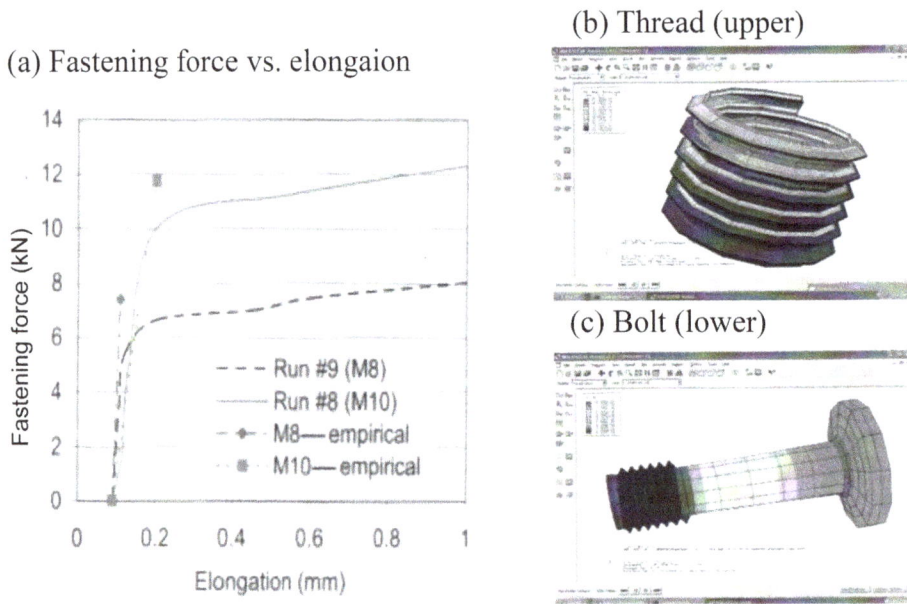

(a) Fastening force vs. elongaion

(b) Thread (upper)

(c) Bolt (lower)

© SAE International.

The validity of the FEA is verified in Figure 12.3.10, and a typical severity comparison of stress levels at bolt thread root, bolt thread flank, and bolt head fillet is given in Figure 12.3.11. The stress behaviors at the three locations developed in the bolt described here are based on run 9 condition (Table 12.3.1). The maximum von Mises stress shows up at the bolt head fillet when the bolt-fastening load is low. As the bolt-fastening force increases, the stress levels at the thread flank and thread root surpass that at the head fillet and eventually the stress level at the thread root becomes the final cause of failure. Effects of individual factors and multifactorial interactions resulting from design contrasts because of parametric variations based on the 2_{IV}^{8-4} design matrix are listed in Table 12.3.2.

FIGURE 12.3.11 Stresses developed at bolt head fillet, thread flank, and thread root (run 9) as functions of fastening force [Chiang et al. 2006].

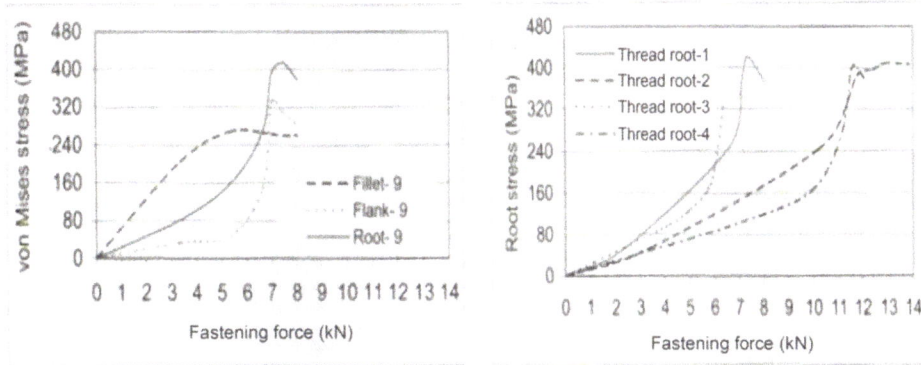

TABLE 12.3.2 Folding 2_{III}^{7-4} over on factor H to obtain 2_{IV}^{8-4} in two blocks (runs 1–8 and runs 9–16) [Chiang et al. 2006].

Run	A	B	C	D	E	F	G	H	Y_{root}	Y_{flank}
1	−1	−1	−1	1	1	1	−1	1	6.68	7.09
2	1	−1	−1	−1	−1	1	1	1	10.1	12.21
3	−1	1	−1	−1	1	−1	1	1	6.0	5.96
4	1	1	−1	1	−1	−1	−1	1	11.0	10.96
5	−1	−1	1	1	−1	−1	1	1	7.26	7.0
6	1	−1	1	−1	1	−1	−1	1	9.84	12.36
7	−1	1	1	−1	−1	1	−1	1	7.0	5.84
8	1	1	1	1	1	1	1	1	11.8	10.98
9	1	1	1	−1	−1	−1	1	−1	6.27	6.88
10	−1	1	1	1	1	−1	−1	−1	10.0	11.68
11	1	−1	1	1	−1	1	−1	−1	6.15	5.82
12	−1	−1	1	−1	1	1	1	−1	10.7	10.92
13	1	1	−1	−1	1	1	−1	−1	7.0	6.8
14	−1	1	−1	1	−1	1	1	−1	10.3	12.4
15	1	−1	−1	1	1	−1	1	−1	6.85	5.75
16	−1	−1	−1	−1	−1	−1	−1	−1	11.3	10.9

Notes:
Y_{root} (kN): Fastening force required to yield the bolt thread root
Y_{flank} (kN): Fastening force required to yield the bolt thread flank

Ambiguities about two-factor interactions resolve the experimenter to take the sense of engineering context and to judge relative magnitudes of individual main effects. When measured by the criterion of statistical significance level of 10%, i.e., t-statistic tail value (or $t_{v,90\%}$), the predictive equations for the yielding resistance of the bolt thread root and for the endurance of the bolt thread flank can be formulated, respectively, as follows:

$$Y_{root,p} = 7.04875 + 4.00375\,a + 0.44375\,b + 0.58125\,c + 0.20375\,d$$

$$+ 0.69625\,a\,b - 0.22125\,a\,c. \qquad (12.3.1)$$

$$\text{and} \quad Y_{\text{flank,p}} = 9.00625 + 5.2275\ a - 0.175\ g - 1.23\ b$$

$$- 0.22125\ a\ c - 0.18\ a\ g. \tag{12.3.2}$$

(a) Stress Concentration at Thread Root: Four factors have significant contributions as main effects on the stress concentration at the thread root, i.e., $Y_{\text{root,p}}$, and they are factors A, B, C, and D. One particular two-factor interaction, i.e., η_{AB}, is proven to have a strong impact on the stress concentration at thread root. The effectiveness of this specific two-factor interaction, resulting from alias equation $\eta_{AB} = A\ B + C\ G + D\ H + E\ F$ based on Equation (3.4.10), comes from interaction AB, i.e., the interaction between the major diameter (factor A) and pitch (factor B). This is due to the fact that factors A and B have great influences individually, while the influences of factors G, H, E, and F are not significant. By the same token, the minor contribution to alias equation $\eta_{AC} = A\ C + B\ G + D\ F + E\ H$ is mainly because of interaction AC.

(b) Stress Concentration at Thread Flank: Only two individual main effects have significant influences on the bolt thread flank, and they are factors A (major diameter) and B (pitch). Their two-factor interaction also appears to be influential on the stress level of the bolt thread flank.

TABLE 12.3.3 Significance levels of effects on yielding resistance of the bolt thread root.

Variable	Effect (kN)	Standard error (t_{11})	t-Ratio	Significance level
Mean	7.04875	0.05788	131	<0.1%
A	4.00375	0.115759	34.6	<0.1%
B	0.44375	0.115759	3.83	<0.1%
C	0.58125	0.115759	5.02	<0.1%
D	0.203765	0.115759	1.76	≈6%
E	−0.08875*	—	—	—
F	0.12625*	—	—	—
G	0.14625*	—	—	—
H	0.11375*	—	—	—
AB	0.69625	0.115759	6.01	<0.1%
AC	−0.22125	0.115759	1.91	<5%
AD	0.08625*	—	—	—
AE	−0.00125*	—	—	—
AF	0.06375*	—	—	—
AG	0.19375*	—	1.674	≈7%
AH	−0.00375*	—	—	—

*: Small effects, assumed to be insignificant.

© SAE International.

In light of the two mean values in Tables 12.3.3 and 12.3.4, one can tell that the yielding endurance of the bolt thread flank is superior to the yielding resistance of the bolt thread root. However, this may vary according to the given treatment, e.g., runs 1 to 16 in Table 12.3.2.

TABLE 12.3.4 Significance levels of effects on yielding endurance of the bolt thread flank.

Variable	Effect (kN)	Standard error (t_{11})	t-Ratio	Significance level
Mean	9.00625	0.0659	136.7	<0.1%
A	5.2275	0.1318	39.7	<0.1%
B	−1.2300	0.1318	9.33	<0.1%
C	0.58125	—	—	—
D	0.203765	—	—	—
E	−0.08875*	—	—	—
F	0.12625*	—	—	—
G	0.14625*	0.1318	—	—
H	0.11375*	—	—	—
AB	0.69625	—	6.01	<0.1%
AC	−0.22125	0.1318	1.328	≈11%
AD	0.08625*	—	—	—
AE	−0.00125*	—	—	—
AF	0.06375*	—	—	—
AG	0.19375*	0.1318	1.366	≈10%
AH	−0.00375*	—	—	—

*: Small effects, assumed to be insignificant.

12.4. **Screwed Joints**

Screws are easy to install and often used to hold a to-be-clamped part to a base (Figure 12.4.1). Special screws and tightening processes are used in automotive mountings with heavy mechanical and vibration resistance such as screws with hexagonal or flanged heads for axle and suspension assemblies. Thread-rolling screws have gained momentum in many automotive applications because of their low field costs and structural robustness.

   With self-aligning and self-tapping capabilities, self-tapping screws are used in instrument panels. When tapping into unthreaded base parts or nut, screws are able to do work hardening at the mating interface and thus offer a torque-prevailing feature with improved vibration resistance. The torque–angle method is a cost-effective method to determine the right installation torque of screwed joints [Vangipuram et al. 2007].

FIGURE 12.4.1 Schematic drawing of a screwed joint.

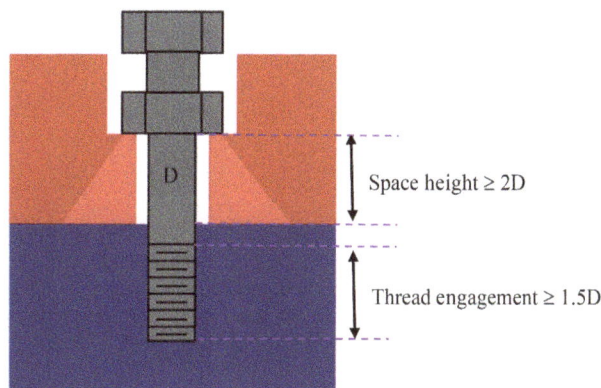

Space height ≥ 2D

Thread engagement ≥ 1.5D

12.4.1. Screw-Tightening Process

The use of automated screw-torquing devices in manufacturing assembly has become a standard practice in various industries, especially for automotive interior systems [Chiang and Barber 1997]. Laboratory instrumentation for an automatic screw-torquing process using small screws (mainly for fastening automotive electronic parts) was successfully set up and tested to examine the factorial trending of a screw-torquing process. Three different sensors were employed to measure the force curve, torque curve, and rotating speed of screwdrivers, respectively. A programmable horizontal servo slide was taught to do the torquing job.

DOE was then set up to identify the significant factors that affect the final torquing process. In the case of closed-loop control, knowing the variation that the system can induce will help to determine the compensation range that the controller has to get consistent final torque levels. Control factors and their design levels selected for this study are listed in Table 12.4.1. Note that the vacuum pressure bushing, i.e., factor H, is installed at the end of the electric driver to suck the screw and align it such that the screw axis is "normally" coincident with the central axis of the hole in the workpiece. The trace of an external manual torque measuring device is applied in order to validate the accuracy of automated screw-torque process.

TABLE 12.4.1 Factors and design levels for experimental tests on screw-tightening process [Malek et al. 1993].

Variables	Level (−)	Level (+)
A. Screw (stainless): size	M2	M3
B. Workpiece: material	Acetal homopolymer	Stainless steel
C. Screw (stainless): head geometry	Button head	Flat head
D. Machine: rotating speed	600 rpm	910 rpm
E. Machine: servo-sliding speed	10 mm/s	20 mm/s
F. Machine: vacuum pressure	10 mm Hg	14 mm Hg
G. Workpiece: tipping angle	1°	3°
H. Workpiece: eccentricity	0.25 mm	0.5 mm

12.4.2. Self-Tapping Screwed Joints

Effects of mechanical and thermal loads on metallic screw-fastened plastic joints in automotive interior trims (plastics) were analyzed using nonlinear finite element methods and statistical DOE [Chiang and Barber 1997]. Temperature-dependent constitutive models were considered to investigate the potential failure of plastic bosses, being fastened by self-tapping double-threaded steel screws with different thread heights. Both geometric and material nonlinearities are considered. A high compressive stress at the inside diameter of the molded hole in the plastic boss can be produced after the tapping and fastening process is completed, when the hole size is big. The compressive stress may result in the loss of fastening force, when the screwed joint is subjected to an elevated temperature.

The fastening force increases from zero to the target value, i.e., factor A, ranging from 502.4 to 628 N, as it is threaded into the plastic boss at room temperature (23°C). Then, the temperature increases from 23 to 98°C. The failure function is defined as the ratio of the induced von Mises effective stress (σ_e, i.e., column Y in Table 12.4.3) calculated from FEA to the yield strength of the plastic boss (σ_y). At 83°C, σ_e/σ_y = 39.9 MPa/39 MPa = 102.3%. The result coincides well with most laboratory observations that the fastening force disappears at 83°C.

Fractional factorial design 2_V^{5-1}, having alias equation E = ABCD, is set up to explore the influences of the five factors of interest, as listed in Table 12.4.2. The design matrix is exhibited in Table 12.4.3. Calculations for main effects and multiple-factor interactions are performed according to the contrast coefficients. The significance levels for main effects and multiple-factor interactions are obtained by checking the t-distribution, of which the error is estimated from pooling the six small interactions together as

$$\text{Error} = [(\eta_{AB}^2 + \eta_{AC}^2 + \eta_{BC}^2 + \eta_{CE}^2 + \eta_{BE}^2 + \eta_{AE}^2)/6]^{1/2}$$

$$= \{[(0.062)^2 + (0.188)^2 + (0.138)^2 + (-0.012)^2 + (0.025)^2 + (0.2)^2]/6\}^{1/2}$$

$$= 0.128.$$

With a parting significance level of 5%, the predictive model is

$$Y_p = 35.775 + 2.000\,a - 0.625\,b + 1.750\,c + 0.963\,d - 0.250\,e$$

$$- 0.725\,a\,d - 0.388\,a\,b\,c - 1.175\,a\,b\,d + 0.313\,a\,c\,d. \tag{12.4.1}$$

It is assumed that interactions between factor E and other factors in the above equation are insignificant, since the main effect of factor E is small relative to other main effects. The predictive values are listed in column Y_p.

Both fastening force (factor A) and screw pitch (factor C) have great impacts on the stress level of the self-tapping screwed joint according to Equation (12.4.1). Let B = 60° (b = 1), D = 1.545 mm (d = −1), and E = 6 mm (e = 1), and Equation (12.4.1) reduces to be

$$Y_p = 33.935 + 3.9\,a + 1.75\,c - 0.701\,a\,c. \tag{12.4.2}$$

The response surface plot of von Mises stress versus fastening force and screw pitch is depicted in Figure 12.4.2.

In summary, a robust design to reduce the stress level can be achieved by combining the following design parameters: large thread angle, short pitch, small inside diameter of the boss hole, and large outside diameter of the boss. Consider the maximum fastening force of 628 N as the applied wrench torque can be controlled up to. The highest von Mises effective stress in the plastic boss can be reduced from 39.9 MPa (with failure function σ_e/σ_y = 39.9 MPa/39 MPa = 102.3% at working temperature 83°C) corresponding to the preliminary design (original design configuration in the existing product) to 33.8 MPa (with failure function σ_e/σ_y = 33.8 MPa/39 MPa = 86.8% at working temperature 83°C) for the suggested new design that B = 60°, C = 3.18 mm, D = 1.545 mm, and E = 6 mm.

TABLE 12.4.2 Factors and design levels for self-tapping screwed joints [Chiang and Barber 1997].

Variables	(−1)	(+1)	Coded variable
A. Fastening force (N)	502.4	628	a = (A − 565.2)/62.8
B. Thread angle of screw (deg)	50	60	b = (B − 55)/5
C. Pitch of screw (mm)	3.18	3.887	c = (C − 3.5335)/0.3535
D. Inside radius of molded hole (mm)	1.545	1.84	d = (D − 1.6925)/0.1475
E. Outside radius of boss (mm)	5.5	6	e = (E − 5.75)/0.25

TABLE 12.4.3 Fractional factorial design 2_V^{5-1} for self-tapping screwed joints [Chiang and Barber 1997].

(I) Main effects							
Run	a	b	c	d	e = abcd	Y (MPa)	Y_p
1	−1	−1	−1	−1	1	30.5	29.6
2	1	−1	−1	−1	−1	36.2	35.4
3	−1	1	−1	−1	−1	30.9	29.5
4	1	1	−1	−1	1	33.8	33.0
5	−1	−1	1	−1	−1	36.4	35.2
6	1	−1	1	−1	1	42.6	41.5
7	−1	1	1	−1	1	34.4	34.1
8	1	1	1	−1	−1	41.4	40.2
9	−1	−1	−1	1	−1	36.8	35.7
10	1	−1	−1	1	1	38.9	37.9
11	−1	1	−1	1	1	34.5	33.3
12	1	1	−1	1	−1	38.3	37.8
13	−1	−1	1	1	1	38.0	37.1
14	1	−1	1	1	−1	39.5	38.8
15	−1	1	1	1	−1	36.4	35.7
16	1	1	1	1	1	39.2	37.6
Contrast	4.000	−1.250	3.500	1.925	−0.500		
Effect	2.000	−0.625	1.750	0.963	−0.250	35.775 (average)	
Error	0.128	0.128	0.128	0.128	0.128		
t_6	15.59	4.872	13.64	7.503	1.949		
$t_{6,5\%}$	1.943	1.943	1.943	1.943	1.943		
Significant?	Yes	Yes	Yes	Yes	Yes		

(II) Two-factor and three-factor interactions									
Run	ab	ac	ad	bc	bd	cd	abc	abd	acd	bcd
1	1	1	1	1	1	1	−1	−1	−1	−1
2	−1	−1	−1	1	1	1	1	1	1	−1
3	−1	1	1	−1	−1	1	1	1	−1	1
4	1	−1	−1	−1	−1	1	−1	−1	1	1
5	1	−1	1	−1	1	−1	1	−1	1	1
6	−1	1	−1	−1	1	−1	−1	1	−1	1
7	−1	−1	1	1	−1	−1	−1	1	1	−1
8	1	1	−1	1	−1	−1	1	−1	−1	−1
9	1	1	−1	1	−1	−1	−1	1	1	1
10	−1	−1	1	1	−1	−1	1	−1	−1	1
11	−1	1	−1	−1	1	−1	1	−1	1	−1
12	−1	−1	1	−1	1	−1	−1	1	−1	−1
13	1	−1	−1	−1	−1	1	1	1	−1	−1
14	−1	1	1	−1	−1	1	−1	−1	1	−1
15	−1	−1	−1	1	1	1	−1	−1	−1	1
16	1	1	1	1	1	1	1	1	1	1
Contrast	0.125	0.375	−1.450	0.275	−0.025	0.050	−0.775	−2.350	0.625	0.400
Effect	0.062	0.188	−0.725	0.138	−0.012	0.025	−0.388	−1.175	0.313	0.200
Error	0.128	0.128	0.128	0.128	0.128	0.128	0.128	0.128	0.128	0.128
t_6	0.487	1.462	5.651	1.072	0.097	0.195	3.020	9.159	2.436	1.559
$t_{6,5\%}$	1.943	1.943	1.943	1.943	1.943	1.943	1.943	1.943	1.943	1.943
Significant?	No	No	Yes	No	No	No	Yes	Yes	Yes	No

where
Y (MPa): von Mises effective stress (σ_e) obtained from FEA
Y_p (MPa): von Mises effective stress (σ_e) predicted by Equation (12.4.1)

FIGURE 12.4.2   Response surface plot of von Mises stress versus fastening force and screw pitch.

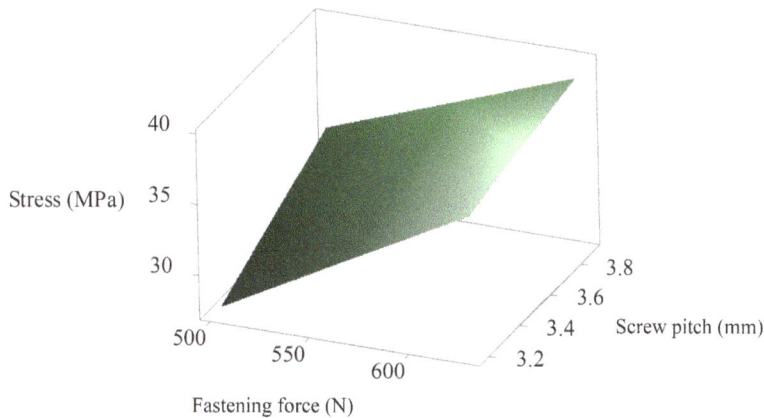

12.5. **Riveted Joints**

Riveted joints are commonly used in aircrafts, on-ground vehicles, buildings, and bridges, to permanently hold thin metallic parts together. As rivets are applied at a lap joint, loads are transferred by rivet shear and through friction between the fastened members. Rivets can be completely solid for the concern of product reliability and safety when applied to on-ground vehicles and aircrafts or designed with a hollow core to reduce the amount of riveting force for light-duty applications.

A rivet shall plastically deform and fill the applied hole and form another head to permanently assemble parts together. The uneven load transfer through in-plane bearing load, clamping friction, and flexible bending of riveted joint leads to a complex 3D stress distribution around the joint [Silva et al. 2000]. There are mainly two kinds of solid rivets used for both on-ground vehicles and aircrafts:

1. Bulky Rivet: Widely used in on-ground vehicles to fasten structural parts. A bulky rivet may have a spherically round head or just a flat head.

2. Slug Rivet: Widely used for assembling aircraft parts because of the need of higher interference fit and long fatigue life expectancy. Rivets applied in aircraft fuselage are shown in Figure 12.5.2.

Advantages of riveted joints are low-profile heads and the low-cost assembly where large numbers of fasteners are required. Rivets are available in high-strength steels and aerospace alloys, and their solid shaft is able to transmit the greatest possible shear force for a given hole size. Furthermore, the resistance to vibration is technically sound.

FIGURE 12.5.2   Rivets applied in aircraft fuselage [Skorupa and Skorupa 2012].

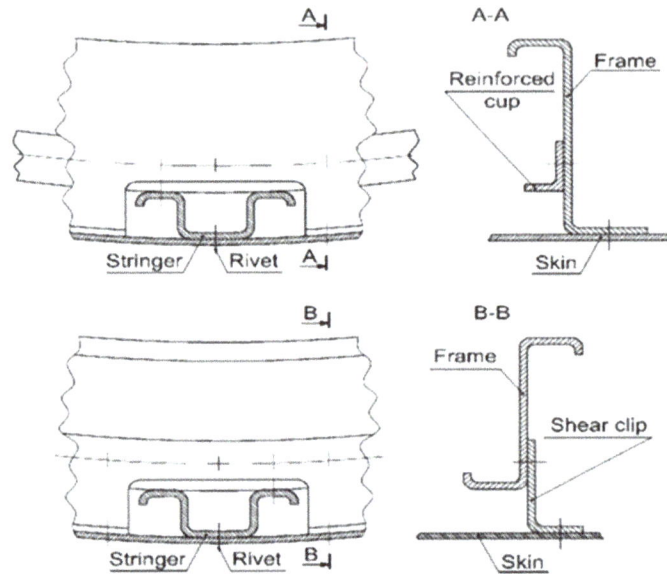

The example application of DOE for examining the reliability of riveted joints addressed hereon is based on slug rivets that are commonly used in aircraft fuselage.

12.5.1. Quality Metrics as Installed

Mating between rivets and fastened parts at installation has been of great concern for the long-term performance. According to Aircraft Standard Handbook (1991), quality metrics for the appearance of riveted joints as installed in light of Figure 12.5.3 can be assessed using the following four measures:

1. D: Diameter of the rivet head as formed, e.g., $0.171875'' \leq D \leq 0.21875''$ for $1/8''$ rivets
2. H: Height of the rivet head as formed, e.g., $0.046875'' \leq H \leq 0.078125''$ for $1/8''$ rivets
3. Gap/clearance: Zero, because a gap between the sheet and rivet leads to a loose rivet
4. Flush: The less the better, e.g., less than $0.01''$ protrusion for $1/8''$ rivets

Sharp corners beneath the head may cause the head to crack. Tearing between the rivet holes, shearing, and crushing between the rivet and the fastened members are considered to be the major in-pull failure modes in the joint.

FIGURE 12.5.3   Quality assessment of the appearance of slug-riveted joints [Bajracharya 2006, Chen et al. 2011].

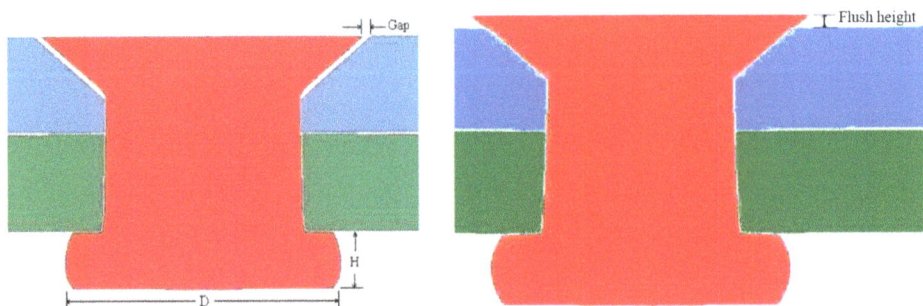

Reprinted with permission. © Springer.

What would be the relationship between the squeeze force and the head diameter as formed? Assume that the rivet head is shaped into cylindrical disk ideally and the volume conservation applies

$$H \, (\pi \, D^2 \, / \, 4) = H_o \, (\pi \, D_o^2 \, / \, 4). \tag{12.5.1}$$

$$\text{Thus,} \quad H \, / \, H_o = (D_o \, / \, D)^2. \tag{12.5.2}$$

An incremental change in the gage length with respect to its length is defined as the strain increment:

$$d\varepsilon = dz \, / \, z. \tag{12.5.3}$$

An integration of the strain increment given above from the original length (H_o) to the current length (H) yields the total strain as

$$\varepsilon = \int d\varepsilon = \int_{H_o}^{H} \left(\frac{dz}{z} \right) = \ln(H \, / \, H_o). \tag{12.5.4}$$

Therefore, the nominal true strain along the vertical axis (z-axis) can be approximated by the above equation. Next, consider the constitutive equation for an elastoplastic material under quasi-static loading [Szolwinski and Farris 2000]:

$$+ \qquad \sigma_{zz} \approx K \, (\varepsilon_{zz})^n, \tag{12.5.5}$$

where
K is the coefficient, e.g., K = 552 MPa for Al 2117(T4) and K = 730 MPa for Al 2024(T3)
n is the hardening exponent, e.g., n = 0.15 for Al 2117(T4) and n = 0.1571 for Al 2024(T3)

The nominal squeeze force, namely F_z, is then

$$F_z \equiv (\pi \, D^2 \, / \, 4) \, \sigma_{zz} = (\pi \, D^2 \, / \, 4) \, K \, (\varepsilon_{zz})^n. \tag{12.5.6}$$

Plugging Equation (12.5.4) into Equation (12.5.6) yields

$$F_z = (\pi \, D^2 \, / \, 4) \, K \, [\ln(H \, / \, H_o)]^n. \tag{12.5.7}$$

Plugging Equation (12.5.2) into the above equation yields

$$F_z = (\pi \, D^2 \, / \, 4) \, K \, \{\ln[(D_o \, / \, D)^2]\}^n. \tag{12.5.8}$$

Equations (12.5.7) and (12.5.8) relate the applied squeeze force (F_z) to the head height (H) and head diameter (D), respectively, as formed under an ideal uniaxial loading condition without any material being pushed into the hole in the fastened member. These two equations can be used as the first approximation to the exploration of quality metrics before applying the FEA and conducting physical tests [Chen et al. 2011].

Subject to the assumption of constant volume, the strain invariant persists as

$$\varepsilon_{xx} + \varepsilon_{yy} + \varepsilon_{zz} = 0. \tag{12.5.9}$$

Since no lateral force in the forming operation is applied, the three strain components are related to each other by

$$\varepsilon_{xx} = \varepsilon_{yy} = -\nu\,\varepsilon_{zz}. \tag{12.5.10}$$

Substitution of the above equation into Equation (12.5.9) yields

$$\nu = 0.5. \tag{12.5.11}$$

This means that the material is in fully plastic deformation. Thus, Poisson's ratio of the rivet (aluminum alloy) goes from 0.33 in the early elastic deformation to 0.5 in the fully plastic deformation range. Since the elastic strain is much smaller than the fully plastic strain, it is reasonable to assume that $\nu = 0.5$ prevails in the forming process.

When the squeeze force is applied rapidly and thus the working temperature is elevated significantly, instead of using Equation (12.5.5), one may use Johnson–Cook constitutive equation [Johnson and Cook 1983], in which strain, strain rate, and temperature are taken into consideration. Johnson–Cook constitutive equation for the nominal flow stress (σ) is portrayed in terms of five material parameters, namely A, B, C, n, and m, as

$$\sigma = [A + B\,(\varepsilon_{eq}{}^p)^n]\left\{1 + C\,\ln\left[\frac{d\varepsilon_{eq}{}^p/dt}{(d\varepsilon_{eq}{}^p/dt)_0}\right]\right\}\left[1 - \left(\frac{T - T_{room}}{T_m - T_{room}}\right)^m\right], \tag{12.5.12}$$

where

T_{room} is the room temperature

T_m is the melting point (temperature) of the material

A is the yield strength at a strain rate of 1 S^{-1} and room temperature (T_{room})

B is the strain hardening coefficient at a strain rate of 1 S^{-1} and room temperature (T_{room})

n is the strain hardening exponent at T_{room}

C is the strain rate hardening coefficient at T_{room}

m is the strain softening exponent with respect to temperature variation

$\varepsilon_{eq}{}^p$ is the equivalent plastic strain, e.g., von Mises equivalent plastic strain

$d\varepsilon_{eq}{}^p/dt$ is the equivalent plastic strain rate, e.g., von Mises equivalent plastic strain rate

$(d\varepsilon_{eq}{}^p/dt)_0$ is the reference equivalent plastic strain rate, usually at 1.0 S^{-1}

$(d\varepsilon_{eq}{}^p/dt)/(d\varepsilon_{eq}{}^p/dt)_0$ is the normalized equivalent plastic strain rate

12.5.2. Structural Performance of Riveted Joints

The inhomogeneity of riveting interference in the thickness direction of the aircraft panels may lead to an inevitable undesired deformation. This phenomenon may significantly degrade the dimensional accuracy of the final product. Luckily, the load distribution in each contact zone as installed reveals the initial fastening quality regarding whether it would lead to any of the potential failure modes. At a high riveting force, there is a severe compressive contact between the rivet and fastened members such that both the rivet and the hole in contact will endure potential plastic deformations. The tangential hoop stresses exerted onto the buck tail, and its surrounding material are usually lower than those at the rivet head. The objective function for studying riveted joints is then generally divided into the following three categories:

1. Potential fretting damage because of vibration [Harish and Farris 1998]
2. Residual stress in the rivet [Li and Shi 2004]
3. Fatigue life (onset of a crack) of the joint [Szolwinski and Farris 2000, Skorupa et al. 2015]

Residual stress is due to highly plastically deformed parts, and fretting is induced to related contact edge effects that act as a stress concentration and even stress-singular sites. In general, the squeeze force exerted onto the rivet is expected to fill up the hole properly and to enlarge the rivet head for increasing the clamping force between the two sheets as fastened. Two major load transfer mechanisms are thus created by the exerted squeeze operation:

(a) Axial Clamping: When a rivet fills the hole, a uniform bearing pressure distribution is anticipated to exist along the hole axis. This is beneficial to enhancing load transfer by the contact friction between the fastened sheets when they are pushed together against each other as an integral joint [Muller 1995].

(b) Radial Constraint: Once the squeeze force increases to a certain extent, the compressive residual stress at the radial interference between the rivet and fastened members grows accordingly. The expansion of the rivet against the hole wall induces a compressive hoop stress that counteracts the crack opening stress in the rivet and consequently retards the potential crack growth [Szolwinski and Farris 2000].

Finite element methods have been proven to be one of the most effective tools for evaluating the structural performance of a riveted joint. The 3D FEA model shown in Figure 12.5.4, based on 8-noded solid elements, is a typical approach.

FIGURE 12.5.4   FEA model of a riveted joint [Bajracharya 2006].

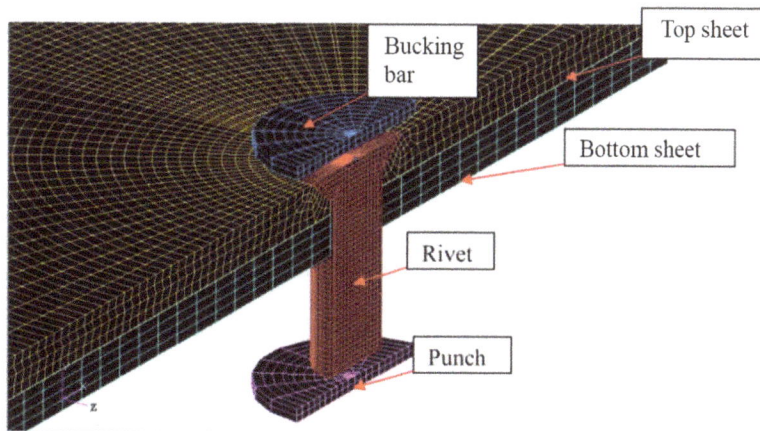

12.5.3. Design and Operating Factors

Riveting is to place a rivet in a hole, apply pressure to and deform the tenon to create the head, and thus make a permanent mechanical joint. Potential design and operating factors for enhancing the structural reliability of riveted joints applied in an aircraft fuselage are summarized in Table 12.5.1.

TABLE 12.5.1 Potential design and operating parameters of 1/8″ rivets as applied in the aircraft fuselage [Bajracharya 2006].

Factors (nominal values)	Level (−)	Level (0)	Level (+)
Rivets			
Rivet length (mm; 0.25–0.32″)	6.35	7.239	8.128
Rivet diameter (mm; 0.125″)	3.0988	3.175	3.2512
Rivet countersink angle (deg; 100°)	—	100°	—
Rivet countersink depth (mm; 0.042″)	—	1.0668	—
Rivet material	—	Al 2117 (T3)	—
Fastened members			
Sheet thickness (mm; 0.064″)	—	1.6256	—
Sheet hole diameter (mm; 0.1285″)	—	3.2639	—
Sheet countersink angle (deg)	—	100°	—
Sheet countersink depth (mm; 0.032″)	—	0.8128	—
Sheet material	—	Al 2024 (T3)	—
Buck cavity	Flat	—	Dome
Forcing functions			
Squeeze force (kN)	6.672	—	13.345
Squeeze speed (mm/s)	—	—	—
Pressing dwell time after squeeze (sec)	—	—	—
Friction between fastened members	—	0.7	—
Friction between rivet and fastened members	—	0.7	—

Note: Level (0)—nominal values

© SAE International.

12.5.4. Friction Riveting

Friction riveting, so-called FricRiveting, is an innovative riveting technique for joining polymeric structure using metallic rivets, developed and patented by the Helmholtz-Zentrum Geesthacht in Germany [Amancio-Filho et al. 2009]. A schematic presentation of friction-riveting process is given in Figure 12.5.5, which shows how to install an aluminum rivet into a thin polycarbonate (PC) part. Joining is achieved by mechanical interference and adhesion between a metallic rivet and polymeric joining partners. It is based on the principles of riveting and friction welding [Bajracharya 2006]. The joining mechanism is activated by the rotation of the metallic rivet that would energize frictional heat in the contact area.

FIGURE 12.5.5   Friction-riveting process—inserting an aluminum rivet into PC part [Blaga et al. 2015].

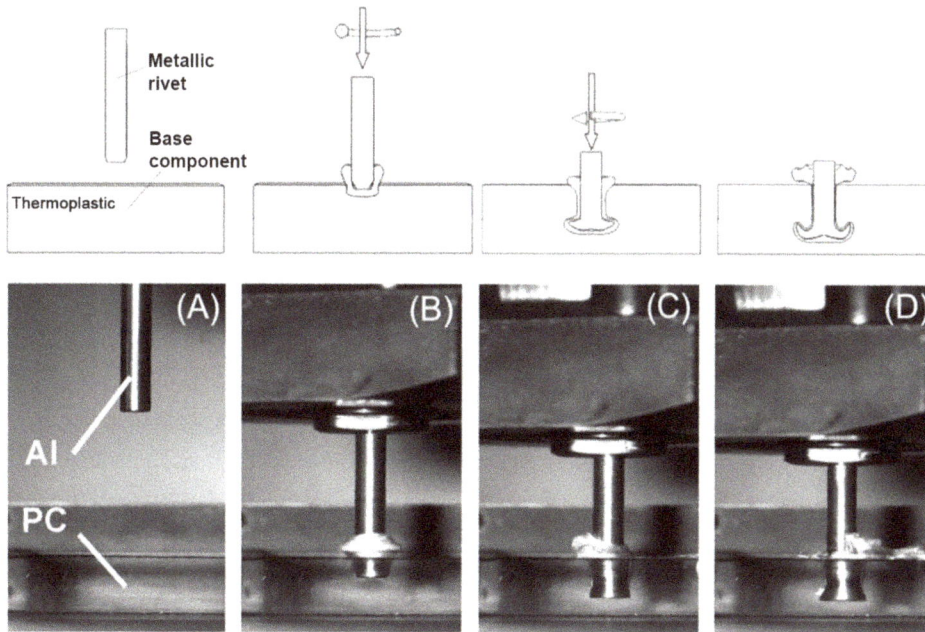

12.5.5. DOE for Riveting

How to relate the design and operating parameters to an objective function, in the category of either quality metrics or structural performance, of a riveted joint can be fulfilled using DOE. The DOE can be performed in combination with FEA and physical tests. Four factors listed in Table 12.5.2 are considered by Bajracharya [2006] for investigating their impacts on the quality metrics of installing the typical 1/8″ rivet used in aircraft fuselage. The design matrix based on full factorial design 2^4 and the corresponding data obtained from elastoplastic FEA are given in Table 12.5.3, including the diameter of formed head, height of formed head, and flush. Note that a net flush means to yield a zero gap.

TABLE 12.5.2 Design and operating factors of 1/8″ rivets applied in DOE [Bajracharya 2006].

Factors	Level (−)	Level (0)	Level (+)
(A) Sheet hole diameter (mm)	3.2639	3.4671	4.0259
(C) Rivet diameter (mm)	3.0988	—	3.2512
(B) Rivet length (mm)	6.35	—	8.128
(D) Pressing force (kN)	6.672	—	13.345

TABLE 12.5.3 2^4 factorial design for evaluating the appearance quality of riveted joints [Bajracharya 2006].

Run	A	B	C	D	Y_d	Y_h	Y_f	Y_g
1	−1	−1	−1	−1	3.785	1.836	0.190	0.0
2	1	−1	−1	−1	3.744	0.963	0.0	0.0
3	−1	1	−1	−1	4.122	1.814	0.254	0.0
4	1	1	−1	−1	4.021	1.151	0.025	0.0
5	−1	−1	1	−1	4.369	2.423	0.19	0.0
6	1	−1	1	−1	4.138	1.824	0.008	0.0
7	−1	1	1	−1	4.397	2.852	0.254	0.0
8	1	1	1	−1	4.054	2.431	0.038	0.0
9	−1	−1	−1	1	4.956	0.950	0.163	0.0
10	1	−1	−1	1	4.140	0.439	0.020	0.0
11	−1	1	−1	1	5.255	1.062	0.216	0.0
12	1	1	−1	1	4.684	0.673	0.030	0.0
13	−1	−1	1	1	5.545	1.410	0.168	0.0
14	1	−1	1	1	5.070	1.085	0.051	0.0
15	−1	1	1	1	5.682	1.572	0.206	0.0
16	1	1	1	1	5.319	1.285	0.028	0.0

where
Y_d (mm): Head diameter as formed, shown in Figure 12.5.2; obtained from FEA
Y_h (mm): Head height as formed, shown in Figure 12.5.2; obtained from FEA
Y_f (mm): Flush, shown in Figure 12.5.2; obtained from FEA
Y_g (mm): Gap, shown in Figure 12.5.2; obtained from FEA

(A) Diameter of Formed Head Y_d

The response surface regression is conducted on head diameter Y_d versus factors A, B, C, and D, without quadratic terms, using Minitab. The ANOVA table is given as follows:

Source	SS_{adj}	DOF	MS_{adj}	F-value	p-Value
Model	6.05366	10	0.60537	36.52	0.0%
Linear	5.69597	4	1.42399	85.91	0.0%
A	0.54071	1	0.54071	32.62	0.2%
B	0.19984	1	0.19984	12.06	1.8%
C	0.93406	1	0.93406	56.36	0.1%
D	4.02135	1	4.02135	242.62	0.0%
Interactions	0.35769	6	0.05961	3.60	9.1%
A B	0.00209	1	0.00209	0.13	73.7%
A C	0.00085	1	0.00085	0.05	82.9%
A D	0.14227	1	0.14227	8.58	3.3%
B C	0.07949	1	0.07949	4.80	8.0%
B D	0.02810	1	0.02810	1.70	25%
C D	0.10488	1	0.10488	6.33	5.3%
Error	0.08287	5	0.01657		
Subtotal	6.13653	15			
Grand average	—	1			
Total	—	16			

The predictive equation of the head diameter as formed, with R = 98.65% and R_{adj} = 95.95%, is given as follows:

$$Y_{d, pred} = 4.5799 - 0.1838\ a + 0.1118\ b + 0.2416\ c + 0.5013\ d$$

$$- 0.0943\ a\ d - 0.0705\ b\ c + 0.0810\ c\ d. \tag{12.5.13}$$

Given that the rivet diameter is 3.175 mm (0.125″) with b = 0 and the rivet height is 63.5 mm (2.5″) with c = −1, the above equation reduces to

$$Y_{d, pred} = 4.3383 - 0.1838\ a + 0.4203\ d - 0.0943\ a\ d. \tag{12.5.14}$$

With the response surface plot given in Figure 12.5.6, it is shown that the small hole of diameter A = 3.2639 mm (a = −1) and high pressing force D = 13.345 kN (d = 1) will produce the largest formed head of diameter Y_d = 5.0367 mm. The larger the hole, the better the joint. However, the final decision has to be also checked against the fatigue life of the joint.

In summary, with a = −1 (A = 3.2639 mm) and d = 1 (D = 13.345 kN), Equation (12.5.14) yields a formed head of diameter = 5.0367 mm, which falls within the acceptable range, i.e., 4.3656 < Y_d < 5.55625 mm for applying standard 0.125 × 2.5″ rivets (b = 0 and c = −1).

FIGURE 12.5.6 Diameter of formed head (Y_d) as a function of sheet hole diameter (factor A) and press force (factor D).

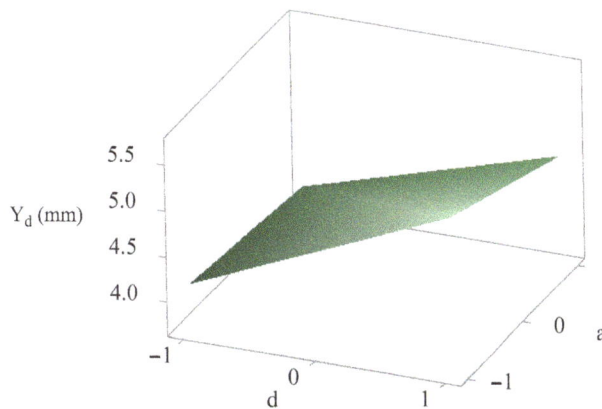

(B) Height of Formed Head Y_h

The response surface regression is conducted on head diameter Y_h versus factors A, B, C, and D, without quadratic terms, using Minitab. The ANOVA table is given as follows:

Source	SS$_{adj}$	DOF	MS$_{adj}$	F-value	p-Value
Model	6.75361	10	0.67536	63.63	0.0%
Linear	6.41344	4	1.60336	151.06	0.0%
A	1.03484	1	1.03484	97.50	0.0%
B	0.22803	1	0.22803	21.48	0.6%
C	2.24580	1	2.24580	211.59	0.0%
D	2.90477	1	2.90477	273.68	0.0%
Interactions	0.34017	6	0.05670	5.34	4.3%
AB	0.01881	1	0.01881	1.77	24.1%
AC	0.04026	1	0.04026	3.79	10.9%
AD	0.06845	1	0.06845	6.45	5.2%
BC	0.04940	1	0.04940	4.65	8.3%
BD	0.01518	1	0.01518	1.43	28.5%
CD	0.14808	1	0.14808	13.95	1.3%
Error	0.05307	5	0.01061		
Subtotal	6.80668	15			
Grand average	—	1			
Total	—	16			

The predictive equation of the head height as formed, with correlation R = 99.22% and adjusted correlation R_{adj} = 97.66%, is given as follows:

$$Y_{h, pred} = 1.4856 - 0.2543\,a + 0.1194\,b + 0.3746\,c - 0.4261\,d$$

$$+ 0.0502\,a\,c + 0.0654\,a\,d + 0.0556\,b\,c - 0.0962\,c\,d. \qquad (12.5.15)$$

Next, plug a = −1 (A = 3.2639 mm) and d = 1 (D = 13.345 kN) into the above equation to calculate the formed head height for standard 0.125 × 0.25″ rivets (b = 0 and c = −1). It yields a formed head height of Y_h = 1.0202 mm, which falls out of the acceptable range, i.e., 1.1906 < Y_h < 1.9844 mm.

(C) Flush Y_f

The response surface regression is conducted on head diameter Y_f versus factors A, B, C, and D, without quadratic terms, using Minitab. The ANOVA table is given as follows:

Source	SS$_{adj}$	DOF	MS$_{adj}$	F-value	p-Value
Model	0.139697	10	0.013970	233.77	0.0%
Linear	0.134414	4	0.033604	562.33	0.0%
A	0.129632	1	0.129632	2169.29	0.0%
B	0.004278	1	0.004278	71.59	0.0%
C	0.000117	1	0.000117	1.95	22.1%
D	0.000387	1	0.000387	6.48	5.1%
Interactions	0.005283	6	0.000881	14.73	0.5%
AB	0.001920	1	0.001920	32.13	0.2%
AC	0.000178	1	0.000178	2.98	14.5%

Source	SS$_{adj}$	DOF	MS$_{adj}$	F-value	p-Value
AD	0.002391	1	0.002391	40.01	0.1%
BC	0.000117	1	0.000117	1.95	22.1%
BD	0.000678	1	0.000678	11.34	2.0%
CD	0.000000	1	0.000000	0.01	93.8%
Error	0.000299	5	0.000060		
Subtotal	0.139996	15			
Grand average	—	1			
Total	—	16			

The predictive equation of the flush induced, with correlation R = 99.79% and adjusted correlation R_{adj} = 99.36%, is given as follows:

$$Y_{f, pred} = 0.11509 - 0.09001\ a + 0.01635\ b - 0.00492\ c$$

$$- 0.01095\ a\ b + 0.01222\ a\ d - 0.00651\ b\ d, \tag{12.5.16}$$

Next, plug a = –1 (A = 3.2639 mm) and d = 1 (D = 13.345 kN) into the above equation to calculate the formed flush for standard 0.125 × 0.25″ rivets (b = 0 and c = –1). It yields a formed flush Y_f = 0.1978 mm, which falls within the acceptable range, i.e., Y_f < 0.254 mm.

12.6. Joining with Adhesives

Adhesive bonding is the process of joining materials together by surface adhesion with the aid of adhesives. Adhesive bonding possesses some attractive advantages such as uniform distribution of stress over the entire bond area of joints, except the free boundaries. It also allows for easier fabrication of smooth and complex contoured structures which exhibit esthetic appearance and reduce the possibility of electrolytic corrosion problems. In addition to temperature and moisture, the effects of porosity in the bond layer can have an influence on ductile response.

Delamination is the weak failure mode of an adhesive joint, and it may occur because of interfacial diffusion as a function of temperature and moisture content [Ocañaa et al. 2015]. Blistering is another potential failure mode, and it may occur because of an open-faced adhesive joint at high relative humidity while expedited by temperature rise. Fick's law is a fairly accurate description of moisture diffusion [Wylde and Spelt 1998]. Waviness of composite adhesive joints is another concern [Zheng and Sun 2004].

12.6.1. Adherence

Adhesive ductility is an important factor in minimizing the adverse effects of shear and peel stress peaks in the bond layer. Ductility has a pronounced influence on mechanical response of adhesive joints, and restricting the design to elastic response deprives the application of a significant amount of additional structural capability. A study on the adherence of adhesive joints using polyvinyl alcohol (PVA) was studied by a group of students to explore more influential factors beyond temperature, moisture, and porosities. Four factors were identified as listed in Table 12.6.1. The experimental tests were conducted at room temperature with a relative humidity of 60% and no airflow (wind) at all. Two runs were recorded for each treatment condition. The measurable is how much force (grams) can be applied to separate the glued parts according to a Chinese Standard (2011). Note that 1000-g force is equal to 9.807 N.

© SAE International.

TABLE 12.6.1 Factors and levels used for the study on adherence in adhesive joints.

Factor	Level (−)	Level (+)
A: Cure time (min)	0	10
B: Glued area (cm·cm)	1.7 × 2.0	2.5 × 3.4
C: Concentration	60%	100%
D: Roughness	Coated plank	Rough wall

The 16 run treatments for full factorial design matrix 2^4 are listed in Table 12.6.2(a). The calculated main effects are given in Table 12.6.2(a), while the multifactor interactions are given in Table 12.6.2(b). Since there are two replications for each run, the sample standard deviation can be calculated as

$$S_i^2 = [\sum_{n=1}^{2} (y_{in} - y_{i,ave})^2] / (2 - 1).$$

(12.6.1)

Since there are two replications for each run, the unbiased estimate of the sample standard deviation for all test data can be obtained from individual sample standard deviations using Equation (2.5.3) as

$$S = (\frac{v_1 S_1^2 + v_2 S_2^2 + \ldots + v_{N-1} S_{N-1}^2 + v_N S_N^2}{v_1 + v_2 + \ldots + v_{N-1} + v_N})^{1/2}$$

$$= (\frac{1 S_1^2 + 1 S_2^2 + \ldots + 1 S_{15}^2 + v_N S_{16}^2}{1 + 1 + \ldots + 1 + 1}) = 20.5.$$

It is shown that three main effects have different levels of influence on the result, while no interactive effects are significant, based on either $t_{16,95\%}$ or $t_{16,90\%}$ given in Table 12.6.2. Thus, the predictive equation can be written as

$$Y_p = 268.13 + 51.25 \, a + 61.88 \, b - 73.13 \, d - 23.8 \, a \, d.$$

It can be seen that the 10-min cure time, larger glue area, and smooth surface do increase the adherence of adhesive joints. Nevertheless, the interaction between curing time and surface roughness has a negative minor effect on the adherence strength, i.e., a long curing time for a rough surface may deteriorate the adherence.

TABLE 12.6.2 The full factorial design matrix of 2^4 for testing adherence of adhesive joints.

(a) Main factors and test results

Run	A	B	C	D	Mean	y_1	y_2	y_{ave}	S_i	S_i^2
1	−1	−1	−1	−1	1	160	140	150	14.14	200
2	1	−1	−1	−1	1	400	360	380	28.28	800
3	−1	1	−1	−1	1	410	400	405	7.071	50
4	1	1	−1	−1	1	470	480	475	7.071	50
5	−1	−1	1	−1	1	200	180	190	14.14	200
6	1	−1	1	−1	1	340	360	350	14.14	200
7	−1	1	1	−1	1	300	340	320	28.28	800
8	1	1	1	−1	1	440	480	460	28.28	800
9	−1	−1	−1	1	1	160	120	140	28.28	800
10	1	−1	−1	1	1	140	180	160	28.28	800
11	−1	1	−1	1	1	200	220	210	14.14	200
12	1	1	−1	1	1	240	280	260	28.28	800
13	−1	−1	1	1	1	120	120	120	0	0
14	1	−1	1	1	1	140	180	160	28.28	800
15	−1	1	1	1	1	200	200	200	0	0
16	1	1	1	1	1	320	300	310	14.14	200
Contrast	102.5	123.8	−8.75	−146.3	—					
Effect	51.25	61.88	4.375	−73.13	268.13					
S	20.5	20.5	20.5	20.5	20.5					
$t_{16,1-\alpha}$	2.50	3.02	0.21	3.57	13.1					
$t_{16,95\%}$	1.746	1.746	1.746	1.746	1.746					
$t_{16,90\%}$	1.337	1.337	1.337	1.337	1.337					

(b) Interactions

Run	AB	AC	AD	BC	BD	CD	ABC	ABD	ACD	BCD	ABCD
1	1	1	1	1	1	1	−1	−1	−1	−1	1
2	−1	−1	−1	1	1	1	1	1	1	−1	−1
3	−1	1	1	−1	−1	1	1	1	−1	1	−1
4	1	−1	−1	−1	−1	1	−1	−1	1	1	1
5	1	−1	1	−1	1	−1	1	−1	1	1	−1
6	−1	1	−1	−1	1	−1	−1	1	−1	1	1
7	−1	−1	1	1	−1	−1	−1	1	1	−1	1
8	1	1	−1	1	−1	−1	1	−1	−1	−1	−1
9	1	1	−1	1	−1	−1	−1	1	1	1	−1
10	−1	−1	1	1	−1	−1	1	−1	−1	1	1
11	−1	1	−1	−1	1	−1	1	−1	1	−1	1
12	1	−1	1	−1	1	−1	−1	1	−1	−1	−1
13	1	−1	−1	−1	−1	1	1	1	−1	−1	1
14	−1	1	1	−1	−1	1	−1	−1	1	−1	−1
15	−1	−1	−1	1	1	1	−1	−1	−1	1	−1
16	1	1	1	1	1	1	1	1	1	1	1
Contrast	−10	10	−47.5	−6.25	−23.8	13.75	22.5	35	10	21.25	−12.5
Effect	−5	5	−23.8	−3.13	−11.9	6.88	11.3	17.5	5	10.6	−6.25
S	20.5	20.5	20.5	20.5	20.5	20.5	20.5	20.5	20.5	20.5	20.5
$t_{16,1-\alpha}$	0.24	0.24	1.16	0.15	0.58	0.34	0.55	0.86	0.24	0.52	0.31
$t_{16,95\%}$	1.746	1.746	1.746	1.746	1.746	1.746	1.746	1.746	1.746	1.746	1.746
$t_{16,90\%}$	1.337	1.337	1.337	1.337	1.337	1.337	1.337	1.337	1.337	1.337	1.337

12.6.2. Relaxation Life Cycle Tests of Adhesive Joints

Accelerated relaxation life cycle tests for examining the strength decay of single-lapped adhesive joints are conducted. The test specimens are 6Al-4V-Ti alloy/AF126-2 single-lap joints. Adhesive F126-2 is a modified epoxy film, of which the material properties in tension are E = 2.03 GPa, σ_{uts} = 41 MPa, and ν = 0.41 as supplied by 3M, UK. The adhesive contains a carrier fabric for controlling the bond-line thickness. Prior to bonding, the adherends were degreased with 1,1,1 trichloroethane and then grit-blasted using 80/120 alumina to produce a uniform matte finish. A pressure of 0.5861 MPa (85 psi) was used to grit blast the areas that are bonded. Any dust remaining after grit blasting was removed with clean compressed air. The grit-blasted/ degreased specimens are then immersed for 30 min in a chromic acid etch solution at a temperature of 60–70°C. The specimens are then cured by means of an extractor fan at 120°C for a few minutes [Broughton and Mera 1999].

Pseudo-full factorial design 3^2 for strength tests of adhesive joints was then conducted [Broughton and Mera 1999]. It is so-named because both design parameters are not really arranged following the standard code (–1, 0, 1) at three levels of the same distance. Tests are performed according to ASTM D3846 under compressive loading at a displacement rate of 1 mm/min as the specimens are exposed to elevated temperatures in humid environments. The test results in terms of load per unit width (N/mm), after 1, 3, 7, and 17 days of environmental exposures, are listed in Table 12.6.3 [Broughton and Mera 1999]. It has been observed that the diffusion coefficient and the saturation moisture content for AF126-2 increase with temperature and relative humidity. The saturation moisture content is approximately 6.0% by weight at 70°C.

TABLE 12.6.3 Pseudo-full factorial design 3^2 for strength tests of adhesive joints [Broughton and Mera 1999].

Run	Temperature (°C)	Humidity (%)	Joint strength (N/mm)			
			Y_1	Y_3	Y_7	Y_{17}
1	25	45	368	368	368	368
2	25	85	281	270	276	187
3	25	96	254	235	241	194
4	40	45	348	300	310	206
5	40	85	276	252	262	173
6	40	96	243	228	247	148
7	70	45	278	248	236	229
8	70	85	273	240	244	195
9	70	96	243	224	128	84

© SAE International.

Y_1, Y_3, Y_7, and Y_{17}: Observations after 1, 3, 7, and 17 days of environmental exposures

Assume that the hazard rate is the ratio of the ultimate strength at day 1 to the ultimate tensile strength at the date of concern. Having the acceleration factor of the load level taking the qth step (q = 1, 3, 7, and 17), as described by Equation (8.6.14), one may assume the following degradation model:

$$\frac{\sigma_{uss}(T_q, H_{Rq})}{\sigma_{uss}(T_0, H_{R0})} = \left(\frac{E_q}{E_0}\right) \exp\left[B\left(\frac{1}{T_0} - \frac{1}{T_q}\right) + C\left(\frac{1}{H_{R0}} - \frac{1}{H_{Rq}}\right)\right], \quad (12.6.1)$$

where

$\sigma_{uss}(T_0, H_{R0})$ is the ultimate shear strength of the joint at day 0

$\sigma_{uss}(T_q, H_{Rq})$ is the ultimate shear strength of the joint at day q, where q = 1, 3, 7, and 17

E_q/E_0 is the ratio of Young's modulus at time t = q to the initial Young's modulus (at t = 0)

Assume that the material follows the first order (i.e., linear solid model) based on the Prony series. Then,

$$E_q / E_0 = 1 - p_1 [1 - \exp(- t / \lambda_1)], \tag{12.6.3}$$

where

p_1 is the first-order coefficient of Prony series for adhesive relaxation
λ_1 is the characteristic time of normal modulus of relaxation

Substituting the above equation into Equation (12.6.2) and taking natural logarithmic transformations, one has

$$\ln[\sigma_{uss}(T_0, H_{R0}) / \sigma_{uss}(T_q, H_{Rq})] = \ln\{1 - p_1 [1 - \exp(- t / \lambda_1)]\}$$
$$+ B (1 / T_0 - 1 / T_q) + C (1 / H_{R0} - 1 / H_{Rq}). \tag{12.6.4}$$

Plugging the four sets of data, i.e., Y_1, Y_3, Y_7, and Y_{17}, as given in Table 12.6.4, into the above equation, one can solve four simultaneous equations for unknown parameters p_1, λ_1, B, and C using multifactorial regression.

12.6.3. Delamination

Delamination is the failure mode because of weak stress intensity factors at the boundaries of an adhesive joint. This stress intensity factor is subject to material properties of joined parts, and it may be worsened owing to interfacial diffusion as a function of temperature and moisture content [Ocañaa et al. 2015]. Energy release rates, i.e., mode I G_I (opening mode) and mode II G_{II} (in-plane shear mode) at the very onset of a crack and during its propagation because of in-plane mixed loading modes, may exceed the threshold of the resistance to delamination failure.

An experiment is designed to investigate how the combined mixed-mode critical energy release rate, $G = f(G_I, G_{II})$, affects the delamination (dis-bond) growth under in-plane mixed crack modes [Li et al. 2015]. It was conducted using a quasi-static test setup on a single-lap joint with two pieces of unidirectional laminae, where pre-cracking was initiated following ASTM D5528 in order to remove the effect of the Teflon insert. The applied shear force is aligned with the fibrous direction. Experimental tests are conducted in accordance with ASTM D 5573-99. It is shown that most specimens fail the adhesive's cohesion to the substrate at even low-mode mixtures, while the crack growth rate increases with respect to increasing in-plane shear strain energy release rate.

12.6.4. Bolted-Bonded Joints

The schematic drawing of a typical bolted-bonded joint is depicted in Figure 12.6.3. Five factors are listed in Table 12.6.3 as proposed for conducting an experimental design by Lopez-Cruz et al. [2017]. The tests are carried out according to the design matrix given in Table 12.6.4, and the loads are applied in accordance with ASTM D3165-07 at 11.7 mm/min until separation. Three replicates for each treatment (run) have been completed, and their averages and standard deviation are listed in Table 12.6.5. A typical failure mode is demonstrated in Figure 12.6.4, in light of the applied shear force versus displacement corresponding to run 5 (Table 12.6.5). The predictive equation based on the obtained data, while drawing the statistical inference from a significance level of 10% ($\alpha = 10\%$), is given as follows:

$$Y_p = 16.91 + 1.42 \, a - 1.095 \, b.$$

It is seen that the adherend thickness (factor A) is the most prominent design factor, followed by adhesive type (factor B). The other three factors, i.e., adhesive thickness, washer clamping area, and clearance between the hole and bolt, are not statistically significant. Note that these conclusions are valid only when the design levels of these factors fall within the ranges given in Table 12.6.4.

FIGURE 12.6.3 Schematic drawing of bolted-bonded joints for composites [Lopez-Cruz et al. 2017].

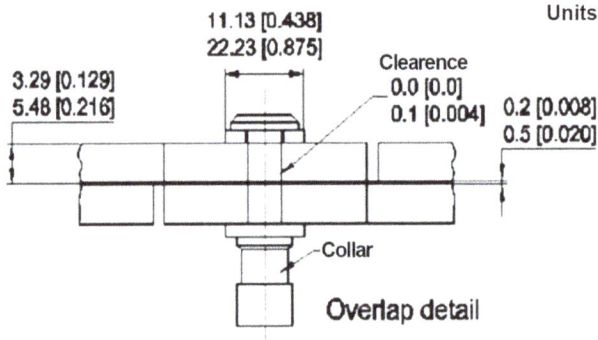

Reprinted with permission. © Elsevier Science & Technology Journals.

TABLE 12.6.4 Factors for study on bolted-bonded joints for composites [Lopez-Cruz et al. 2017].

Factor	Level (−)	Level (+)	Normalized variable
A: Adherend thickness (mm)	3.3	5.5	$a = (A - 4.4)/1.1$
B: Adhesive type	EA9361	FM300-2M	$b = B_-$ or B_+
C: Adhesive thickness (mm)	0.2	0.5	$c = (C - 0.35)/0.15$
D: Washer clamping area (mm²)	387	97	$d = -(D - 242)/145$
E: Clearance—bolt to hole (mm)	0	0.1	$e = (E - 0.05)/0.05$

FIGURE 12.6.4 Failure mode of a bolted-bonded joint for composites: run 5 in Table 12.6.1 [Lopez-Cruz et al. 2017].

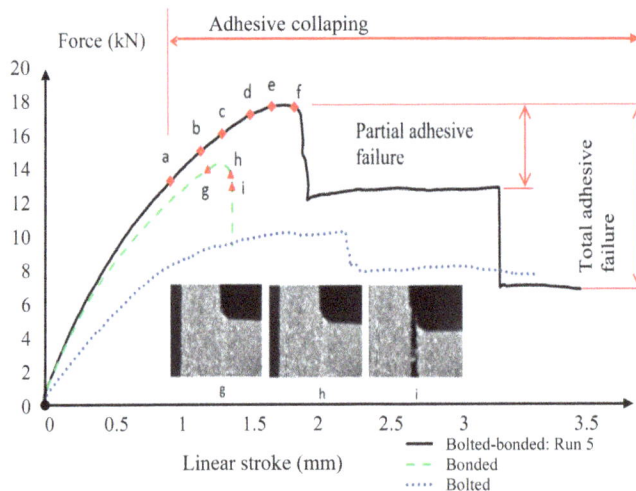

Reprinted with permission. © Elsevier.

TABLE 12.6.5 Experimental matrix and test data applying 2_{III}^{5-2} to bolted-bonded joints.

Run	A	B	C	D = AB	E = AC	Y_bolted-bonded (kN)
1	−1	−1	−1	1	1	14.40 ± 0.32
2	1	−1	−1	−1	−1	20.35 ± 0.97
3	−1	1	−1	−1	−1	14.17 ± 0.45
4	1	1	−1	1	1	16.98 ± 0.78
5	−1	−1	1	−1	1	17.93 ± 1.22
6	1	−1	1	1	−1	19.34 ± 0.25
7	−1	1	1	−1	−1	15.46 ± 1.08
8	1	1	1	1	1	16.65 ± 0.91
Contrast	2.84	−2.19	0.87	−0.84	−1.54	
Effect	1.42	−1.095	0.435	−0.42	−0.77	16.91 (average)
Error	0.8216	0.8216	0.8216	0.8216	0.8216	
$|t_{16}|$	1.7284	1.3328	0.5295	0.5112	0.9372	
$T_{16,90\%}$	1.337	1.337	1.337	1.337	1.337	
Significant?	Yes	Minor	No	No	No	

where Y (kN): Standard deviations for bolted-bonded joints from three replicated tests [Lopez-Cruz et al. 2017]

12.7. Spot Welding

A body in white is assembled from preformed pieces of sheet metals and composites on the shop floor of an automotive body plant. A shop may be divided into various workstations, ranging from 12 to 18 blocks, each representing a welding complex that covers many welding operations (mainly spot welding) at different stations [Spieckermann et al. 2000]. To buffer the production process, decoupling is usually introduced between two subsequent blocks forming a structure of blocks and buffers. After the elements of the underbody have been completed in respective blocks, they are assembled and welded [mainly metal inert gas (MIG)] in the follow-up blocks. Then, the side frames and the roof are attached to the underbody forming the body frame. Brazing and grinding activities are then carried out before the doors, the trunk lid, and the hatchback are attached in the finish area. The finished body in white then leaves the body shop and is conveyed to the paint shop. It could take more than 100 welding robots and other various equipment to complete the body in white before it is conveyed to the next assembly process.

12.7.1. Resistance Spot Welding (RSW)—2^3 Design

RSW is an effective welding process for joining automotive sheet metals, of which two contacting metal surfaces are joined by the heat that is converted from electric resistance. Typically, the thickness of sheet metals as applied ranges from 0.6 to 3 mm. Spot welding involves the following three steps (Figure 12.7.2):

1. The electrodes are brought into contact with the surface of the sheet metals, and a small amount of pressure is exerted.

2. The current from the electrodes is then applied briefly. Weld time ranges from 0.01 to 0.63 sec depending on the sheet material, thickness of the sheet metal, electrode force, and diameter of electrodes.

3. After the current is removed, the electrodes remain in place in order for the material to cool down gradually.

One of the attractive features of spot welding is that a lot of energy can be delivered to the spot in a very short time. That permits the welding to occur without excessive heating of the remainder of the sheet. Spot welding is very material-dependent [Karci et al. 2009], and electrode sticking during micro-resistance welding of thin metal sheets can be a concern [Zhou et al. 2000]. Potential factors that affect the sticking include welding current, weld time, electrode tip coating, electrode force, and electrode spacing. Reducing welding current and weld time and increasing electrode force and electrode spacing were found to reduce electrode sticking [Dong et al. 2002].

FIGURE 12.7.2   Schematic drawing of a RSW process.

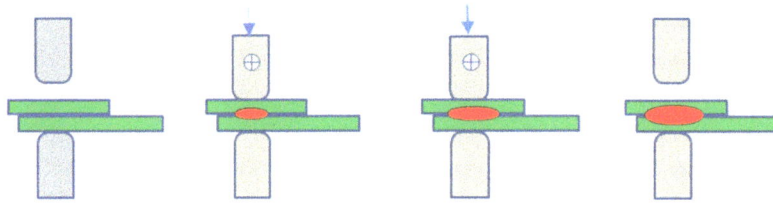

© SAE International.

DOE is intended to examine the robustness of a given grade of high-strength metal (mainly steel) when subjected to variations in key process variables, as argued by the American Welding Society [AWS D8.9M:2012 2012], although the long current level/range testing and weld property testing together generally provide a good indication of the RSW behavior [Rowlands and Antony 2003]. The top five viable factors evolving spot welding are the applied weld current, weld time, squeeze time (hold time), exerting pressure (or electrode force), material of metal sheet, and thickness of metal sheet.

One DOE is set up to study the influences of these three processing parameters on spot welding of automotive metal sheets made of SAE 1008 [Pashazadeh et al. 2016]. Their design levels are listed in Table 12.7.1. In general, the weld strength is a function of nugget size, the larger the better. Thus, the objective function is to maximize the nugget diameter, of which the data obtained from experiments are listed in columns Y_1 and Y_2 in Table 12.7.2. The value of Y is the average of Y_1 and Y_2 that are obtained from two replicated runs on each treatment.

TABLE 12.7.1 Design factors for spot welding of automotive sheet steels [Pashazadeh et al. 2016].

Factor	Level (−)	Level (+)	(−1, 1), Equation (2.4.5)
A: Exerted pressure (MPa)	0.294	0.4312	a = (A − 0.3626)/0.0686
B: Electric current (kA)	5	6.5	b = (B − 5.75)/0.75
C: Welding time (cycles)	6	12	c = (C − 9)/3

TABLE 12.7.2 2^3 factorial design for spot welding of automotive sheet steels.

Treatment	a	b	c	ab	ac	bc	abc	Y_1	Y_2	Y	Y_p
1	−1	−1	−1	1	1	1	−1	2.68	2.53	2.605	2.71
2	1	−1	−1	−1	−1	−1	1	3.45	3.36	3.405	3.307
3	−1	1	−1	−1	1	1	1	3.91	3.98	3.945	4.046
4	1	1	−1	1	−1	−1	−1	3.53	3.57	3.55	3.449
5	−1	−1	1	1	−1	−1	1	3.57	3.43	3.5	3.496
6	1	−1	1	−1	1	1	−1	3.79	3.7	3.745	3.749
7	−1	1	1	−1	−1	−1	−1	4	4.2	4.1	4.096
8	1	1	1	1	1	1	1	3.79	3.89	3.84	3.844
Contrast	0.098	0.545	0.42	−0.425	−0.105	−1.98	0.1725			—	
Effect	0.049	0.273	0.21	−0.213	−0.053	−0.099	0.0863			3.586 (average)	
Error	0.0847	0.0847	0.0847	0.0847	0.0847	0.0847	0.0847	0.042			
t-Ratio	0.576	3.217	2.479	2.509	0.62	1.166	1.018	—	—	84.68	
$t_{8,1-\alpha} = t_{8,90\%}$	1.397	1.397	1.397	1.397	1.397	1.397	1.397				
$t_{8,1-\alpha} = t_{8,85\%}$	1.108	1.108	1.108	1.108	1.108	1.108	1.108				
Significant?	No	Yes	Yes	Yes	No	Minor	Minor				

Y (mm): Diameter, average of two test results, i.e., Y_1 and Y_2 [Pashazadeh et al. 2016]
Y_p (mm): Predicted value of Y, based on the criterion that $\alpha = 15\%$

There are two runs for each treatment. Thus, $\nu_1 = \nu_2 = \nu_3 = \nu_4 = \nu_5 = \nu_6\ \nu_7 = \nu_8 = 1$. Based on Equation (2.5.3), individual sample variances associated with the eight treatments can be used for computing the sample variance for each individual effect, main or interactive, as

$$S = \left(\frac{\nu_1 S_1^2 + \nu_2 S_2^2 + \ldots + \nu_8 S_8^2}{\nu_1 + \nu_2 + \ldots + \nu_8} \right)^{1/2}$$

$$= \left[\frac{1 \times 1.0113 + 1 \times 0.0041 + 1 \times 0.0024 + 1 \times 0.0008 + 1 \times 0.0098 + 1 \times 0.004 + 1 \times 0.02 + 1 \times 0.005}{8} \right]^{1/2}$$

$$= 0.0847,$$

where, for example, sample variance $S_1^2 = [(2.68 − 2.605)^2 + (2.53 − 2.605)^2]/(\nu_1 − 1) = 1.0113$ and $\nu_1 = 2 − 1$ for treatment 1 that has two individually and independently replicated data available. Similarly, other sample variances for the other seven treatments can be obtained.

FIGURE 12.7.3   Pareto plot of all the effects on the weld diameter.

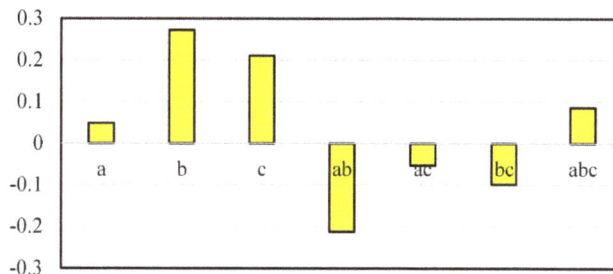

The Pareto plot of all the effects on the resultant weld diameter (given in Table 12.7.2) is exhibited in Figure 12.7.3. Based on the parting line of significance at $\alpha = 10\%$ that $t_{8,1-\alpha} = t_{8,90\%} = 1.397$, one has the following regression model for this case study on the weld diameter:

$$Y_p = 3.5863 + 0.2725\,b + 0.21\,c - 0.213\,a\,b, \tag{12.7.1}$$

$$= 3.5863 + 0.2725\,[(B - 5.75) / 0.75] + 0.21\,[(C - 9) / 3]$$

$$- 0.213\,[(A - 0.3626) / 0.0686]\,[(B - 5.75) / 0.75]$$

$$= -7.744 + 23.755\,A + 1.861\,B + 0.07\,C - 4.130\,A\,B. \tag{12.7.2}$$

The predicted values (Y_p) based on the above equation for all eight treatments (runs) are plotted against the test values (Y) in Figure 12.7.4, which shows that the correlation is $R = 93.8\%$ when the correlation level is set at $\alpha = 10\%$; the correlation is $R = 98.6\%$ when the confidence level is set at $\alpha = 17\%$. According to Equation (12.7.2), the more the weld time (cycles), the higher the weld diameter (mm). Given that weld time is set at 12 cycles, the response surface plot of the weld diameter versus the response surface of weld diameter versus the exerted pressure and electric current is shown in Figure 12.7.5.

FIGURE 12.7.4   Correlation between test data Y (vertical axis) and predicted values Y_p (horizontal axis).

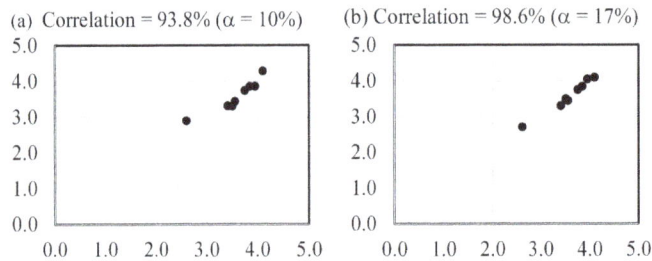

© SAE International.

FIGURE 12.7.5   Response surface plot of weld diameter versus exerted pressure and electric current at welding time = 12 cycles.

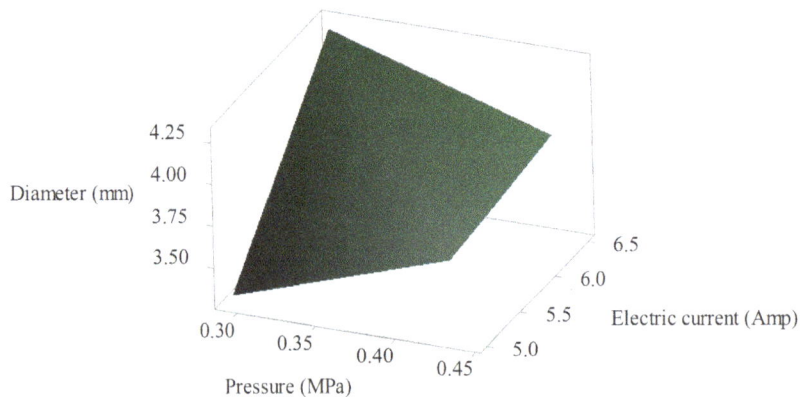

© SAE International.

However, the ratio of the three-factor interaction to the sample mean is 0.0863/3.5863 = 2.4% and the ratio of AC interaction to the sample mean (grand average) is 0.0525/3.5863 = 1.5%. Making engineering sense of the approach, one may assume that 2% (i.e., corresponding to $t_{8,1-\alpha} \approx t_{8,83\%}$) is accepted. More aggressively by engineering sense, based on the parting line of significance $\alpha = 17\%$ that $t_{8,1-\alpha} = t_{8,83\%} = 2\%$ (from Table 12.7.2), one has another regression model that

$$Y_p = 3.5863 + 0.2725\,b + 0.21\,c - 0.213\,a\,b - 0.099\,b\,c + 0.0863\,a\,b\,c, \qquad (12.7.3)$$

or $Y_p = -17.8063 + 52.7345\,A + 2.9646\,B + 1.4887\,C$

$$- 5.2651\,A\,B - 3.215\,A\,C - 0.2467\,B\,C + 0.5591\,A\,B\,C. \qquad (12.7.4)$$

In light of the above equation based on the confidence interval of 83% ($\alpha = 17\%$), it can be concluded that both electric current (factor B) and welding time (factor B) have strong individual main effects on the nugget's size. The impacts of two two-factor interactions are also significant. The interaction of high electric current and high exerted pressure (interaction AB) will damage the size of nugget, so will the interaction of the high electric current and high welding time (interaction BC). However, the three-factor interaction helps grow the nugget size, if all of them operate at high levels.

The predicted values (Y_p) based on Equation (12.7.3) for all eight treatments (runs) are plotted against the test values (Y) in Figure 12.7.3(b), which shows that the predictive correlation is R_{pred} = 98.6% at 83% confidence. The great predictive correlation (R = 98.6%) shows that the DOE modeling is valid, and thus, conclusions drawn from Equation (12.7.3) can be used as the quality assessment guideline for nugget's size of a spot weld within the operating ranges of the relevant parameters given in Table 12.7.1.

For clear thinking and pragmatism, one may be curious about which equation should be applied in practice. The difference between Equations (12.7.3) and (12.7.1) seems rather dubious to experimenters. One comes with a correlation of 93.8% at 90% confidence, while the other comes with a correlation of 98.6% at 83% confidence. Both are correct when banking on statistical inferences. The decision to choose the "right one" will depend on how much these processing factors can be controlled by the manufacturer, as measured by cost, timing, and reliability. Regardless of how successfully one is in constructing an experimental design that conserves natural relationships, one is still faced with scaling artifacts of enclosure. Dimensional considerations can also potentially be used to minimize these artifacts and to maximize comparability among experiments that differ in scale [Gardner 2001].

12.7.2. Nut Welding

Weld nuts are automatically fed to four lower alignment pins for an automotive seat-bottom frame with a pneumatic feed system. Each upper cylinder has an attached electrode that mates with the seat pan, weld nut, and lower electrode. When the cylinder is excited, an electrical circuit delivers a current that is received from a transformer via a copper shunt to the electrode. This current passes through the nut and seat pan, generating heat to weld the nut to the seat pan. The four spot profusions on individual nuts are melted and fused with the sheet metal of the seat-bottom frame, as shown in Figure 12.7.6. A full factorial design 2^3 augmented with four center points [Gore 2019] is employed to examine the influence on the break torque strength (Nm), as listed in column Y in the table, of the welded nuts by the following three factors:

- Factor A: Current (kA), that is controlled by a transformer and weld timer
- Factor B: Time (cycles) for an electrode to make contact before the current is through
- Factor C: Contact pressure (kPa) between the electrode and sheet metal

The time (factor B) expressed in terms of cycles and squeeze time, hold time, and cool time is included in each cycle. The minimum requirement of breakaway torque is 40 Nm, and the automotive seat manufacturer used to have an original defect rate of 150 ppm (parts per million). What would be the expected defect rate after the DOE study?

FIGURE 12.7.6 Seat-bottom frame with four aligned fastening nuts [Gore 2019].

© Stat-Ease Inc.

Based on the ANOVA with the F-distribution, the overall contributions of individual factors and their interactions to the torque strength, with the given experimental conditions given in Table 12.7.3, are presented in Table 12.7.4. The analysis is summarized in the following predictive equation:

$$Y_p = 42.40 - 1.133\ B + 0.2122\ A^2 + 0.1101\ A\ B - 0.006674\ A\ C + 0.001225\ B\ C$$

$$(R = 99.6\%,\ R_{adj} = 99.3\%,\ and\ R_{pred} = 97.5\%). \tag{12.7.5}$$

The above predictive equation is effective with high correlations, including unadjusted model correlation, adjusted model correlation, and predictive correlation, while its model adequacy is evidenced by the insignificance of lack of fit.

TABLE 12.7.3 Full factorial design 3^2 augmented with four center points on torque strength of spot-welded nuts in automotive seat-bottom frame [Gore 2019].

(i) Design factors and levels							
Factor	**(−1)**		**(+1)**			**Unit variate**	
A: Current (kA)	10		15			$a = (A − 12.5)/2.5$	
B: Time (cycles)	4		30			$b = (B − 17)/13$	
C: Contact pressure (kPa)	240		480			$c = (C − 360)/120$	

(ii) Design matrix							
Run	**a**	**b**	**c**	**A**	**B**	**C**	**Y (Nm)**
1	−1	−1	−1	10	4	240	48.0
2	1	−1	−1	15	4	240	70.6
3	−1	1	−1	10	30	240	56.5
4	1	1	−1	15	30	240	90.4
5	−1	−1	1	10	4	480	35.6
6	1	−1	1	15	4	480	46.0
7	−1	1	1	10	30	480	48.1
8	1	1	1	15	30	480	76.3
9	0	0	0	12.5	17	360	54.2
10	0	0	0	12.5	17	360	56.5
11	0	0	0	12.5	17	360	57.7
12	0	0	0	12.5	17	360	56.5

Y (Nm): Break torque

TABLE 12.7.4 ANOVA for torque strength of spot-welded nuts in automotive seat-bottom frame.

Source	SS_{adj}	DOF	MS_{adj}	$F_{u,v}$	p-Value
Regression	2391.00	5	478.200	164.48	0.0%
B	50.48	1	50.480	17.36	0.6%
A^2	396.61	1	396.611	136.41	0.0%
AB	102.943	1	102.94	35.41	0.1%
AC	328.83	1	328.833	113.10	0.0%
BC	31.17	1	31.170	10.72	1.7%
Error	17.44	6	2.907		
Lack of fit	11.02	3	3.672	1.71	33.4%
Pure error	6.43	3	2.142		
Subtotal	2408.45	11			
Grand average		1			
Total	—	12			

Time is money in an assembly process. Let B = 4 cycles, i.e., the minimum time applied in the specified range in the experimental design. Then, Equation (12.7.5) for predicting the break torque reduces to

$$Y_p = 37.868 + 0.2122\ A^2 + 0.4404\ A + 0.0049\ C - 0.00674\ A\ C. \tag{12.7.6}$$

The response surface of torque strength expressed in terms of factor A (current) and factor C (pressure) corresponding to the above equation is presented in Figure 12.7.7. It can be seen the amount of current is the deciding factor for increasing the torque strength with the experimental conditions given in Table 12.7.3.

FIGURE 12.7.7 Response surface of break torque strength of spot-welded nuts in automotive seat-bottom frame.

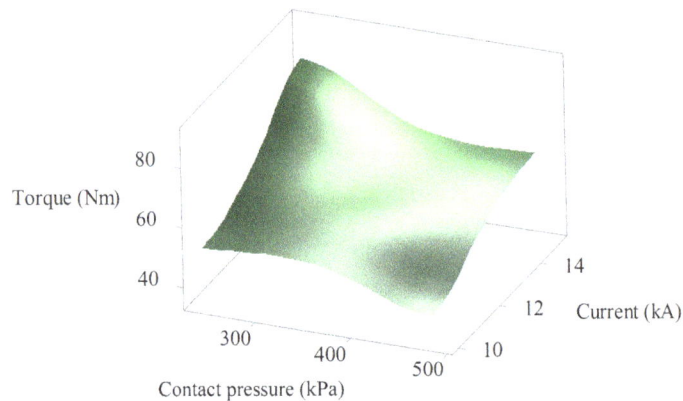

© SAE International.

12.8. High-Strength Welding

Besides spot welding, welding techniques for high-strength joints such as MIG welding, ultrasonic welding (USW), and magnetic pulse welding are also applied as permanent structural joints in vehicles.

12.8.1. MIG Welding

MIG welding, also called gas metal arc welding (GMAW), comprises heating, melting, and solidification of base metals and filler material in localized fusion zone by a transient heat source to form a joint. GMAW is one of the most applied welding methods in industrial environments because of its versatility, fast welding speed, and adaptability for process automation (e.g., robotics). MIG welding is widely used for welding a variety of ferrous and nonferrous materials.

It has been applied in automotive body in white for assembling most structural parts made of carbon steels, low-alloy high-strength steels, ultimate high-strength steels and stainless steels, and some aluminum alloys. One essential feature of MIG welding is the electrode wire that is fed continuously into the arc from a coil. These metals can be welded in all positions with this process by choosing the appropriate shielding gas, electrode, and welding variables. MIG process makes use of inert gas to shield the area being welded from the contamination caused by air, which may slow down the welding process and lead to poor-quality welds. Thus, it is also called submerged arc welding (SAW).

Because of the complexity of a MIG welding process, DOE and the related optimization techniques are often employed to unveil the influences of controllable variables on the structural integrity. The quality of MIG weld can be characterized by the resultant bead size, dilution, distortion, voids, and heat-affected zone (HAZ) including its depth. The bead geometry can be measured using its penetration, width, and over-thickness (protrusion). Factors of concern include electric voltage (V), electric current (A), wire feed rate (m/min or mm/s), nozzle-to-plate distance (mm), angle of electrode to workpiece (deg; normally 90°), welding (travel) speed (mm/s), and gas flow rate (L/min).

An approach to getting familiar with the impacts of MIG welding parameters on the joint strength and depth of penetration of die steel (carbon (C): 0.32–0.45%, silicon (Si): 0.80–1.2%, molybdenum (Mo): 1.1–1.75%, vanadium (V): 0.8–1.2%, sulfur (S) $\leq$ 0.3%, chromium (Cr): 4.75–5.5%, phosphorus (P) $\leq$ 0.3%, manganese (MN): 0.25–0.5%) via fractional factorial design 33-1 was conducted by Kumar and Goyal [2018]. Design factors for MIG welding of structural Y (MPa) and depth of penetration D (mm), are listed in Table 12.8.2.

TABLE 12.8.1 Design factors for MIG welding of structural parts [Kumar and Goyal 2018].

Factor	(−1)	(0)	(+1)	Coded variable
A: Electric current (A)	180	190	200	$a = (A − 190)/10$
B: Electric voltage (V)	21	24	27	$b = (B − 24)/3$
C: Nozzle-to-plate distance (mm)	12	16	20	$c = (C − 16)/4$
D: Angle of electrode to workpiece (deg)	90° (nominal)			
E: Welding (travel) speed (mm/s)	Fixed			
F: Wire feed rate (m/min)	Fixed			
G: Gas flow rate (L/min)	Fixed			
H: Electrode	Mild steel wire coated with copper			
I: Workpiece (mm)	Die steel (200 × 100 × 5)			

TABLE 12.8.2 Fractional factorial design 3_{III}^{3-1} for MIG welding [Kumar and Goyal 2018].

Run	a	b	c	A	B	C	Y (MPa)	D (mm)
1	−1	−1	−1	180	21	12	284.6	2.32
2	−1	0	0	180	24	16	300.4	2.08
3	−1	1	1	180	27	20	302.8	2.20
4	0	−1	0	190	21	16	298.4	2.80
5	0	0	1	190	24	20	328.7	2.60
6	0	1	−1	190	27	12	340.6	3.02
7	1	−1	1	200	21	20	333.2	2.98
8	1	0	−1	200	24	12	360.4	3.20
9	1	1	0	200	27	16	390.8	3.08

Y (MPa): Ultimate tensile strength
D (mm): Depth of penetration

By locating Minitab → Stat → Regression → Regression → Fit Regression Model and having the "most insignificant" factor or interaction that comes with the largest p-value weeded out one by one until all p-values are less than 10%, the final regression models for ultimate tensile strength (MPa) and depth of penetration (mm) are given as follows:

(a) Ultimate Tensile Strength (MPa)

$$Y_p = 3122.4 - 29.633\ A - 18.60\ B - 19.783\ C + 0.06133\ A^2 - 0.7653\ B^2$$

$$+\ 0.32944\ A\ B + 0.10604\ A\ C \qquad (R = 100\%\ \text{and}\ R_{adj} = 100\%).$$

It can be seen that the influence of each individual factor (A, B, or C) on the ultimate tensile strength is highly nonlinear. According to the p-values listed in the ANOVA table, i.e., Table 12.8.3(a), the input power (interaction AB) has the greatest impact on the ultimate tensile strength within the ranges of input data.

(b) Depth of Penetration (mm)

$$D_p = -57.97 + 0.6800\ A - 0.664\ B - 0.001633\ A^2 + 0.01394\ B^2 + 0.00458\ C^2$$

$$-\ 0.000939\ A\ C \qquad (R = 99.94\%,\ R_{adj} = 99.76\%,\ \text{and}\ R_{pred} = 98.61\%).$$

It can be seen that the influence of each individual factor (A, B, or C) on the depth of penetration is highly nonlinear. According to the p-values listed in the ANOVA table, i.e., Table 12.8.3(b), the input current (factor A) has the greatest impact on the depth of penetration within the ranges of input data.

Predictive equations given above are effective with high unadjusted model correlations, adjusted model correlations, and predictive correlations. The correlation between the ultimate tensile strength and depth of penetration is moderate (R = 78.4%) as shown in Figure 12.8.2.

TABLE 12.8.3 ANOVA for MIG welding of die steel.

(a) Strength	Source of variance	SS_{adj}	DOF	MS_{adj}	F-value	p-Value
	Regression	9320	7	1331.44	31,954.63	0.4%
	A	117.4	1	117.37	2816.92	1.2%
	B	7.16	1	7.16	171.93	4.8%
	C	27.15	1	27.15	651.55	2.5%
	A^2	75.24	1	75.24	1805.65	1.5%
	B^2	72.29	1	72.29	1734.86	1.5%
	AB	195.4	1	195.36	4688.65	0.9%
	AC	28.79	1	28.79	690.88	2.4%
	Error	0.04	1	0.04		
	Subtotal	9320	8			
	Grand average	—	1			
	Total	—	9			

(b) Depth	Source of variance	SS_{adj}	DOF	MS_{adj}	F-value	p-Value
	Regression	1.372	6	0.228717	276.6	0.4%
	A	0.064	1	0.063926	77.31	1.3%
	B	0.03	1	0.029709	35.93	2.7%
	A^2	0.053	1	0.053356	64.53	1.5%
	B^2	0.030	1	0.030324	36.67	2.6%
	C^2	0.0127	1	0.012653	15.30	6.0%
	AC	0.019	1	0.018644	22.55	4.2%
	Error	0.0017	2	0.000827		
	Subtotal	1.374	8			
	Grand average	—	1			
	Total	—	9			

© SAE International.

FIGURE 12.8.2 Correlation between ultimate tensile strength (MPa) and depth of penetration (mm) for MIG welding.

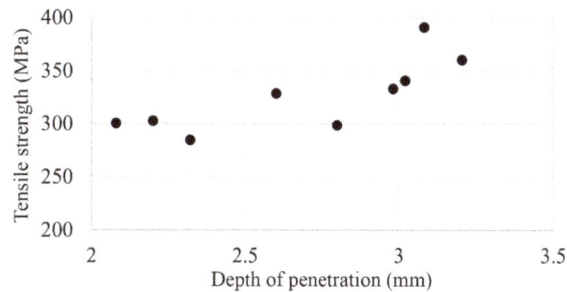

© SAE International.

12.8.2. **USW**

USW is a welding process in solid state by applying high-frequency vibrations under pressure. The high-frequency relative motion between the parts to be joined facilitates the solid progressive shearing and plastic deformation between their mating interfaces and produces an atomic bond at an elevated temperature, i.e., typically at 0.3 to 0.5 times the absolute melting temperature (°K) of the substrate materials. It takes two-way clamping access, as shown in Figure 12.8.3. On the bottom, there is an anvil to support the parts to be joined, and on top of the parts, there is a sonotrode. A sonotrode is a tool, typically tuned to act as a half-wavelength resonant device, that creates ultrasonic vibrations and applies the vibrational energy to parts to fabricate the joint [Gallego-Juárez and Graff 2015].

FIGURE 12.8.3 Schematic drawing of an USW device [Norouzi et al. 2012].

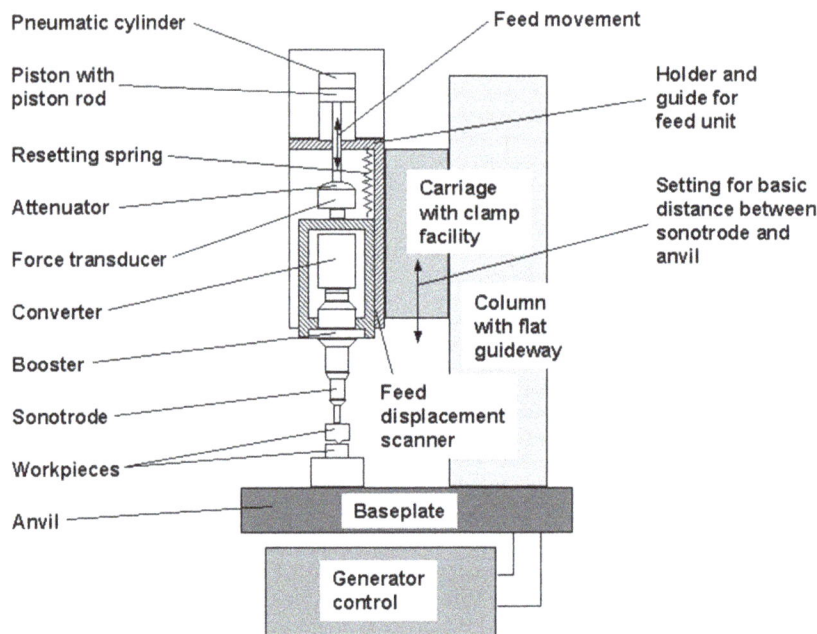

Reprinted with permission. © Elsevier.

It is a joining method that has been applied to a wide range of metals, thin sheet metals, and foils, especially used for manufacturing of battery systems [Aland et al. 2018]. Copper wires that connect the equipment like alternators, rectifiers, and batteries of electric vehicles (EVs) have been applied in GM Chevrolet Volt, Spark, and Bolt, as well as Nissan LEAF. Plastic parts and plastic-based composites used in the interior of vehicles can be joined together using USW. The heat is built up in the vibration mode upon the application of pressure with a vibration frequency ranging from 10 to 70 kHz and a vibration amplitude ranging from 10 to 250 μm [Grewell et al. 2003].

Parts made of composite materials are difficult to weld together because of the significant differences in their material properties between melting temperatures of reinforcing elements and matrix, such as melting point, thermal conductivity, and specific heat of the components that would lead to incompatibility in thermal fields and crystallization conditions. Design matrix 2^3 was employed by Pop-Calimanu and Fleser [2012] to investigate the effect of process variables and energy input on joint formation between two metallic composite disks, i.e., Al/20% silicon carbide (SiC), of 1 mm in thickness and 50 mm in diameter. The process parameters and their design levels are given in Table 12.8.4. The objective is to find out the influences of these process parameters on the net joining area and the input energy on joint formation, respectively, denoted by Y (mm^2) and J (Joule). Design matrices and related calculations based on Student's t-distribution are given in Table 12.8.5. Following the calculations based on t-distribution (Chapter 2), one can have the following two predictive equations for the area of weld and energy to fail the weld, respectively,

$$Y_p = 26.91 - 3.638\,a + 2.5275\,b - 3.815\,c + 6.975\,a\,b - 1.573\,a\,c$$

$$+ 4.545\,a\,b\,c \qquad\qquad \text{(Area of Weld)}, \qquad\qquad (12.8.1)$$

and
$$J_p = 541 - 68.25\,a + 38\,b + 31\,c + 46.75\,a\,b + 20.25\,a\,c - 21\,b\,c$$

$$+ 47.25\,a\,b\,c \qquad \text{(Energy of Weld Energy)}. \qquad\qquad (12.8.2)$$

The predicted values using the above two equations are also listed in Table 12.8.2 in columns Y_p and J_p, respectively.

TABLE 12.8.4 Design factors for ultrasonic welds of Al/20%SiC composite disks [Pop-Calimanu and Fleser 2012].

Factor	(−1)	(+1)	Unit variate
A: Pressure (MPa)	0.14	0.17	a = (A − 0.155)/0.015
B: Weld time (sec)	1.2	2	b = (B − 1.6)/0.4
C: Vibration amplitude of sonotrode (μm)	70	85	c = (C − 77.5)/7.5

TABLE 12.8.5 Full factorial design 2^3 for evaluating strength of ultrasonic welds [Pop-Calimanu and Fleser 2012].

(i) Net joining area Y (mm²)									
Run	**a**	**b**	**c**	**ab**	**ac**	**bc**	**abc**	**Y**	**Y$_p$**
1	−1	−1	−1	1	1	1	−1	32.68	32.69
2	1	−1	−1	−1	−1	1	1	23.69	23.70
3	−1	1	−1	−1	1	−1	1	32.90	32.89
4	1	1	−1	1	−1	−1	−1	33.63	33.62
5	−1	−1	1	1	−1	−1	1	37.31	37.30
6	1	−1	1	−1	1	−1	−1	3.850	3.84
7	−1	1	1	−1	−1	1	−1	19.30	19.31
8	1	1	1	1	1	1	1	31.92	31.93
Contrast	−7.275	5.055	−7.63	13.95	−3.145	***−0.025***	9.09		
Effect	−3.638	2.5275	−3.815	6.975	−1.573	***−0.0125***	4.545	26.91 (average)	
Error*	0.0125	0.0125	0.0125	0.0125	0.0125	0.0125	0.0125		
t-Ratio	291	202.2	305.2	558	125.8	—	363.6		
t$_{1,95\%}$	6.314	6.314	6.314	6.314	6.314	6.314	6.314		
Significant**	Yes	Yes	Yes	Yes	Yes	No	Yes		
(ii) Input energy on joint formation J (Joule)									
Run	**a**	**b**	**c**	**ab**	**ac**	**bc**	**abc**	**J**	**J$_p$**
1	−1	−1	−1	1	1	1	−1	539	539
2	1	−1	−1	−1	−1	1	1	363	363
3	−1	1	−1	−1	1	−1	1	658	658
4	1	1	−1	1	−1	−1	−1	480	480
5	−1	−1	1	1	−1	−1	1	697	697
6	1	−1	1	−1	1	−1	−1	413	413
7	−1	1	1	−1	−1	1	−1	543	543
8	1	1	1	1	1	1	1	635	635
Contrast	−136.5	76	62	93.5	40.5	−42	94.5		
Effect	−68.25	38	31	46.75	20.25	−21	47.25	541 (average)	

* : 0.538, i.e., 2% of 26.91 (grand average), is here employed as the parting value for efficacy of each effect, since there is no replicated run to generate the statistical error. This leads to the conclusion that error = −0.0125 (interactive effect of factors B and C)

** : 10.82, i.e., 2% of 541 (grand average), is here employed as the parting value for efficacy of each effect, since there is no replicated run to generate the statistical error. This leads to the conclusion that all main and interactive effects are valid

12.8.3. Magnetic Pulse Welding

Automotive steering wheel driveshaft is a prominent component which transfers the rotating movement from the steering wheel to tire wheels. Eccentric vibrations could happen in a mixture of torsional and bending that may result in undesired noise, vibration, and harshness (NVH) and thus shorten the fatigue life, when the driveshaft is designed using all aluminum for the purpose of weight reduction. Joining with dissimilar materials for a driveshaft that ends up with aluminum for the pipe and steel for the yoke is a necessity. A feasibility study on joining of an aluminum pipe (Al alloy 1070) to a steel yoke (SM 45C) is conducted by Shim et al. [2011], and the joint strength because of magnetic pulse welding has been investigated using full factorial design 2^3 augmented with axial and center points. Design factors and the related design matrix are listed in Table 12.8.6.

The objective is to reduce the angle of distortion (Y). By locating Minitab → Stat → Regression → Regression → Fit Regression Model and having the "most insignificant" factor or interaction that comes with the largest p-value weeded out one by one until all p-values are less than 10%, the final regression model is

$$Y_p = -6.61 + 1.326 \, A + 22.30 \, B - 3.46 \, C - 16.81 \, B^2$$

$$(R^2 = 98.8\%, \ R^2_{adj} = 98.2\%, \ \text{and} \ R^2_{pred} = 97.1\%). \tag{12.8.3}$$

The above predictive equation is effective with high unadjusted model correlation, adjusted model correlation, and predictive correlation. The corresponding ANOVA used for justifying Equation (12.8.3) is given in Table 12.8.7. The predicted angle of distortion for each treatment (run condition) is also given in Table 12.8.6 as column Y_p.

Let A = 8.5 kV. Then, Equation (12.8.3) reduces to

$$Y_p = 4.661 + 22.30 \, B - 3.46 \, C - 16.81 \, B^2. \tag{12.8.4}$$

The reduction in angular distortion becomes a problem of dimensional control. The angle of distortion can be mitigated dramatically by the increase in gap between the pipe and yoke rod (factor B) and reduction in the thickness of outer pipe, as shown in Figure 12.8.4. However, these decisions need to be checked against the required break torque strength of the joint.

TABLE 12.8.6 Full factorial design 2^3 augmented with axial and center points for magnetic pulse welding of automotive driveshaft made of dissimilar materials [Shim et al. 2011].

(I): Factors and their design levels			
Factor	**(−1)**	**(0)**	**(+1)**
A: Applied electric voltage (kV)	8.0	8.5	9.0
B: Gap between pipe and yoke rod (mm)	0.8	1.0	1.2
C: Thickness of outer pipe (mm)	0.6	0.8	1.0

(II): Design matrix and experimental results								
Run	**a**	**b**	**c**	**A**	**B**	**C**	**Y**	**Y_p**
1	−1	−1	−1	8.0	0.8	0.6	9.3	9.00
2	1	−1	−1	9.0	0.8	0.6	9.8	10.33
3	−1	1	−1	8.0	1.2	0.6	3.9	4.48
4	1	1	−1	9.0	1.2	0.6	5.8	5.80
5	−1	−1	1	8.0	0.8	1.0	7.8	7.62
6	1	−1	1	9.0	0.8	1.0	8.8	8.95
7	−1	1	1	8.0	1.2	1.0	3.4	3.09
8	1	1	1	9.0	1.2	1.0	3.9	4.42
9	−1.6	0	0	7.7	1.0	0.8	5.8	6.32
10	1.6	0	0	9.3	1.0	0.8	8.8	8.42
11	0	−1.5	0	8.5	0.7	0.8	9.3	9.27
12	0	1.5	0	8.5	1.3	0.8	2.9	2.47
13	0	0	−1.5	8.5	1.0	0.5	8.8	8.42
14	0	0	1.5	8.5	1.0	1.1	6.3	6.35
15	0	0	0	8.5	1.0	0.8	7.8	7.38

where
Y (MPa): Angle of distortion, measured
Y_p (MPa): Angle of distortion, predicted

TABLE 12.8.7 ANOVA for magnetic pulse welding of automotive steering wheel driveshaft.

Source of variance	SS_{adj}	DOF	MS_{adj}	F-value	p-Value
Regression	79.311	4	19.8279	98.26	0.0%
A	5.769	1	5.7691	28.59	0.0%
B	1.524	1	1.5241	7.55	2.1%
C	5.986	1	5.9858	29.66	0.1%
B^2	3.485	1	3.4854	17.27	0.2%
Error	2.018	10	0.2018	—	—
Subtotal	81.329	14	—	—	—
Grand average	—	1			
Total	—	15			

FIGURE 12.8.4 Response surface of joining distortion angle of an automotive steering wheel driveshaft as a function of gap and thickness of outer pipe.

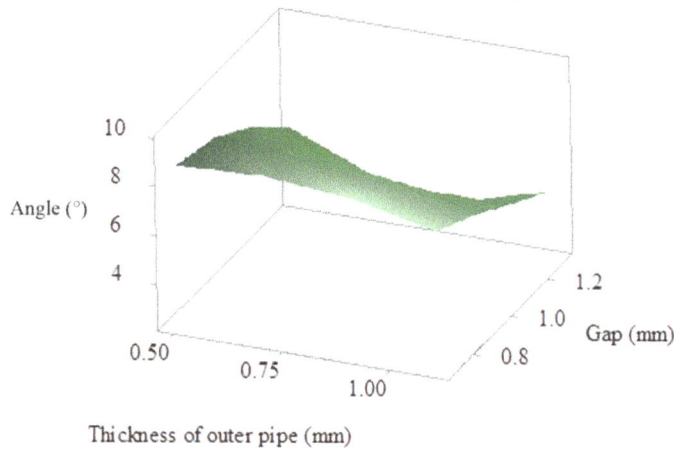

12.9. Ball Joints

Ball joints have been widely applied in automotive steering mechanisms and suspension systems. For example, a ball joint makes a connection between the knuckle and the lower control arm in a steering system. A general ball joint is made of a socket, bearing, plug, and ball stud, as demonstrated in Figure 12.9.1(a). The process of assembling these parts together, involving spinning and pressing, to produce such a ball joint is known as a caulking process, as shown in Figure 12.9.1(b).

FIGURE 12.9.1 Socket-ball joint and caulking machine [Sin and Lee 2014].

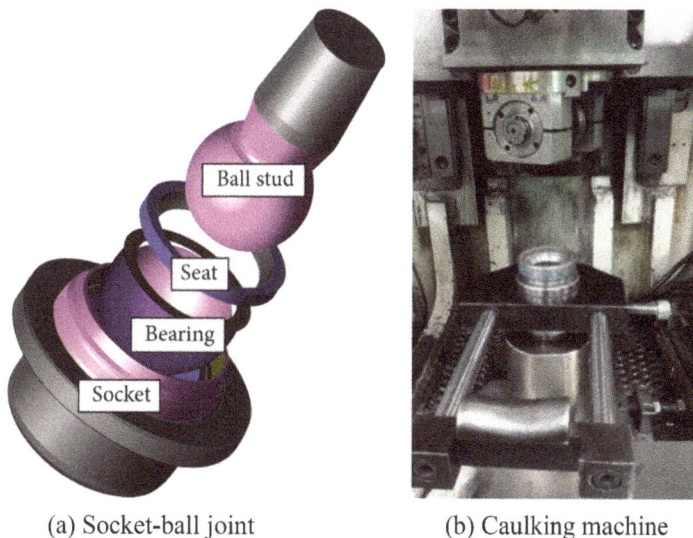

(a) Socket-ball joint　　　　(b) Caulking machine

A DOE has been conducted on the pull-out test of ball joints using 3_{III}^{3-1} [Sin and Lee 2014]. Three applied design levels of the three selected process parameters are listed in Table 12.9.1. Three distinct performance responses, i.e., pull-out force Y_{pull} (kN), caulking depth Y_c (mm), and maximum stress Y_s (MPa), are recorded against the process factors listed in Table 12.9.1. The caulking depth is defined as the deformed depth (bent length) of the socket. The pull-out force and maximum stress are evaluated at the moment of separation, as an indicator of the structural integrity in operation. The data of these three performance responses were obtained based on implicit multibody FEA according to fractional factorial design 3_{III}^{3-1}, and they are listed in Table 12.9.2. It was noticed that the pull-out strength is well correlated to the caulking depth (R = 93.14%). The maximum stress at the final stage is 840.7 (±5.5%) MPa, which means that the failure occurs approximately at the same stress level as expected.

TABLE 12.9.1 Process factors for ball joints in automotive steering systems [Sin and Lee 2014].

Factor	Level (−1)	Level (0)	Level (1)
A: Spinning (rpm)	200	300	400
B: Roller—axial travel distance (mm)	2.7	2.9	3.1
C: Roller—pressing time (sec)	0.1	0.5	0.9

TABLE 12.9.2 Pull-out test of ball joints.

Run	a	b	c	A	B	C	Y_{pull}	Y_c	Y_s
1	1	1	1	200	2.7	0.1	22.5	0.96	866
2	1	2	2	200	2.9	0.5	19.3	0.87	847
3	1	3	3	200	3.1	0.9	8.1	0.51	868
4	2	1	3	300	2.7	0.9	1.9	0.26	795
5	2	2	1	300	2.9	0.1	31.5	1.04	823
6	2	3	2	300	3.1	0.5	29	0.95	833
7	3	1	2	400	2.7	0.5	4.1	0.68	843
8	3	2	3	400	2.9	0.9	6.5	0.38	846
9	3	3	1	400	3.1	0.1	35	1.21	845

where
Y_{pull} (kN): Pull-out strength; data obtained from Sin and Lee [2014]
Y_c (mm): Caulking
Y_s (MPa): Maximum stress

The ANOVA of the pull-out strength (Y_{pull}) is carried out using the data, as given in Table 12.9.2. The objective is to increase the pull-out strength (Y_{pull}). By locating Minitab → Stat → Regression → Regression → Fit Regression Model and having the "most insignificant" factor or interaction that comes with the largest p-value weeded out one by one until all p-values are less than 10%, the final regression model for pull-out force is

$$Y_{pull} = -179.61 + 0.08117\ A + 74.151\ B + 84.82\ C + 0.11854\ A\ C - 51.93\ B\ C$$

$$- 0.000281\ A^2$$

$$(R = 100\%, R_{adj} = 99.99\%, \text{ and } R_{pred} = 99.94\%). \tag{12.9.1}$$

High correlation, adjusted correlation, and predictive correlation given above can be used for justifying the effectiveness and adequacy of the above equation.

TABLE 12.9.3 ANOVA of pull-out strength of socket-ball joint.

Source	SS_{adj}	DOF	MS_{adj}	F-value	p-Value (α)
Regression	1294.97	6	215.829	13,936.91	0.0%
A	2.29	1	2.288	147.76	0.7%
B	139.14	1	139.135	8984.52	0.0%
C	10.07	1	10.068	650.16	0.2%
AC	44.97	1	44.967	2903.69	0.0%
BC	27.61	1	27.611	1782.98	0.1%
A^2	12.00	1	12.000	774.89	0.1%
Error	0.03	2	0.015	—	—
Subtotal	1295.00	8			
Grand average	—	1			
Total	—	9			

The maximum value of pull-out strength can be obtained by taking first differentiations of Y_{pull} with respect to factors A, B, and C and setting them to zero, respectively:

$$\partial Y_{pull} / \partial A = 0.08117 + 0.11854\ C - 0.000562\ A = 0$$

$$\partial Y_{pull} / \partial B = 74.151 - 51.93\ C = 0$$

and $$\partial Y_{pull} / \partial C = 84.82 + 0.11854\ A - 51.93\ B = 0.$$

These three equations can be solved for three unknowns: A = 445.61, B = 2.645, and C = 1.4279, which would yield the minimum pull-out strength since $\partial^2 Y_{pull}/\partial A^2 = -0.000562 < 0$.

In order to maximize the torque within the treatments (run conditions) for these three factors, one may choose A = 200, B = 3.1, and C = 0.1.

According to the MS_{adj} (adjusted mean squares) given in Table 12.9.3, factor B is the most significant contributor to the pull-out strength. Assume that factor A (spin speed) is set at 200 rpm, i.e., operated at level −1. Then, Equation (12.9.1) becomes

$$Y_{pull} = -174.616 + 74.151\ B + 108.528\ C - 51.93\ B\ C. \qquad (12.9.2)$$

The corresponding 3D surface plot is given in Figure 12.9.2. The pull-out strength (kN) can be improved dramatically by the increase in axial travel distance of the roller (factor B) and reduction in pressing time. However, the decision has been checked against the required caulking depth of the joint.

FIGURE 12.9.2 Pull-out force (N) of a ball joint as a function of axial travel distance (mm) and pressing time (sec) of the roller.

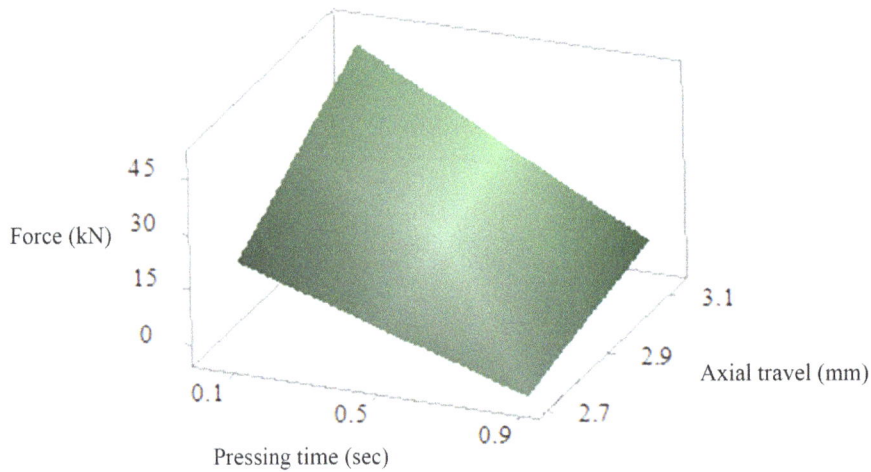

12.10. Mechanical Seals

A mechanical seal is a device used to seal fluids or soft material within a limited space. The sealing mechanism consists of two main components with flat or circular faces in contact with each other. Typical types of mechanical seals include

1. Laminated seals for flexible packaging [Dixon et al. 2006, Jang and Ahn 2016, Ilhan et al. 2021]
2. O-rings [Koga et al. 2011, Biredar et al. 2021]
3. Radial lip seals for rotating shafts [Simons 1966]
4. Gaskets

On the macro-scale, two faces in contact may have a relative motion as designed (e.g., radial lip seals for rotating shafts) or remain relatively stationary (e.g., gaskets between a cylinder head and an engine block). Nevertheless, micro-relative motions between any two surfaces in contact may result from dynamic operations, even if they are fastened together by bolts.

12.10.1. Laminated Seals for Flexible Packaging

A product, or material, is inserted into the pouch, and then, it is sealed by heat or other mechanisms (e.g., ultrasonic devices) across sides to create a sealed package. Applications include, but are not limited to,

1. Battery pouches [Jang and Ahn 2016]
2. Food pouches [Ilhan et al. 2021]
3. Medical pouches [Dixon et al. 2006]

For evaluating the performance and reliability of a two-layered laminate, i.e., polyethylene terephthalate (PET) (24 µm)/low-density polyethylene (LDPE)-metallocene linear LDPE (mLLDPE)-C4 (40 µm) for sealing flexible material in a package, an experimental design is carried out by D'huys et al. [2019]. The laminate is selected for the study because LDPE is a sealing medium that is commonly used in commercial packaging films and the PET outer layer enhances the mechanical and barrier properties. In this process of flexible packaging, seal settings include

- A (sec): Time duration, when the ultrasonic device is loaded
- B (N/mm): Line force applied at sealing jaw
- C (μm): Amplitude of the applied ultrasonic wave

The objective aims at high seal strength, limited ultrasonic horn displacement, and low energy consumption. After the USW is completed, a holding force of 2 N/mm is applied for 0.5 sec to finalize the sealing process. The design matrix and test results are listed in Table 12.10.1.

TABLE 12.10.1 ANOVA for laminated seals in flexible packaging [D'huys et al. 2019].

Run	A	B	C	S	H	E
1	0.2	2	36	1.31	10	27.75
2	0.3	4	18	0.30	10	21.93
3	0.1	4	36	1.76	40	17.76
4	0.2	2	18	0.05	10	7.50
5	0.2	4	27	2.90	30	21.05
6	0.2	4	27	2.43	30	20.10
7	0.1	6	27	2.90	80	16.30
8	0.2	4	27	2.09	30	17.65
9	0.3	6	27	2.11	60	47.62
10	0.2	6	18	1.18	40	22.58
11	0.1	2	27	0.96	10	7.55
12	0.2	6	36	3.52	70	59.64
13	0.3	2	27	0.74	10	15.61
14	0.3	4	36	3.22	50	55.86
15	0.1	4	18	0.52	20	5.89

where
A (sec): Time duration, when the ultrasonic device is loaded, $0.1 \leq A \leq 0.3$
B (N/mm): Line force applied at sealing jaw, $2 \leq B \leq 4$
C (μm): Amplitude of the applied ultrasonic wave, $18 \leq C \leq 36$
S (N/mm): Strength of seal, the larger the better
H (mm): Horn displacement, the smaller the better
E (J): Energy consumption, the smaller the better

The data analysis is carried out using Minitab (Stat → Regression → Regression → Fit Regression Model). The "most insignificant" factor or interaction that comes with the largest p-value is weeded out one by one until all p-values are less than 10%, which is here defined as the parting significance level based on the F-distribution. In light of the ANOVA presented in Table 12.10.2, the final predictive equations for the seal strength, ultrasonic horn displacement, and energy consumption are, respectively,

$$S_p = -6.04 + 0.4156 B + 0.379 C - 34.1 A^2 - 0.00692 C^2 + 0.514 A C$$

$$(R = 94.4\%, R_{adj} = 91.1\%, \text{ and } R_{pred} = 81.0\%), \tag{12.10.1}$$

$$H_p = -11.32 + 0.629 B^2 + 0.3079 B C$$

$$(R = 95.2\%, R_{adj} = 94.4\%, \text{ and } R_{pred} = 91.4\%), \tag{12.10.2}$$

and $$E_p = 90.0 - 164.9 A - 12.95 B - 5.23 C + 0.790 B^2 + 0.0833 C^2$$

$$+ 29.07 A B + 6.13 A C + 0.2335 B C$$

$$(R = 99.3\%, R_{adj} = 98.3\%, \text{ and } R_{pred} = 92.8\%). \tag{12.10.3}$$

These three predictive equations are effective and adequate with high correlations, including unadjusted model correlation, adjusted model correlation, and predictive correlation.

The criterion space for the laminated seal for flexible packaging corresponding to the design space presented by the design space given in Table 12.10.1 is depicted in Figure 12.10.1. A high seal strength can be achieved with a low horn displacement at the expense of high energy. It can also be achieved with a high horn displacement at low energy consumption.

TABLE 12.10.2 ANOVA tables for laminated seals in flexible packaging.

Y_s: Source	SS_{adj}	DOF	MS_{adj}	$F_{u,v}$	p-Value
Regression	15.4199	5	3.0840	14.68	0.0%
B	5.5278	1	5.5278	26.32	0.1%
C	1.1340	1	1.1340	5.40	4.5%
A^2	1.2181	1	1.2181	5.80	3.9%
C^2	1.1706	1	1.1706	5.57	4.3%
AC	1.2895	1	1.2895	6.14	3.5%
Error		9	1.8902	0.21	
Lack of fit	1.5594	7	0.2228	1.35	49%
Pure error	0.3309	2	0.1654		
Subtotal	17.3101	14			
Grand average		1			
Total	—	15			

Y_d: Source	SS_{adj}	DOF	MS_{adj}	$F_{u,v}$	p-Value
Regression	6826.7	2	3413.34	57.96	0.0%
BB	293.3	1	293.34	4.98	4.5%
BC	1168.7	1	1168.70	19.85	0.1%
Error	706.7	12	58.89		
Lack of fit	406.7	6	67.78	1.36	36.1%
Pure error	300.0	6	50.00		
Subtotal	7533.3	14			
Grand average		1			
Total	—	15			

Y_e: Source	SS_{adj}	DOF	MS_{adj}	$F_{u,v}$	p-Value
Regression	3907.80	8	488.475	50.63	0.0%
A	80.52	1	80.520	8.35	2.8%
B	87.37	1	87.375	9.06	2.4%
C	187.27	1	187.270	19.41	0.5%
B^2	37.08	1	37.076	3.84	9.8%
C^2	169.20	1	169.203	17.5	0.6%
AB	135.26	1	135.257	14.02	1%
AC	121.66	1	121.661	12.61	1.2%
BC	70.64	1	70.644	7.32	3.5%
Error	57.88	6	9.647		
Lack of fit	51.73	4	12.932	4.20	20.1%
Pure error	6.16	2	3.078		
Subtotal	3965.69	14			
Grand average		1			
Total	—	15			

FIGURE 12.10.1 Criterion space for the laminated seal for flexible packaging.

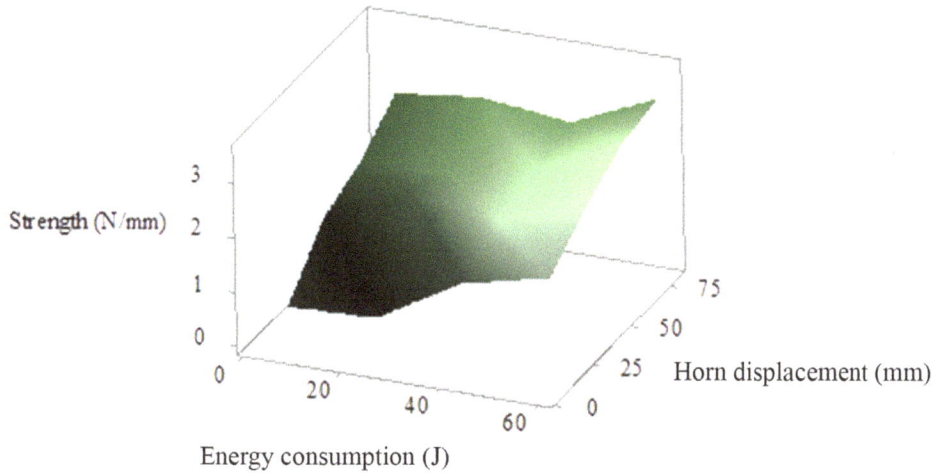

12.10.2. Sustainable O-rings for Sealing Hydrogen in Storage

Rubber O-rings for sealing hydrogen in storage are another major concern in the highly pressurized hydrogen environment. An experimental design was set up to investigate the potential blister fracture of rubber O-rings for sealing highly pressurized hydrogen. Blister fracture, also called explosive decompression failure (XDF), within an O-ring may occur when the sealed highly pressurized hydrogen is decompressed suddenly. As listed in Table 12.10.3, eight potential packaging parameters that may have influences on blistering fracture of rubber O-rings subjected to highly pressurized hydrogen have been investigated by Koga et al. [2011]. A cyclic pressurization cycle test is defined as follows:

(a) The rubber O-ring is pressurized at the upper pressure limit (factor E) until the amount of gas permeation is in the stable condition with the assigned holding time (factor F).

(b) Then, it is released to the lower pressure limit (factor A) with the assigned holding time (factor G).

The cyclic pressurization test is then repeated ten times for each experimental treatment (run) prescribed in Table 12.10.4. After the cyclic pressurization tests, the mechanical strength of each rubber O-ring was assessed using a tensile test.

A blister is defined as a cavity filled with hydrogen with sizes varying in several sub-micrometers that are hardly observed by optical microscopy. A crack is defined as a cavity filled with hydrogen with sizes of more than 1 μm, which can easily be observed by optical microscopy. Because there exist some invisible cracks and blisters as initiated from the interior of a rubber O-ring when subjected to the cyclic pressurization test, the strength reduction ratio will be used as an indirect index for generalizing the severity of embrittlement:

$$Y = (F_{fracture,\ original} - F_{fracture,\ after\text{-}test}) / F_{fracture,\ original},$$

where

$F_{fracture,\ original}$ (N) is the force required to break the O-ring before the embrittlement test

$F_{fracture,\ after\text{-}test}$ is the force required to break the O-ring after the embrittlement test

Strength degradation Y is attributed partially to the volume increase in the rubber O-ring because swelling may extrude the O-ring and initiate surface crack. All in all, Y follows the rule that the small the better.

An experimental design was set up to investigate the potential blister fracture of rubber O-rings for sealing highly pressurized hydrogen. Blister fracture, also called XDF, within an O-ring may occur when the sealed highly pressurized hydrogen is decompressed suddenly. Eight factors that have potential influences on the blister fracture of rubber O-rings are identified in Table 12.10.3 [Koga et al. 2011]. Factorial design $2 \times 3^{7-5}_{III}$ is employed for assessing the potential blister fracture, and test results are listed in Table 12.10.4. The loss of O-ring sealing is defined as

$$Y = (F_0 - F) / F_0, \qquad\qquad (12.10.4)$$

where

  Y is the O-ring degradation, i.e., loss of O-ring sealing capacity; the smaller the better
  F is the force that causes the O-ring to fracture after the pressure cycle test
  F_0 is the force that causes the new O-ring to fracture

Two replicated runs are conducted for each treatment, as listed in columns Y_1 and Y_2 of Table 12.10.4.

TABLE 12.10.3 Design factors assigned for assessing potential failure of rubber O-rings subject to highly pressurized hydrogen [Koga et al. 2011].

Factor	Level 1	Level 2	Level 3
A (MPa): Lower pressure limit	8	1	
B (Durometer hardness A80): material	EPDM	HNBR	VMQ
C (°C): Ambient temperature	100	30	0
D (%): O-ring filling ratio (FR)	86	77	67
E (MPa): Upper pressure limit	90	35	10
F (sec): Holding time at upper pressure limit	120	60	30
G (sec): Holding time at lower pressure limit	120	60	30
H (sec): Decompression time	60	10	3

where
EPDM: Ethylene-propylene-diene monomer rubber
VMQ: Silicone rubber
HNBR: Hydrogenated acrylonitrile butadiene rubber

TABLE 12.10.4 Factorial design $2 \times 3^{7-5}_{III}$ for assessing potential failure of rubber O-rings subject to highly pressurized hydrogen [Koga et al. 2011].

Run	A	B	C	D	E	F	G	H	Y_1	Y_2	Y
1	8	EPDM	100	86	90	120	120	60	0.615	0.688	0.638
2	8	EPDM	30	77	35	60	60	10	0.150	0.100	0.124
3	8	EPDM	0	67	10	30	30	3	0.238	0.250	0.236
4	8	VMQ	100	86	35	60	30	3	0.436	0.443	0.444
5	8	VMQ	30	77	10	30	120	60	0.000	0.007	0.026
6	8	VMQ	0	67	90	120	60	10	0.043	0.100	0.114
7	8	HNBR	100	77	90	30	60	3	0.477	0.502	0.507
8	8	HNBR	30	67	35	120	30	60	0.174	0.138	0.140
9	8	HNBR	0	86	10	60	120	10	0.116	0.089	0.090
10	1	EPDM	100	67	10	60	60	60	0.075	0.160	0.161
11	1	EPDM	30	86	90	30	30	10	0.160	0.255	0.224
12	1	EPDM	0	77	35	120	120	3	0.140	0.243	0.178
13	1	VMQ	100	77	10	120	30	10	0.043	0.043	0.035
14	1	VMQ	30	67	90	60	120	3	0.093	0.129	0.089
15	1	VMQ	0	86	35	30	60	60	0.021	0.057	0.025
16	1	HNBR	100	67	35	30	120	10	0.045	0.167	0.090
17	1	HNBR	30	86	10	120	60	3	0.147	0.128	0.180
18	1	HNBR	0	77	90	60	30	60	0.031	0.016	0.032

The data analysis is carried out using Minitab (Stat → Regression → Regression → Fit Regression Model). The "most insignificant" factor or interaction that comes with the largest p-value is weeded out one by one until all p-values are less than 10%, which is here defined as the parting significance level based on the F-distribution. In light of the ANOVA given in Table 12.10.6, the regression models of O-ring degradation, with three different correlations, are given as follows:

$$Y_{EPDM} = 3.329 + 0.02074\ A - 0.0884\ D + 0.001826\ E - 0.00282\ F - 0.02825\ H$$

$$+ 0.000020\ C^2 + 0.000622\ D^2 + 0.000021\ F^2 + 0.000420\ H^2 . \tag{12.10.5}$$

$$Y_{HNBR} = 3.242 + 0.02074\ A - 0.0884\ D + 0.001826\ E - 0.00282\ F - 0.02825\ H$$

$$+ 0.000020\ C^2 + 0.000622\ D^2 + 0.000021\ F^2 + 0.000420\ H^2. \tag{12.10.6}$$

And $\quad Y_{VMQ} = 3.191 + 0.02074\ A - 0.0884\ D + 0.001826\ E - 0.00282\ F - 0.02825\ H$

$$+ 0.000020\ C^2 + 0.000622\ D^2 + 0.000021\ F^2 + 0.000420\ H^2. \tag{12.10.7}$$

($R = 97.8\%$, $R_{adj} = 96.8\%$, and $R_{pred} = 95.0\%$).

Predictive equations given above are effective and adequate with high correlations, including unadjusted model correlation, adjusted model correlation, and predictive correlation, while the model adequacy is evidenced by the insignificance of lack of fit (Table 12.10.5). The predicted values for these 18 experimental treatments are listed in column Y_p of Table 12.10.4.

TABLE 12.10.5 ANOVA for sealing H_2 by rubber O-rings.

Source	SS_{adj}	DOF	MS_{adj}	$F_{u,v}$	p-Value
Regression	1.02774	11	0.093431	48.99	0.0%
A	0.18966	1	0.189660	99.45	0.0%
B	0.11723	2	0.058617	30.74	0.0%
D	0.02165	1	0.021655	11.36	0.3%
E	0.13404	1	0.134037	70.29	0.0%
F	0.00816	1	0.008160	4.28	5.0%
H	0.15155	1	0.151555	79.47	0.0%
C^2	0.28971	1	0.289710	151.92	0.0%
D^2	0.02507	1	0.025073	13.15	0.1%
F^2	0.01078	1	0.010784	5.65	2.6%
H^2	0.14502	1	0.145016	76.04	0.0%
Error	0.04577	24	0.001907		
Lack of fit	0.01632	6	0.002721	1.66	18.8%
Pure error	0.02945	18	0.001636		
Subtotal	1.07350	35			
Grand average	—	1			
Total	—	36			

Conclusions drawn from the study on sustainability of O-rings for sealing hydrogen in storage are addressed as follows:

1. Strength reduction ratios of these three different kinds of rubber differ from one another, as implied by the p-value of factor B, which is statistically significant (Table 12.10.5). VMQ performs better than HNBR; EPDM is the worst among these three.

2. The strength reduction ratio increases linearly with increasing (factor A) and increasing upper pressure limit (factor E), respectively.

3. The strength reduction ratio varies quadratically with rubber O-ring FR (factor D), which is related to the material properties. Taking $\partial Y_{VMQ}/\partial D = 0$, one would obtain that D = 71. Since $\partial^2 Y_{VMQ}/\partial D^2 = 0.001244 > 0$, the rubber O-ring causes the least strength reduction at the FR of 71%.

4. The strength reduction ratio also varies quadratically with increasing ambient temperature (factor C), hold time at upper pressure limit (factor F), and decompression time (factor H), respectively. However, the influence of upper pressure limit (factor F) is minor in light of its MS_{adj} value.

According to the MS_{adj} data (Table 12.10.5), factor A (power pressure limit), factor C (ambient temperature), and factor H (decompression time) are the top three factors that have tremendous impact on the strength reduction. Since the influence of power pressure limit (factor A) on the strength reduction is linear (the less the better), focus will be on how the strength gets reduced by factor C and factor H. Assume that A = 4.5 MPa, B = VMQ, D = 77%, E = 35 MPa, F = 50 sec, and G = 60 sec. Then, Equation (12.10.7) reduces to

$$Y_{VMQ} = 0.16249 - 0.02821\,H + 0.00002\,C^2 + 0.000419\,H^2. \qquad (12.10.8a)$$

The response surface plot corresponding to the above equation is demonstrated in Figure 12.10.2. It can be seen that the impact of ambient temperature on the strength reduction is accelerated beyond a critical temperature. The impact of decompression time on the strength reduction reaches the minimum at 33.9 sec, as derived from $\partial Y_{VMQ}/\partial H = 0$. It suggests that there might be two or more driving mechanisms behind the decompression time (factor H). In conclusion, strength reduction can be mitigated by varying the decompression time, while the ambient temperature should be well controlled in real-world applications such as fuel-cell-based EVs.

© SAE International.

FIGURE 12.10.2 Strength reduction (Y) as a function of ambient temperature and decompression time.

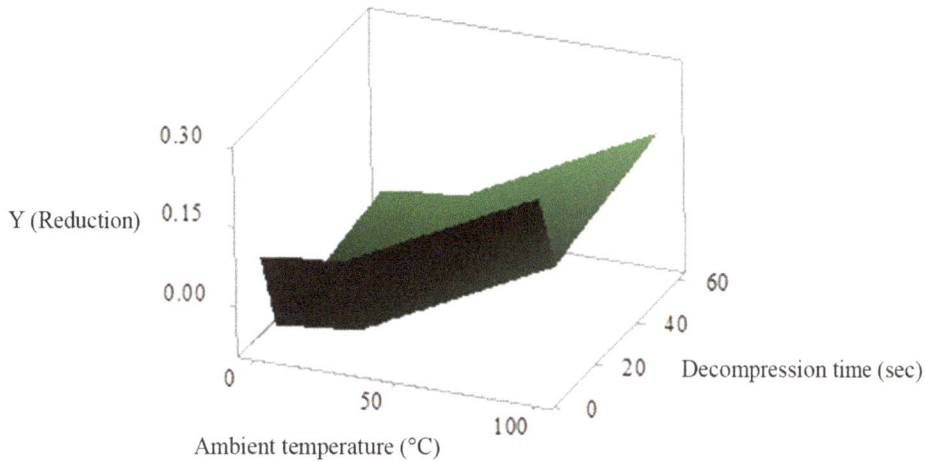

12.10.3. Radial Lip Seals for Rotating Parts

Proper functioning of rotating parts in a vehicle (e.g., shafts in automotive transmissions) relies on their related mechanical seals to keep fluids, such as lubricants, confined in the compartment.

A schematic drawing of a radial lip seal joint is shown in Figure 12.10.3 [Sinzara et al. 2018].

Reproduced with permission from Sinzara, Sherrington, Smith, Brooks, & Onsy, (2018). Effects of Eccentric Loading on Lip Seal Performance.

FIGURE 12.10.3 Schematic drawing of a radial lip seal joint [Sinzara et al. 2018].

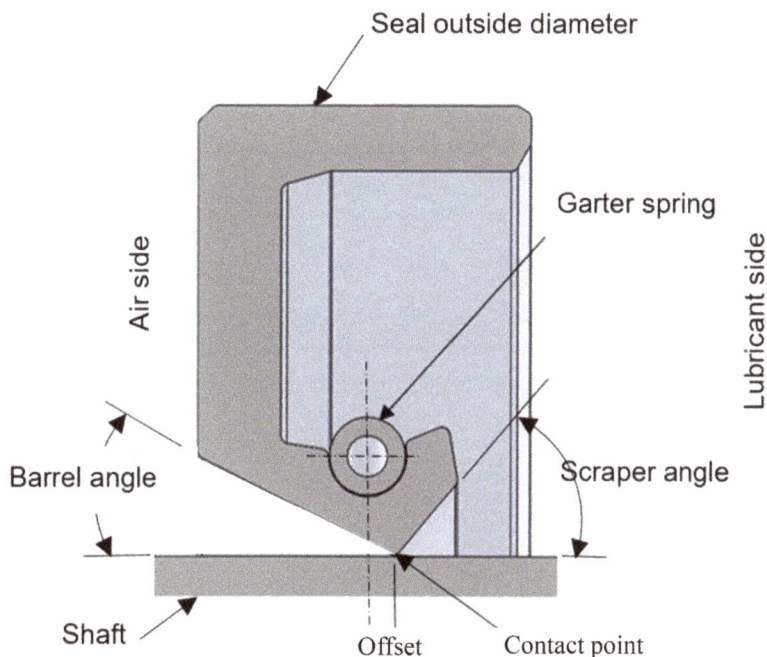

An experimental design is set up to investigate the effects of seven design parameters of a radial lip-type oil seal by Simons [1966]. Design parameters and their related design levels are listed in Table 12.10.6. Fractional factorial design 2_{IV}^{7-3}, as shown in Table 12.10.7, is employed to explore the main effects of individual factors and their related two-factor interactions.

TABLE 12.10.6 Design factors of a radial lip-type oil seal [Simons 1966].

Factors	Level (−)	Level (+)	Coded variable
A (mm): Trim diameter	46.736	47.12	a = (A − 47.12)/0.191
B (N/mm): Radial line force at lip	0.070	0.128	b = (B − 0.099)/0.029
C (mm): Shaft-to-bore offset	0.3048	0.635	c = (C − 0.470)/0.165
D (mm): Spring position (offset)	1.016	1.27	d = (D − 1.143)/0.127
E (mm): Contact path width	0.127	0.254	e = (E − 0.191)/0.064
F (mm): Flex section thickness	0.889	1.143	f = (F − 1.016)/0.127
G (MPa): Seal modulus (93.3°C)	0.1689	0.296	g = (G − 0.233)/0.064

Note: The trim diameter (factor A) refers to the free lip identity (ID) without a garter spring

© SAE International.

TABLE 12.10.7 Factorial design 2_{IV}^{7-3} (E = ABC, F = BCD, and G = ABD) for evaluating the leakage of a radial lip-type oil seal [Simons 1966].

Run	a	b	c	d	e	f	g	Y	Y_p
1	−1	−1	−1	−1	−1	−1	−1	5.726	5.71
2	1	1	1	1	1	1	1	1.682	1.63
3	1	1	1	1	−1	−1	−1	0	0.06
4	−1	−1	−1	−1	1	1	1	6.09	6.11
5	−1	−1	1	1	−1	1	−1	0	−0.06
6	1	1	−1	−1	1	−1	1	1.908	1.89
7	1	1	−1	−1	−1	1	−1	7.715	7.73
8	−1	−1	1	1	1	−1	1	1.147	1.20
9	1	−1	1	−1	1	−1	−1	33.136	33.12
10	−1	1	−1	1	−1	1	1	0.006	−0.05
11	−1	1	−1	1	1	−1	−1	0.094	0.15
12	1	−1	1	−1	−1	1	1	1.465	1.48
13	1	−1	−1	1	1	1	−1	0.005	−0.05
14	−1	1	1	−1	−1	−1	1	0.004	−0.01
15	−1	1	1	−1	1	1	−1	25.372	25.39
16	1	−1	−1	1	−1	−1	1	0.003	0.06

© SAE International.

The data analysis is carried out using Minitab (Stat → DOE → Factorial → ···). After insignificant effects and interactions that come with less than 5% (which is here regarded as the engineering significance) of the average are removed, the final predictive equation for the rate of leakage is

$$Y_p = 5.272 + 0.4672\,a - 0.6744\,b + 2.579\,c - 4.905\,d + 3.407\,e - 3.734\,g$$

$$- 2.239\,a\,b + 0.7528\,a\,c - 0.4118\,a\,d - 3.042\,a\,f - 0.7408\,a\,g + 4.076\,b\,f$$

$$+ 0.8184\,a\,b\,f. \tag{12.10.8b}$$

The predicted values of these 16 treatments (Table 12.10.9) are listed in column Y_p with a great correlation between Y_p and Y (R = 99.9990%).

The Pareto chart of effects and interactions corresponding to Equation (12.10.8a) is displayed in Figure 12.10.4. The main effects of factors A, B, C, D, E, and G are statistically significant, and so are some interactions. The finding resembles what was presented by Simons [1966]. Because the design matrix (Table 12.10.9) comes with resolution IV, these main effects may come true with the assumption that the existence of each three-factor interaction is remote. Should it be true, one may reduce the leakage by the following setting (within the test ranges listed in Table 12.10.8):

- Factor D: Spring position (offset) = 1.27 mm; the larger the better
- Factor G: Seal modulus (93.3°C) = 0.296 MPa; the larger the better
- Factor E: Contact path width = 0.127 mm; the smaller the better
- Factor C: Shaft-to-bore offset = 0.3048 mm; the smaller the better
- Factor B: Radial line force at lip = 0.128 N/mm; the larger the better
- Factor A: Trim diameter 46.736 mm; the smaller the better
- Factor F: Flex section thickness is not statistically significant.

As a DOE with resolution IV, the confounding patterns include the following terms:

$$AB \rightarrow AB + CD + EG$$
$$AC \rightarrow AC + BD + FG$$
$$AD \rightarrow AD + BC + EF$$
$$AE \rightarrow AE + BG + DF$$
$$AF \rightarrow AF + CG + DE \text{ (Not statistically significant)}$$
$$AG \rightarrow AG + BE + CF$$
$$\text{and} \quad BF \rightarrow BF + CE + DG.$$

Two-factor interaction AF of all the two-factor interactions is the only one that is not statistically significant. More treatments (tests) are required for resolving significant confounded two-factor interactions.

FIGURE 12.10.4 Pareto chart of effects and interactions.

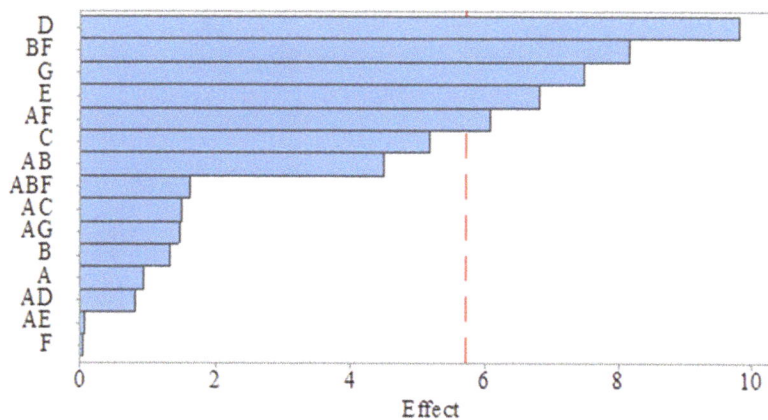

12.10.4. Expansion Joints

Flexible bellows are often used as expandable/compressible components for mechanical sealings in a pipeline network to balance axial movements and retaining forces. The convolution portion of a metal bellow acts like an axial spring, which can extend or contract axially with appropriate forces. The number of convolutions required for a flexible bellow, made of AISI 304 austenitic stainless steel, depends on the operating stress level that eventually affects its fatigue life. According to Expansion Joints Manufacturing Association (EJMA) standards, the maximum meridional stress for each individual convolution subjected to an axial displacement is

$$Y_m = \left(\frac{5\,C\,E}{3\,A^2\,K}\right) D, \qquad (12.10.9)$$

where

$\quad$ Y_m (MPa) is the maximum meridional stress, according to EJMA
$\quad$ A (mm) is the corrugation pitch (length)
$\quad$ C (mm) is the wall thickness
$\quad$ D (mm) is the axial displacement
$\quad$ E (MPa) is the Young's modulus
$\quad$ K is the dimensionless constant

The relationship between the maximum meridional stress and axial displacement of a stainless bellow (AISI 304) is disclosed using finite element models loaded at an axial displacement of 1 mm (D = 1 mm) per one pitch according to the design matrix given in Table 12.10.8 [Kumar et al. 2017, Table 2]. The design configuration with nominal dimensions of the stainless bellow is depicted in Figure 12.10.5. Test results, read from Fig. 8 of Kumar et al. [2017], are recorded in column Y of Table 12.10.8.

FIGURE 12.10.5 Nominal dimensions of the stainless bellow, where A = 15 mm, B = 3.3 mm, and C = 0.5 mm [Kumar et al. 2017, Fig. 2].

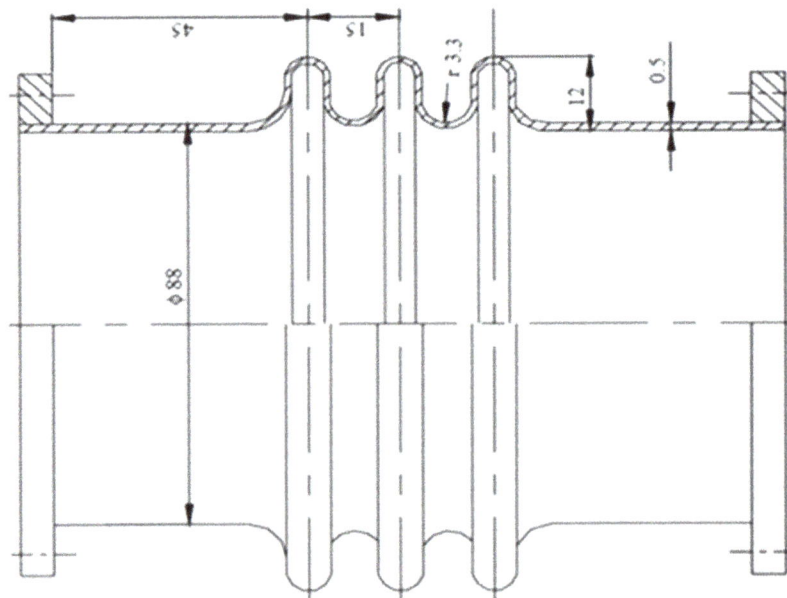

TABLE 12.10.8 Composite design for examining the maximum meridional stress (MPa) of a stainless bellow joint with various dimensions at 1-mm axial displacement per pitch.

Run	A	B	C	Y	Y_p
1	10.8	2.97	0.50	235	230.4
2	13.2	2.97	0.50	170	163.5
3	10.8	3.63	0.50	175	180.3
4	13.2	3.63	0.50	145	159.4
5	10.8	3.30	0.45	180	183.4
6	13.2	3.30	0.45	145	139.5
7	10.8	3.30	0.55	225	229.6
8	13.2	3.30	0.55	180	185.7
9	12.0	2.97	0.45	174	181.8
10	12.0	3.63	0.45	172	154.7
11	12.0	2.97	0.55	235	228.0
12	12.0	3.63	0.55	215	201.0
13	12.0	3.30	0.50	185	190.2

where
Y (MPa): Maximum meridional stress from FEA
Y_p (MPa): Predicted maximum meridional stress using Equation (12.10.10)
B (mm): Convolution radius

The data analysis is carried out using Minitab (Stat → Regression → Regression → Fit Regression Model). The "most insignificant" factor or interaction that comes with the largest p-value is weeded out one by one until all p-values are less than 10%, which is here defined as the parting significance level based on the F-distribution. In light of the ANOVA presented in Table 12.10.9, the regression model is identified as follows:

$$Y_p = 894 - 4.75\,A^2 - 389\,B + 462.1\,C^2 + 29.0\,A\,B$$

$$(R = 95.4\%, R_{adj} = 93.0\%, R_{pred} = 85.7\%). \tag{12.10.10}$$

The predictive equation given above is effective and adequate with good correlations, including unadjusted model correlation, adjusted model correlation, and predictive correlation. The predicted values for these 13 experimental treatments are put in column Y_p of Table 12.10.8. Some new discoveries due to this study are summarized as follows:

1. According to Equation (12.10.10), the wall thickness (factor C) has the most influence on the stress level, followed by corrugation pitch (factor A). The maximum meridional stress increases nonlinearly with respect to an increasing wall thickness, but it reduces nonlinearly with respect to the increasing corrugation pitch.

2. The effect of the interaction between wall thickness (factor C) and corrugation pitch (factor A) on the stress level is also statistically significant.

The findings given above differ very much from what was published in Kumar et al. [2017].

Conclusions drawn from this study are valid only for the factors that fall within the ranges listed in Table 12.10.8. The optimal solution was attained using the least wall thickness, i.e., C = 0.45 mm. Then, Equation (12.10.10) reduces to

$$Y_p = 987.575 - 4.75\,A^2 - 389\,B + 29.0\,A\,B. \tag{12.10.11}$$

In light of the above equation (wall thickness = 0.45 mm), a surface plot of the maximum meridional stress versus factor A (corrugation pitch) and factor B (convolution radius) is exhibited in Figure 12.10.6. The maximum meridional stress decreases with respect to the convolution radius (factor B) very much when the corrugation pitch is short (e.g., A = 10.8 mm). Nevertheless, the influence of the convolution radius (factor B) is statistically insignificant when the corrugation pitch is long (e.g., A = 13.2 mm). In other words, Equation (12.10.9) bestowed by EJMA standards is accurate on the condition that the corrugation pitch is large enough.

Given that the wall thickness C = 0.45 mm, the optimal solution can be calculated using

$$\partial Y_p/\partial A = 0 \text{ and } \partial Y_p/\partial B = 0, \tag{12.10.12}$$

which lead to A = 13.41 mm and B = 4.39 mm. Substituting these data into Equation (12.10.11) minimizes the maximum meridional stress as

$$Y_p = 132.91 \text{ MPa}. \tag{12.10.13}$$

In the meanwhile, a similar solution can be obtained from Equation (12.10.9). Substituting A = 13.41 mm, B = 4.39 mm, and C = 0.45 mm into Equation (12.10.9) with constant K = 2π yields

$$Y_m = 132.76 \text{ MPa}. \tag{12.10.14}$$

The difference between Y_m = 132.76 MPa and Y_p = 132.91 MPa falls within the numerical error. It elucidates that Equations (12.10.10) and (12.10.9) lead to the same optimal solution as long as the corrugation pitch is long (e.g., A = 13.2 mm) and the bellow wall is thin (e.g., C = 0.45 mm).

TABLE 12.10.9 ANOVA for examining the relationship between the maximum meridional stress and axial displacement of a stainless bellow joint.

Source	SS_{adj}	DOF	MS_{adj}	$F_{u,v}$	p-Value
Regression	10,270	4	2568	20.12	0.0%
B	780.3	1	780.3	6.11	3.9%
A^2	890.6	1	890.6	6.98	3.0%
C^2	4275.3	1	4275	33.5	0.0%
AB	629.8	1	629.8	4.93	5.7%
Error	1021	8	127.6		
Subtotal	11,291	12			
Grand average	—	1			
Total	—	13			

© SAE International.

FIGURE 12.10.6 Surface plot of the maximum meridional stress versus factor A (corrugation pitch) and factor B (convolution radius).

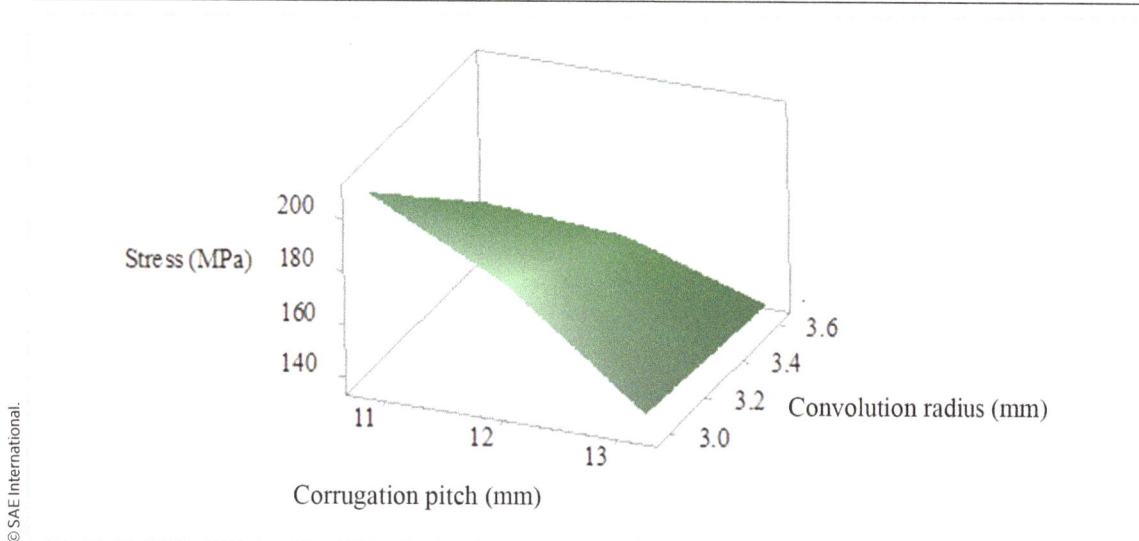

© SAE International.

References

Aland, K., Elangovan, S., and Rathinasuriyan, C. (2018), "Modeling and Prediction of Weld Strength in Ultrasonic Metal Welding Process Using Artificial Neural Network and Multiple Regression Method," *Material Science & Engineering International Journal*, 2(2), pp. 40-47.

Amancio-Filho, S. T., Beyer, M., and dos Santos, J. (2009), "Method for Connecting a Metallic Bolt to a Plastic Piece," US 7.575.149 B2.

Archer, D. (2009), "Fastening Threads: Torque Control vs. Angle Control," Assembly, December 23, 2009.

Ashely, S. (1993), "Failure Analysis Beats Murphy's Law," Mechanical Engineering, ASME.

ASTM F959/F959M-17a (2022), "Standard Specification for Compressible-Washer-Type Direct Tension Indicators for Use with Structural Fasteners, Inch and Metric Series," May 27, 2022.

AWS D8.9M:2012 (2012), *Test Methods for Evaluating the Resistance Spot Welding Behavior of Automotive Sheet Steel Materials*, 3rd Edition, American Welding Society, Approved by the ANSI.

Bajracharya, B. (2006), "Effects of Variations of Riveting Process on the Quality of Riveted Joints," MS Thesis, Wichita State University, WA, December 2006.

Bhattacharya, A., Sen, A., and Das, S. (2010), "An Investigation on the Anti-Loosening Characteristic of the Threaded Fasteners under Vibratory Conditions," *Mechanism and Machine Theory*, 45, pp. 1215-1225.

Bickford, J. H. (1995), *Introduction to the Design and Behavior of Bolted Joints*, Marcel Dekker, New York.

Bijak-Zochowski, M. et al. (1991), "Reduction of Contact Stress by Use of Relief Notches," *Experimental Mechanics*, 31(3), pp. 271-275.

Biredar, A. et al. (2021), "Design of Experimental Setup for Investigation of Leakage in O-Rings," in *Recent Advances in Applied Mechanics, VSAM 2021*, Virtual, pp. 521-534.

Biron, G., Vadean, A., and Tudose, L. (2012), "Optimal Design of Interference Fit Assemblies Subjected to Fatigue Loads," *Structural and Multidisciplinary Optimization*, 29(3), pp. 232-243.

Blaga, L. et al. (2015), "Friction Riveting (FricRiveting) as a New Joining Technique in GFRP Lightweight Bridge Construction," *Construction and Building Materials*, 80, pp. 167-179.

Broughton, W. A. and Mera, R. D. (1999), "Environmental Degradation of Adhesive Joints Accelerated Testing," NPL Report CMMT(A) 197, National Physical Laboratory Teddington, Middlesex, UK, August 1999.

Chen, N., Ducloux, R., Pecquet, C., Malrieux, J., Thonnerieux, M., Wan, M., and Chenot, J. L. (2011), "Numerical and Experimental Studies of the Riveting Process," *International Journal of Material Forming*, 4(1), pp. 45-54.

Chiang, Y. J. (2019), *Mechanics and Design for Product Life Prediction*, Chongqing University Press; ISBN: 978-7-5689-1917-6.

Chiang, Y. J., Nassar, S. A., and Barber, G. C. (2006), "Multi-Variable Effects of Fastening Parameters on Stress Development in Bolt Threads," SAE Technical Paper 2006-01-1253, doi: https://doi.org/10.4271/2006-01-1253.

Chiang, Y. J. and Barber, G. C. (1997), "Self-Threading Bolts Tapped into Temperature-Dependent Plastic Bosses," *International Journal of Materials and Product Technology*, 212(2-3), pp. 49-66.

Chiang, Y. J. and Rowlands, R. E. (1991), "Finite Element Analysis for Mixed-Mode Fracture of Bolted Composite Joints," *Journal of Composites Technology and Research*, 13(4), pp. 227-235.

Chiang, Y. J. and Rowlands, R. E. (1987), "Fracture Analysis of Cracks Emanating from a Pin-Loaded Hole in Composites," *Developments in Mechanics*, 14(b), pp. 581-586.

Davet, G. P. (2023), "Using Belleville Springs to Maintain Bolt Preload," Retrieved January 29, 2023, https://www.aspseal.com/Valve-Packing/pdf/Using_Bellville_springs_to_maintain_bolt_preload.pdf.

Devore, J. L. (2016), *Probability and Statistics for Engineering and the Science*, 9th Edition, Cengage Learning; ISBN: 978-1305251809; Chapter 11, End of Chapter, Supplementary Exercises, Exercise 59.

Dixon, D. et al. (2006), "Application of Design of Experiment (DOE) Techniques to Process Validation in Medical Device Manufacture," *Journal of Validation Technology*, 12(2), pp. 1-9.

Dong, S., Kelkar, G., and Zhou, Y. (2002), "Electrode Sticking during Micro-Resistance Welding of Thin Metal Sheets," *IEEE Transactions on Electronics Packaging Manufacturing*, 25(4), pp. 355-361.

D'huys, K. et al. (2019), "Multi-Criteria Evaluation and Optimization of the Ultrasonic Sealing Performance Based on Design of Experiments and Response Surface Methodology," *Packaging Technology and Science*, 32(1), pp. 165-174.

Eccles, B. (2009), "The Loosening of Prevailing Torque Nuts," *Fastener + Fixing Magazine*, Issue 60, November 2009.

Gallego-Juárez, J. A. and Graff, K. F. (2015), *Power Ultrasonics*, Woodhead Publishing.

Gardner, R. H., Kemp, W. M., Kennedy, V. S., and Petersen, J. E. (2001), "Scaling Relations in Experimental Ecology," Columbia University Press, New York, NY, USA.

Goodier, J. N. and Sweeney, R. J. (1945), "Loosening by Vibration of Threaded Fastenings," *Mechanical Engineering*, 67, pp. 798-802.

Gore, D. (2019), "Design of Experiments Helps Reduce Welding Defects from 24 to 0 per Month," Engineering Technology Department, Middle Tennessee State University, TN, Retrieved May 5, 2019.

Grabon, W., Osetek, M., and Mathia, T. (2018), "Friction of Threaded Fasteners," *Tribology International*, 18, pp. 408-420.

Grewell, A. D., Benatar, A., and Park, J. B. (2003), *Plastics and Composites Welding Handbook*, Hanser Gardner, Munich, Volume 10, 407 pages; ISBN: 978-1569903131.

Grudzinski, K. and Kostek, R. (2007), "An Analysis of Nonlinear Normal Contact Microvibrations Excited by a Harmonic Force," *Nonlinear Dynamics*, 50(4), pp. 809-815.

Harish, G. and Farris, T. N. (1998), "Shell Modeling of Fretting in Riveted Lap Joints," *AIAA Journal*, 36(6), pp. 1087-1093.

Haviland, G. (1981), "Unraveling the Myths of the Fastener World," SAE Technical Paper 810509, doi: https://doi.org/10.4271/810509.

Hüyük, H., Music, O., Koç, A., Karadogan, C., and Bayram, Ç. (2014), "Analysis of Elastic-Plastic Interference-Fit Joints," *Proceedings of the 11th International Conference on Technology of Plasticity*, Nagoya, Japan, October 19–24, 2014, pp. 2030-2035.

Hwang, H. Y. (2013), "Bolted Joint Torque Setting Using Numerical Simulation and Experiments," *Journal of Mechanical Science and Technology*, 27(5), pp. 1361-1371.

Ilhan, I. et al. (2021), "Understanding the Factors Affecting the Seal Integrity in Heat Sealed Flexible Food Packages: A Review," *Packaging Technology and Science*, 34, pp. 321-337.

Ito, Y., Toyoda, J., and Nagata, S. (1979), "Interface Pressure Distribution in a Bolt-Flange Assembly," *Journal of Mechanical Design*, 101(2), pp. 330-337.

Jang, J. H. and Ahn, S. H. (2016), "Numerical and Experimental Analysis of Heat Sealing of Multi-Layered Laminate Films Used in Lithium Polymer Battery Packaging," *Journal of Plastic Film & Sheeting*, 33(2), pp. 142-167.

Jiang, Y., Zhang, M., and Lee, C. (2003), "A Study of Early Stage Self-Loosening of Bolted Joints," *Journal of Mechanical Design*, 125, pp. 518-526.

Jiang, Y., Zhang, M., Park, T., and Lee, C. (2004), "Experimental Study of Self-Loosening of Bolted Joints," *Journal of Mechanical Design*, 126(5), pp. 925-931.

Johnson, G. R. and Cook W. H. (1983), "A Constitutive Model and Data for Metals Subjected to Large Strains, High Strain Rates and High Temperatures," *Proceedings of the 7th Int'l Symposium on Ballistics*, The Hague, the Netherlands, 1983.

Karci, F. et al. (2009), "The Effect of Process Parameters on the Properties of Spot Welded Cold Deformed AISI 304 Grade Austenitic Stainless Steel," *Journal of Materials Processing Technology*, 209, pp. 4011-4019.

Koga, A., Uchida, K., Yamabe, J., and Nishimura, S. (2011), "Evaluation on High-Pressure Hydrogen Decompression Failure of Rubber O-Ring Using Design of Experiments," *International Journal of Automotive Engineering*, 2, pp. 123-129.

Kumar, J. et al. (2017), "Effect of Design Parameters on the Static Mechanical Behavior of Metal Bellows Using Design of Experiment and Finite Element Analysis," *International Journal on Interactive Design and Manufacturing*, 11, pp. 535-545.

Kumar, V. and Goyal, N. (2018), "Parametric Optimization of Metal Inert Gas Welding for Hot Die Steel by Using Taguchi Approach," *Material Science Research India*, 15(1), pp. 100-106.

Lee, Y. L., Chiang, Y. J., and Wong, H. H. (1995), "A Constitutive Model for Estimating Multi-Axial Notch Strains," *Journal of Engineering Materials and Technology*, 117(1), pp. 33-40.

Li, C. et al. (2015), "Long-Term Durability of Adhesively Bonded Composite Joints under Quasi-Static and Fatigue Loading," *Proceedings of the 34th Conference and the 28th Symposium of the ICAF*, Helsinki, Finland, June 3–5, 2015, pp. 819-829.

Li, G. and Shi, G. (2004), "Effect of the Riveting Process on the Residual Stress in Fuselage Lap Joints," *Canadian Aeronautics and Space Journal*, 50(2), pp. 91-105.

Loc, N. H. and Phong, L. V. (2000), "Study of Interference Fit between Steel and Brass Parts," *EUREKA: Physics and Engineering*, 5, pp. 140-149.

Lopez-Cruz, P. et al. (2017), "Investigation of Bolted-Bonded Composite Joint Behavior Using Design of Experiments," *Composite Structures*, 170(3), pp. 192-201.

Madej, J. and Sliwka, M. (2021), "Analysis of Interference-Fit Joints," *Applied Sciences*, 11, p. 11428.

Malek, S. S., Chiang, Y. J., and Mason, J. F. (1993), "Multivariable Effects on an Automatic Screw-Torqueing Process," *Journal of Manufacturing Systems*, 12(6), pp. 457-462.

Marshall, M. B. et al. (2011), "Ultrasonic Measurement of Railway Wheel Hub-Axle Press-Fit Contact Pressures," *Journal of Rail and Rapid Transit*, 225(3), pp. 287-298.

MaxBolt (n.d.), "MaxBolt-Trademark for Load-Indicating Fasteners with Direct Tension Indicator."

Mekid, S. et al. (2019), "Batteryless Wireless Remote Bolt Tension Monitoring System," *Mechanical Systems and Signal Processing*, 128, pp. 572-587.

Muller, R. P. G. (1995), "An Experimental and Analytical Investigation on the Fatigue Behavior of Fuselage Riveted Lap Joints," PhD Thesis, Delft University of Technology, the Netherlands.

Norouzi, A., Hamedi, M., and Adineh, V. R. (2012), "Strength Modeling and Optimizing Ultrasonic Welded Parts of ABS-PMMA Using Artificial Intelligence Methods," *Int'l Journal of Advanced Manufacturing Technology*, 61(1-4), pp. 135-147.

Ocañaa, R. et al. (2015), "Evaluation of Degradation of Structural Adhesive Joints in Functional Automotive Applications," *Procedia Engineering*, 132, pp. 716-723.

Pai, N. G. and Hess, D. P. (2003), "Influence of Fastener Placement on Vibration-Induced Loosening," *Journal of Sound and Vibration*, 268(3), pp. 617-626.

Parsons, B. and Wilson, E. A. (1970), "A Method for Determining the Surface Contact Stresses Resulting from Interference Fits," *Journal of Engineering for Industry*, 92(1), pp. 208-218.

Pashazadeh, H. et al. (2016), "Statistical Modeling and Optimization of Resistance Spot Welding Process Parameters Using Neural Networks and Multi-Objective Genetic Algorithm," *Journal of Intelligent Manufacturing*, 27, pp. 549-559.

Pérez-Cerdán, J. C., Lorenzo, M., and Blanco, C. (2014), "Interference Fit Stress Concentrations with Full Chamfered Hub," *Key Engineering Materials*, 572, pp. 209-212.

Pop-Calimanu, M. and Fleser, T. (2012), "The Increasing of Weld Strength by Parameters Optimization of Ultrasonic Welding for Composite Material Based on Aluminum Using Design of Experiments," *NANOCON*, Brno, Czech Republic, October 23–25, 2012.

Rahman, M. U., Chiang, Y. J., and Rowlands, R. E. (1991), "Stress and Failure Analysis of Double-Bolted Joints in Douglas Fir/Sitka Spruce," *Wood and Fiber Science*, 23(4), pp. 567-589.

Rajakumar, D. R. (2012), "Response of Shrink Fitted Assemblies to the Dynamic Torsion," *Journal of Physics: Conference Series*, 364, p. 012124; *the 25th International Congress on Condition Monitoring and Diagnostic Engineering*, IOP Publishing.

Ranjan, B. et al. (2013), "A Novel Prevailing Torque Thread Fastener and Its Analysis," *Journal of Mechanical Design*, 135(10), p. 101007.

Rowlands, H. and Antony, J. (2003), "Application of Design of Experiments to a Spot-Welding Process," *Assembly Automation*, 23(3), pp. 273-279.

Rowlands, R. E., Rahman, M. U., Wilkinson, T. L., and Chiang, Y. J. (1982), "Single- and Multiple-Bolted Joints in Orthotropic Materials," *Composites*, 13(3), pp. 273-279.

Satoh, Y. et al. (1997), "An Evaluation Test for Influences of the Paint-Film on Self-Loosening of Fasteners," Quarterly Reports, Railway Technical Research Institute, pp. 61-65.

Segalman, D. J. (2001), "An Initial Overview of Iwan Modeling for Mechanical Joints," Sandia Report SAND2001-0811, March 2001.

Shim, J. Y. et al. (2011), "Joining of Aluminum to Steel Pipe by Magnetic Pulse Welding," *Materials Transactions*, 52(5), pp. 999-1002.

Shin, J. et al. (2014), "Using Stress Relief Grooves to Reduce Stress Concentration on Axle Drive Shaft," *Journal of Mechanical Science and Technology*, 28(6), pp. 2121-2127.

Silva, L. F. M. et al. (2000), "Multiple Site Damage in Riveted Lap-Joints: Experimental Simulation and Finite Element Prediction," *International Journal of Fatigue*, 22, pp. 319-338.

Simons, J. D. (1966), "Optimum Lip Seal Design by Fractional Factorial Experimentation," SAE Technical Paper 660380, doi: https://doi.org/10.4271/660380.

Sin, B. S. and Lee, K. H. (2014), "Process Design of a Ball Joint, Considering Caulking and Pull-Out Strength," *The Scientific World Journal*, 2014, Article ID 971679.

Sinzara, W. et al. (2018), "Effects of Eccentric Loading on Lip Seal Performance," *Conference Lubmat '18*, 2018.

Skorupa, M., Machniewicz, T., Skorupa, A., Schijve, J., and Korbel, A. (2015), "Fatigue Life Prediction Model for Riveted Lap Joints," *Engineering Failure Analysis*, 53, pp. 111-123.

Skorupa, A. and Skorupa, M. (2012), *Riveted Lap Joints in Aircraft Fuselage*, Springer, Dordrecht, 332 pages; ISBN: 978-9-4007-4281-9.

Spieckermann, S. et al. (2000), "Simulation-Based Optimization in the Automotive Industry - A Case Study on Body Shop Design," *Simulation*, 75(5), pp. 276-286.

Szolwinski, M. P. and Farris, T. N. (2000), "Linking Riveting Process Parameters to the Fatigue Performance of Riveted Aircraft Structures," *Journal of Aircraft*, 37(1), pp. 130-135.

TurnaSure, LLC (2004), Instruction Manual for Installing High-Strength Bolts with Direct Tension Indicators (ASTM F959M) Metric Series Edition, Retrieved November 18, 2019; https://turnasure.com/pdf/USMetric-3016.pdf.

Uehara, G. A. et al. (2015), "Recycling Assessment of Multilayer Flexible Packaging Films Using Design of Experiments," *Polímeros*, 25(4), pp. 371-381.

Vangipuram, R., Valasin, A., and Squires, R. (2007), "Torque Angle Signature Analysis of Joints with Thread Rolling Screws and Unthreaded Weld Nuts," SAE Technical Paper 2007-01-1665, doi: https://doi.org/10.4271/2007-01-1665.

William, E. (2010), "Tribological Aspects of the Self-Loosening of Treaded Fasteners," PhD Dissertation, University of Central Lancashire.

Wylde, J. W. and Spelt, J. K. (1998), "Measurement of Adhesive Joint Fracture Properties as a Function of Environmental Degradation," *Int'l Journal of Adhesion and Adhesives*, 18, pp. 237-246.

Zeng, D., Zhang, Y., Lu, L., Zou, L., and Zhu, S. (2018), "Fretting Wear and Fatigue in Press-Fitted Railway Axle: A Simulation Study of the Influence of Stress Relief Groove," *International Journal of Fatigue*, 118, pp. 225-236.

Zeng, Q. and Sun, C. T. (2004), "Fatigue Performance of a Bonded Wavy Composite Lap Joint," *Fatigue & Fracture of Engineering Materials & Structures*, 27(5), pp. 413-422.

Zhou, Y., Gorman, P., Tan, W., and Ely, K. J. (2000), "Weldability of Thin Sheet Metals During Small-Scale Resistance Spot Welding Using an Alternating-Current Power Supply," *Journal of Electronic Materials*, 29, pp. 1090-1099.

Problems

P12.1: The goal of the experiment is to maximize the bond strength (MPa) amid mounting an integrated circuit (IC) chip on a metallized glass substrate. Design factors and their design levels are given as follows [Devore 2016]:

Factor	Level (−)	Level (+)
A: Adhesive type	D2A	H-I-E
B: Conductor material	Copper	Nickel
C (min): Cure time (at 90°C)	90	120
D: I. C. post coating	Tin	Silver

The design matrix and 40 responses (five test responses for each treatment) are listed as follows:

Run	A	B	C	D	Y (MPa)
1	−	−	−	−	73.0, 73.2, 72.8, 72.2, 76.2
2	−	−	+	+	87.7, 86.4, 86.9, 87.9, 86.4
3	−	+	−	+	80.5, 81.4, 82.6, 81.3, 82.1
4	−	+	+	−	79.8, 77.8, 81.3, 79.8, 78.2
5	+	−	−	+	85.2, 85.0, 80.4, 85.2, 83.6
6	+	−	+	−	78.0, 75.5, 83.1, 81.2, 79.9
7	+	+	−	−	78.4, 72.8, 80.5, 78.4, 67.9
8	+	+	+	+	90.2, 87.4, 92.9, 90.0, 91.1

Please identify the effects that have influences on the bond strength, i.e., Y (MPa).

P12.2: A study on yield strength of multilayer flexible packaging films using DOE was conducted for the purpose of recycling. The design matrix "$(2^2 + \text{center point}) \times 2$" and test results are given as follows [Uehara et al. 2015]:

Run	A	b	c	A	B	C	Y (MPa)
1	−1	−1	−1	25	5	PE-g-MA	22.0
2	1	−1	−1	75	5	PE-g-MA	37.3
3	−1	1	−1	25	15	PE-g-MA	24.1
4	1	1	−1	75	15	PE-g-MA	27.2
5	0	0	−1	50	10	PE-g-MA	24.2
6	0	0	−1	50	10	PE-g-MA	23.8
7	0	0	−1	50	10	PE-g-MA	25.4
8	−1	−1	1	25	5	E-GMA	22.0
9	1	−1	1	75	5	E-GMA	24.0
10	−1	1	1	25	15	E-GMA	21.7
11	1	1	1	75	15	E-GMA	39.1
12	0	0	1	50	10	E-GMA	21.4
13	0	0	1	50	10	E-GMA	20.9
14	0	0	1	50	10	E-GMA	23.2

where
A (% by weight): Percentage of PET content in a PET/PE film
B (% by weight): Percentage of compatibilizer content of the total weight
C: Compatibilizer material, i.e., either PE-g-MA or E-GMA
Y (MPa): Yield strength

How do factors A, B, and C affect the yield strength of the packaging film?

P12.3: The factorial design 2^4, as applied for evaluating the appearance quality of riveted joints of the size of $0.125 \times 2.5''$, is given in Table 12.5.3. As formed, head height Y_h falls out of the desired range, i.e., $1.1906 < Y_h < 1.9844$ mm. How would parameters A and D be selected to have a properly formed head height?

P12.4: An experimental design is set up to investigate the effects of the rotating speed of the shaft and offset of the contact point from the garter center of a radial lip-type oil seal [Sinzara et al. 2018]. Design parameters and factorial design 3×6 are shown as follows:

Run	A	B	P	L
1	500	0	2.866	0
2	500	0.2	2.245	0
3	500	0.4	2.707	0
4	500	0.6	2.150	0
5	500	0.8	2.598	7.0
6	500	1.0	2.373	0
7	800	0	2.315	0
8	800	0.2	2.018	0
9	800	0.4	1.970	49.2
10	800	0.6	1.653	42.2
11	800	0.8	2.194	224.8
12	800	1.0	2.107	0
13	1000	0	1.995	0
14	1000	0.2	1.995	0
15	1000	0.4	1.855	14.1
16	1000	0.6	1.720	63.2
17	1000	0.8	1.953	63.2
18	1000	1.0	1.886	344.3

where
A (rpm): Rotating speed of shaft
B (mm): Offset, horizontal distance between garter center and contact point
P (Nm): Peak torque in the transient response
L (mL/h): Rate of leakage

The ambient temperature is 21.5°C. How do the rotating speed of the shaft and the offset of the contact point affect the peak torque and rate of leakage?

13

Mechanical Fabrication

The reliability of a manufactured product such as a passenger vehicle is the probability that it will perform satisfactorily for a specified period of time under the given environmental conditions. It is set to engineer key manufacturing process parameters to ascertain the product robustness, i.e., reduction in product variation, poor quality or line equipment reliability, low process availability, unexpected major outages, and even low productivity. Manufacturing reliability is deemed necessary to warrant the ability to make highly sustainable and individually customized vehicles at a low cost. The central theme of DOE, optimization theories, Weibull statistics, and Six-Sigma techniques is determined to control the manufacturing process and improve manufacturing reliability over time.

13.1. Manufacturing Reliability Using Weibull Statistics

During the product development phase, manufacturing operations perform product planning that includes product design and fabrication process verification, supplier selection, manpower analysis, equipment needs, and master scheduling. Once a product is committed to the production department, production control monitors material suppliers and work-in process to ensure that any issues affecting the delivery commitments are addressed and corrected immediately. The output based on quantifiable production data is the principal measure of a manufacturing process. Weibull analysis looks at the production data in a random manner where the time sequence is not so important, while Six-Sigma techniques are concerned about the time sequence. The methods are different but complementary to each other [Barringer 2000]. The basic idea of either Six-Sigma techniques or Weibull analysis is to control and reduce variations. Reliability of manufacturing processes can be obtained from daily production data after process failure criteria are established. Results of the analysis are displayed on statistical (i.e., Weibull) probability plots. Traditional normal distribution curves for a Six-Sigma process are usually bell-shaped and symmetrical, while Weibull curves for production data are skewed such that less variation renders steeper curves on Weibull probability plots, as shown in Figure 13.1.1.

These nonsymmetrical curves of the Weibull probability density function meet the emerging science of complexity explaining many self-organizing events such as manufacturing processes. According to [Barringer 2000], based on the Weibull shape parameter β, manufacturing processes can be divided into the following three generalized groups:

1. When $5 \leq \beta \leq 30$, the manufacturing process presents a great opportunity for improvement.

2. When $30 \leq \beta \leq 75$, the manufacturing process is less prevalent while having some opportunities for innovation.

3. When $\beta \geq 100$, the manufacturing process can be regarded as a successful business practice.

Furthermore, a large value of another Weibull parameter (η), which locates the relative volume of production, is more valued over a small one for the same manufacturing process.

FIGURE 13.1.1   Different daily production data shaping perspective Weibull curves [Barringer 2000].

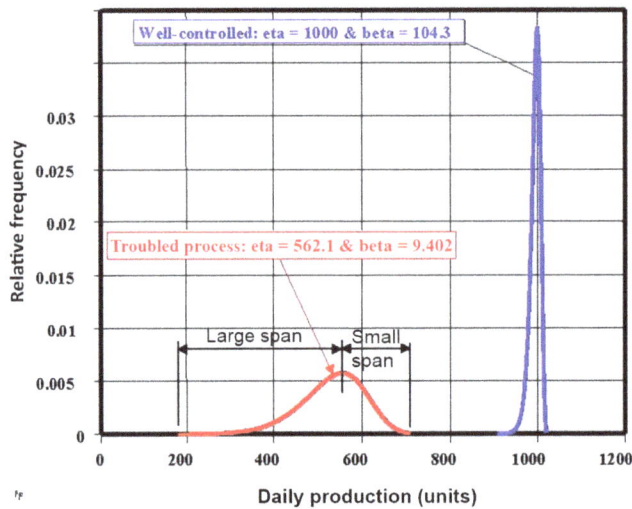

Excellent reliability renders higher utilization of the asset, lower maintenance costs, fewer overall people required to run the facility, better safety performance, and better energy efficiency.

Nowadays, DOE has gained great attention among manufacturing and quality engineers as the viably essential technique that can be employed to develop reliable products. Many successful applications of DOE to various manufacturers for improving fabrication reliability and process robustness have been reported as exhibited in the references. Typical case studies on fabrication are hereupon recapitulated.

13.2. **Injection Molding**

Injection molding is the process whereby a quantity of plastic material or metal of low melting points in solid or granular form is heated and softened in one part of the machine and then forced under pressure into a mold where the material is cooled and hardened so that the plastic material retains the shape imparted to it [Philip 1996]. The schematic drawing of a typical injection molding machine is shown in Figure 13.2.1. It involves the following process parameters:

1. Rheological behavior of material blend, e.g., viscosity, as the temperature varies
2. Cavity balance including shot size
3. Pack pressure, back pressure, and pressure drop rate
4. Gate sealing
5. Pack timing and cooling timing
6. Injection speed and screw speed

   Potential failure modes induced by injection molding include warpage (contour distortions), shrinkage, wear mass loss, flash, wet lines, rough surface, color fading, and low material strength of molded parts, but these can be tuned up by conditioning process parameters given above [Philip 1996].

FIGURE 13.2.1   Schematic drawing of a typical injection molding machine [Mukras 2020].

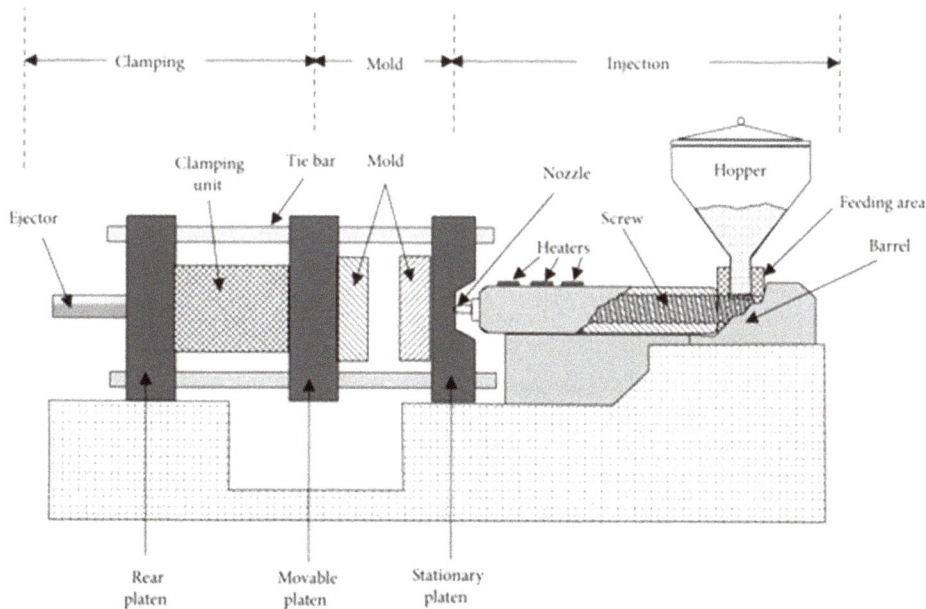

13.2.1. Injection Molding Process Window

Injection molding is the most popular molding process for plastics. The four basic functional components in an injection molding machine are:

1. Injection Unit: It melts the polymer resin and injects the polymer melt into the mold via the gating network.

2. Gating Network: This is composed of sprue, sprue cup, runners, ingates, and overflows, which are designed to guide liquid metal filling. A gating network for a PE container is exhibited in Figure 13.2.1.

3. Mold: A hollow form or cavity is which molten plastic material is solidified to yield the required part shape.

4. Clamping System: It includes fixtures and ejectors that open, close, hold, and eject the finished part.

As injection molding of plastic parts is a near net-shape-forming process for thermoplastic materials, it is the most applied manufacturing process for interior trims and coffers for automobiles [Chiang and Barber 2002]. A process window for injection molding can be defined in terms of injection velocity and pack pressure as shown in Figure 13.2.2, of which the associated boundaries are set up for different applications as follows [MKS 2012]:

(a) Boundaries 1, 2, and 5 form the basis for the machine's injection velocity and pack pressure limits.

(b) Boundary 3 gives the combined upper limits of injection velocity and pack pressure, beyond which unacceptable levels of flash are produced. Flash means that a thin layer of plastic flows outside of the cavity where the two halves of the injection mold meet.

(c) Boundary 4 determines the lower limits of pack pressure and injection velocity beyond which undesirable short shots and sink marks occur. A short shot means that the molded part is not completely filled up.

Besides the process window involving pack pressure and injection velocity, other process parameters including molding temperature, flow rate, and cooling rate are also viable for injection molding.

FIGURE 13.2.2   Process window of an injection molding process and an example molded part.

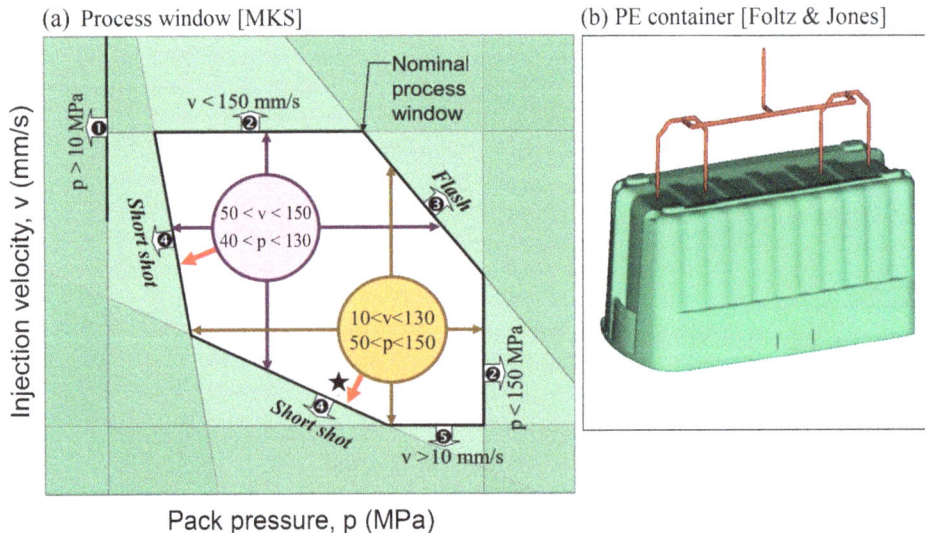

(a) Process window [MKS]

(b) PE container [Foltz & Jones]

On-Line Optimization of Injection Molding, C. McCready, D. Hazen: MKS Instruments, Andover MA; S. Johnston, D. VanDerwalker, D.O. Kazmer: Plastics Engineering Department, University of MA Lowell.

Process setpoints that might be chosen by a plastic manufacturer are shown as circles in the process window. Ranges of process parameters are defined around these settings to encompass the expected long-term variation for, in this example, ten important process factors as given in Table 13.2.1 [MKS 2012]. These factors are independent of each other and controllable during injection molding. DOE for plastics molding have been in use for a long time. It resolves multiple issues and leads to production improvements in the following areas:

(a) Lessening warpage

(b) Enhancing material strength

(c) Minimizing shrinkage

(d) Mitigating flash

(e) Removing air traps and porosities

(f) Eliminating sink marks

(g) Relocating weld lines

(h) Reducing cycle time

(i) Preventing photodegradation

A valid production process lies within the process window boundaries of a process window, while the key performance index of an injection molding process for checking molded plastic parts lies in whether it produces acceptable molded parts with repeatable part weights and thicknesses [Kazmer et al. 2008] without the problems itemized above.

13.2.2. Flash Reduction in Injection Molding Process

An experimental design was conducted for the purpose of reducing the flash of an automotive part by Lin and Chananda [2003] via varying the four process parameters that are listed in Table 13.2.1. Each molded part is measured for flash (mm) at five locations, and the average of the recorded values for each part is considered as the undesired flash. Full factorial design 2^4 with four replications for each treatment was conducted by Lin and Chananda [2003], but the data within one treatment are missing because of broken parts.

TABLE 13.2.1 Process factors and design levels for reducing flash in injection molding of plastics [Lin and Chananda 2003].

Variables	Level (−)	Level (+)	Coded variable
A. Pack pressure (kPa)	1	3	$a = (A − 2)/1$
B. Pack time (sec)	1	5	$b = (B − 3)/2$
C. Injection speed (mm/s)	12.7	50.4	$c = (C − 31.55)/18.85$
D. Screw speed (rpm)	100	200	$d = (D − 150)/50$

Therefore, the data obtained from Lin and Chananda [2003] are rearranged into two blocks of fractional factorial design 2_{IV}^{4-1} without knowing the sample standard deviation (not given in the published paper) and the data in the block corresponding to D = −ABC will be used for assessing the design contrasts, as shown in Table 13.2.2. Following the DOE-t technique presented in Volume I, Chapter 2, one can obtain the contrasts and effects as shown in Table 13.2.3. Since no replicated runs are available for generating the experimental error, it is reasonable to take an engineering measure that any effect under 0.10575, i.e., 2% of 5.2875 (grand average), is considered as a random error. It means that the default main effect of factor B is here regarded as a random error. Then, the predictive equation for flash Y (mm) is formulated as

$$Y_p = 5.2875 + 1.431\,a + 2.681\,c + 0.619\,d$$

$$- 0.963\,a\,b + 0.444\,a\,c - 0.413\,b\,c. \tag{13.2.1}$$

Parameters a, b, c, and d are the four dimensionless coded variables derived from variables A, B, C, and D. Note that $D = -ABC$, which shows the following three alias equations:

$$A\,B = -\,C\,D$$
$$A\,C = -\,B\,D$$
$$\text{and} \quad B\,C = -\,A\,D.$$

Since the main effect of factor B is not insignificant, it is reasonable to assume that the interactions between factor B and the other factors are so "weak" that the presence of interactions AB, BD, and BC are because of the corresponding aliased interactions $-CD$, $-AC$, and $-AD$, respectively. Thus, Equation (13.2.1) is rewritten as

$$Y_p = 5.2875 + 1.431\,a + 2.681\,c + 0.619\,d$$

$$+ 0.963\,c\,d + 0.444\,a\,c + 0.413\,a\,d. \tag{13.2.2}$$

Predicted values (Y_p) for all eight treatments based on the above equation are listed in Table 13.2.3. The strong correlation between the predicted values and the physical test data ($R = 99.997\%$) proves that the predictive equation is effective. Let $b = 0$ (i.e., $B = 3$ sec) and $d = 0$ ($D = 150$ rpm), and Equation (13.2.2) reduces to

$$Y_p = 5.2875 + 1.431\,a + 2.681\,c + 0.444\,a\,c$$

$$= 5.2875 + 1.431\,(A - 2) + 2.681\,(C - 31.55) / 18.85$$

$$+ 0.444\,(A - 2)\,(C - 31.55) / 18.85$$

$$= -0.576 + 0.668\,A + 0.095\,C + 0.024\,A\,C. \tag{13.2.3}$$

The process window corresponding to the experiment done above is shown in Figure 13.2.3. It can be seen that an increase in the pack pressure (factor A) or injection speed (factor C) results in the flash. Their interactive effect (AC) is also statistically significant.

TABLE 13.2.2 Fractional factorial design matrix for 2_{IV}^{4-1} ($D = -ABC$).

Run	A	B	C	D = −ABC	AB	AC	BC	Y (mm)	Y_p
1	−1	−1	−1	1	1	1	1	0.425	0.45
2	1	−1	−1	−1	−1	−1	1	5.625	5.65
3	−1	1	−1	−1	−1	1	−1	0.000	−0.03
4	1	1	−1	1	1	−1	−1	4.375	4.35
5	−1	−1	1	−1	1	−1	−1	6.000	6.03
6	1	−1	1	1	−1	1	−1	9.000	9.03
7	−1	1	1	1	−1	−1	1	9.000	8.98
8	1	1	1	−1	1	1	1	7.875	7.85
Contrast	2.863	0.05	5.363	−1.238	−1.925	0.888	−0.825		
Effect	1.431	_0.025_	2.681	−0.619	−0.963	0.444	−0.413	_5.2875_ (average)	

Y (mm): Flash; test data obtained from Lin and Chananda [2003]
Y_p (mm): Flash; predicted values by the DOE model

FIGURE 13.2.3   Process window for flash reduction in the injection molding process.

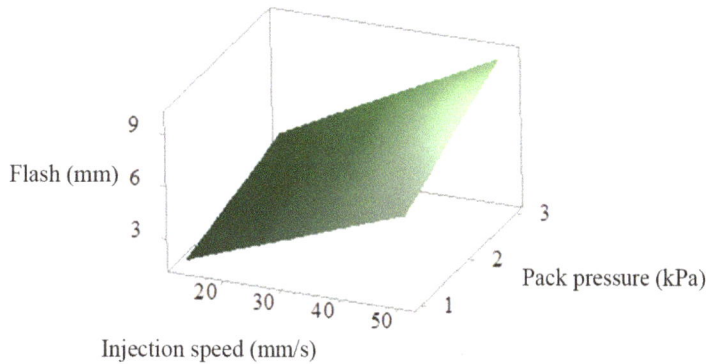

13.2.3.   Identification of Mechanical Properties of Injection-Molded High-Density (HD) Polyethylene (HDPE)/TiO$_2$ Composites

In the fabrication of HDPE/titanium oxide (TiO$_2$) nanocomposites, the impact of injection molding parameters, including concentration of TiO$_2$ (factor A), barrel temperature (factor B), residence time (factor C), and holding time (factor D) on mechanical properties (i.e., yield strength, Young's modulus, and elongation), is under investigation. The corresponding design levels are listed in Table 13.2.3. Fractional factorial design 3^{4-2} is applied, and test results are given in Table 13.2.4.

TABLE 13.2.3   Process factors and design levels for injection molding of HDPE/TiO$_2$ nanocomposites [Pervez et al. 2016].

Factor	Level 1	Level 2	Level 3
A (%): Concentration of nano-TiO$_2$	1	5	10
B (°C): Barrel temperature	200	225	250
C (min): Residence time	30	50	70
D (sec): Holding time	16	18	20

TABLE 13.2.4   Fractional factorial design 3^{4-2} for injection molding of HDPE/TiO$_2$ nanocomposites [Pervez et al. 2016].

Run	a	b	c	d	A	B	C	D	σ_y	E	δ (%)
1	1	1	1	1	1	200	30	16	20.9	384.3	576.9
2	1	2	2	2	1	225	50	18	20.5	364.6	580.8
3	1	3	3	3	1	250	70	20	19.3	363.2	598.6
4	2	1	2	3	5	200	50	20	22.1	481.8	492.5
5	2	2	3	1	5	225	70	16	22.6	346.0	646.8
6	2	3	1	2	5	250	30	18	21.3	401.1	622.0
7	3	1	3	2	10	200	70	18	20.8	438.5	412.0
8	3	2	1	3	10	225	30	20	21.5	425.9	628.6
9	3	3	2	1	10	250	50	16	19.9	422.9	559.3

Notes:
σ_y (MPa): Yield strength
E (GPa): Young's modulus
δ (%): Stretch

By locating Minitab → Stat → Regression → Regression → Fit Regression Model and having the "most insignificant" factor or interaction that comes with the largest p-value weeded out one by one until all p-values are less than 10%, the final regression models for yield strength in tension (MPa) are

$$\sigma_y = -41.3 + 0.905\ A + 0.566\ B - 0.0772\ A^2 - 0.001307\ B^2$$

$$(R = 97.7\%,\ R_{adj} = 85.4\%,\ R_{pred} = 87.5\%).$$

The predictive equation given above is effective with high unadjusted model correlation, adjusted model correlation, and predictive correlation. The response surface plot of yield strength as a function of concentration of nano-TiO_2 (factor A) and barrel temperature (factor B) is displayed in Figure 13.2.4.

It can be seen that the influence of factor A (concentration of nano-TiO_2) or factor B (barrel temperature) on the yield strength is highly nonlinear. Factor C (residence time) and factor D (holding time) have no influence on the yield strength of molded parts. The maximum yield strength can be obtained using the following two equations:

$$\partial\sigma_y\ /\ \partial A = 0.905 - 0.1544\ A = 0$$

$$\text{and} \quad \partial\sigma_y\ /\ \partial B = 0.566 - 0.002614\ B = 0.$$

Solving the above two equations leads to the maximum yield strength ($\sigma_{y,max} = 22.63$ MPa) at A = 5.8614% and B = 216.53°C.

The influences of these four factors on Young's modulus and stretch (columns E and δ in Table 13.2.4) can be resolved by the same token.

FIGURE 13.2.4   Response surface plot of yield strength as a function of concentration of nano-TiO_2 (factor A) and barrel temperature (factor B).

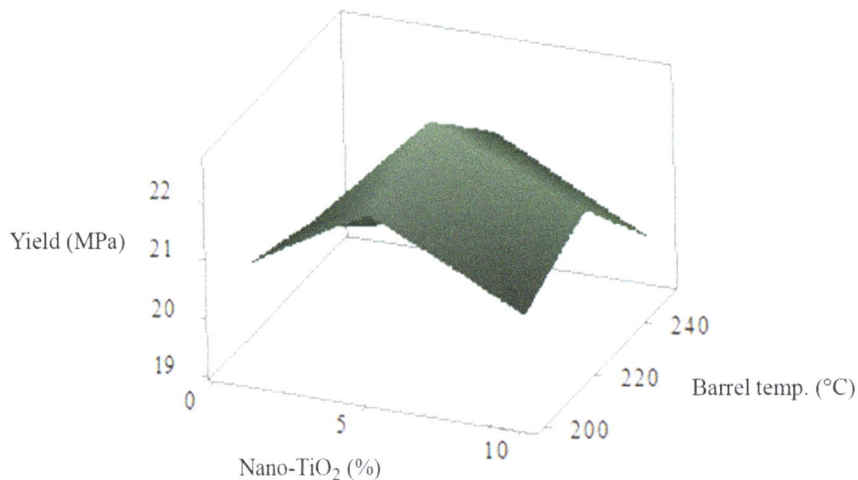

© SAE International.

13.2.4. Shrinkage Reduction in Injection Molding Process

Various robust engineering techniques based on DOE were applied to reducing shrinkage in injection-molded parts, such as Taguchi method [Alton 2010], ANOVA [Mukras 2020], and invasive-weed optimization (IWO) [Akbarzadeh and Sadeghi 2011]. Data obtained from [Gadiyar 2001] will be here examined by traditional full factorial design 3^3 on shrinkage reductions in both parallel and perpendicular directions. Molding parameters selected for the DOE are listed in Table 13.2.5, and the physical test results of injection-molded polystyrene (PS) strips (12.7 × 1.27 mm) are given in Table 13.2.6.

Shrinkage is here defined as the difference between the dimensions of a specimen and the mold cavity as it was molded. Some polymeric molecules, such as PS [Svec et al. 1990], have a strong tendency to become oriented in the direction of melt flow (Table 13.2.7). Both across-the-flow and along-the-flow shrinkages, i.e., $Y_\perp$ (%) and $Y_\parallel$ (%), are represented here by the percentage of linear dimensional changes that occur across and parallel to the direction of the melt flow, respectively.

TABLE 13.2.5 Factors and their design levels for evaluating shrinkages of injection-molded PS strips [Gadiyar 2001].

Factor	Level (0)	Level (1)	Level (2)
A: Melt temperature (°C)	218.3	245.0	270.6
B: Mold temperature (°C)	27.2	45.0	62.2
C: Holding pressure (MPa)	1.724	2.241	2.758
D: Back pressure (MPa)	—	0.6895	—
E: Clamp pressure (MPa)	—	16.0	—
F: Holding time (sec)	—	15	—
G: Cooling time (sec)	—	30	—
H: Cycle time (sec)	—	52.3	—
I: Injection speed (% of maximum)	—	98%	—
J: Rotating speed (% of maximum)	—	50%	—
K: Material (BASF PS 2124)	—	PS	—

Note:
PS has a melting temperature around 218.3°C, although some reported data go higher than that

TABLE 13.2.6 The factorial design matrix of 3^3 for as-molded shrinkages of PS strips [Gadiyar 2001].

Run	a	b	c	A	B	C	$Y_\perp$ (%)	$Y_{\parallel}$ (%)
1	0	0	0	218.3	27.2	1.724	1.477	0.891
2	0	0	1	218.3	27.2	2.241	1.455	0.8088
3	0	0	2	218.3	27.2	2.758	1.38	0.7841
4	0	1	0	218.3	45	1.724	1.437	0.82
5	0	1	1	218.3	45	2.241	1.388	0.7896
6	0	1	2	218.3	45	2.758	1.37	0.7302
7	0	2	0	218.3	62.22	1.724	1.364	0.722
8	0	2	1	218.3	62.22	2.241	1.368	0.7456
9	0	2	2	218.3	62.22	2.758	1.332	0.6968
10	1	0	0	245	27.2	1.724	1.387	0.7774
11	1	0	1	245	27.2	2.241	1.361	0.7169
12	1	0	2	245	27.2	2.758	1.319	0.6841
13	1	1	0	245	45	1.724	1.359	0.7108
14	1	1	1	245	45	2.241	1.284	0.659
15	1	1	2	245	45	2.758	1.312	0.6658
16	1	2	0	245	62.22	1.724	1.329	0.6898
17	1	2	1	245	62.22	2.241	1.278	0.6475
18	1	2	2	245	62.22	2.758	1.243	0.6312
19	2	0	0	270.6	27.2	1.724	1.37	0.7569
20	2	0	1	270.6	27.2	2.241	1.354	0.7075
21	2	0	2	270.6	27.2	2.758	1.295	0.6599
22	2	1	0	270.6	45	1.724	1.318	0.6828
23	2	1	1	270.6	45	2.241	1.326	0.6856
24	2	1	2	270.6	45	2.758	1.265	0.6336
25	2	2	0	270.6	62.22	1.724	1.317	0.6744
26	2	2	1	270.6	62.22	2.241	1.265	0.618
27	2	2	2	270.6	62.22	2.758	1.213	0.6105

$Y_\perp$ (%): Shrinkage perpendicular to the flow direction
$Y_{\parallel}$ (%): Shrinkage along the flow direction

By locating Minitab → Stat → Regression → Regression → Fit Regression Model and having the "most insignificant" factor or interaction that comes with the largest p-value weeded out one by one until all p-values are less than 10%, the final regression models for shrinkages perpendicular to and along the melt flow direction are, respectively,

$$Y_\perp (\%) = 5.759 - 0.02764\ A - 0.0309\ B - 0.566\ C + 0.000043\ A^2$$

$$+ 0.000116\ A\ B + 0.00200\ A\ C + 0.01232\ B\ C - 0.000050\ A\ B\ C$$

$$(R = 97.5\%,\ R_{adj} = 96.4\%,\ \text{and}\ R_{pred} = 94.6\%)$$

and $$Y_{\parallel} (\%) = 4.850 - 0.02796\ A - 0.00987\ B - 0.1294\ C + 0.000051\ A^2$$

$$+ 0.000018\ A\ B + 0.001380\ B\ C$$

$$(R = 97.8\%,\ R_{adj} = 967.1\%,\ \text{and}\ R_{pred} = 95.8\%).$$

The predictive equations given above are effective with high unadjusted model correlation, adjusted model correlation, and predictive correlation.

All main effects have direct impacts on the shrinkages in both directions but are complicated by multi-factorial interactions. A plot of predicted values of $Y_\perp$ (%) versus those of $Y_\parallel$ (%) for the 27 treatments is shown in Figure 13.2.5. It can be clarified that increases in the melt temperature, mold temperature, and holding pressure, within the experimental conditions given in Table 13.2.5, will reduce both shrinkages simultaneously.

TABLE 13.2.7 ANOVA for as-molded shrinkages of PS strips.

$Y_\perp$:	Source of variance	SS_{adj}	DOF	MS_{adj}	$F_{u,v}$	p-Value
	Regression	0.095189	8	0.011899	43.80	0.0%
	A	0.006889	1	0.006889	25.36	0.0%
	B	0.001363	1	0.001363	5.02	3.8%
	C	0.001076	1	0.001076	3.96	6.2%
	AA	0.005244	1	0.005244	19.30	0.0%
	AB	0.001154	1	0.001154	4.25	5.4%
	AC	0.000814	1	0.000814	3.00	10.0%
	BC	0.001129	1	0.001129	4.16	5.6%
	ABC	0.001106	1	0.001106	4.07	5.9%
	Error	0.004889	18	0.000272		
	Subtotal	0.100079	26			
	Grand average	—	1			
	Total	—	27			
$Y_\parallel$:	Source of variance	SS_{adj}	DOF	MS_{adj}	$F_{u,v}$	p-Value
	Regression	0.114964	6	0.019161	72.22	0.0%
	A	0.009080	1	0.009080	34.23	0.0%
	B	0.003349	1	0.003349	12.62	0.2%
	C	0.007446	1	0.007446	28.07	0.0%
	A^2	0.007399	1	0.007399	27.89	0.0%
	AB	0.000811	1	0.000811	3.06	9.6%
	BC	0.001873	1	0.001873	7.06	1.5%
	Error	0.005306	20	0.000265		
	Subtotal	0.120270	26			
	Grand average	—	1			
	Total	—	27			

FIGURE 13.2.5 Along-melt flow shrinkage $Y_\parallel$ (%) versus across-melt flow shrinkage $Y_\perp$ (%) for the 27 treatments (runs).

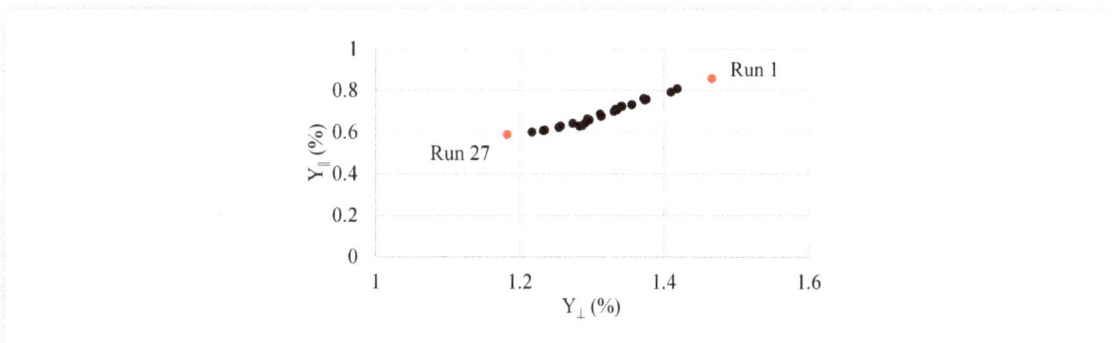

13.2.5. Comparison of DOE Variants

Comparison among different DOE approaches to injection molding is made by MKS [2012]. The combined extrema (minima and/or maxima) of the comprehensive set of process factors listed in Table 13.2.8, as coded −1 and +1 per factor, are used to set up experimental designs for injection molding. Different designs of experiments, i.e., 2_{III}^{10-6} with 16 run, D-optimal with 11 runs, oversaturated design I with 8 runs [MKS 2012], and oversaturated design II with 6 runs [MKS 2012], were set to find out the influence of methodology on the analysis quality. Based on the tests done by MKS [2012], it is concluded that fractional factorial design 2_{III}^{10-6} (16 runs) is able to generate absolutely correct contrasts for the ten factors and the result based on the D-optimal method is acceptable for capturing the trending (more qualitatively than quantitatively), while the oversaturated design matrix with either 8 runs or 6 runs is not so plausible.

TABLE 13.2.8 List of ten factors that may influence injection molding of plastics [MKS 2012].

Variables	Level (−)	Level (+)
A. Pack time (sec)	2	4
B. Material blend	0%	40%
C. Barrel temperature (°C)	195	205
D. Coolant temperature (°C)	25	35
E. Injection speed (mm/s)	60	80
F. Pack pressure (kPa)	60	80
G. Shot size (mm)	20.75	20.75
H. Back pressure (kPa)	20	26
I. Cooling time (sec)	6	10
J. Screw speed (rpm)	100	150

On-Line Optimization of Injection Molding, C. McCready, D. Hazen: MKS Instruments, Andover MA; S. Johnston, D. VanDerwalker, D.O. Kazmer: Plastics Engineering Department, University of MA Lowell.

Predictive equations in terms of multivariable effects explored by a series of DOE that count for warpage, material strength, shrinkage, air traps, sink mark, weld line, and cycle time can be combined for optimization with a multi-objective goal for calibrating, even tuning up, the process window. On the other hand, an individual DOE renders a more reliable molding process that would be customized to meet each individual service need.

13.2.6. Photodegradation of Automotive Plastics

Another unreliability of plastics of concern is photodegradation. Photodegradation means the degradation of materials caused by the absorption of photons, particularly those wavelengths found in sunlight, including infrared radiation, visible light, and ultraviolet (UV) light. Many organic compounds degrade chemically, especially when exposed to UV radiation. The energy of absorbed photons is transferred to electrons in the material and agitates them to reach another excitation state—usually kinematically unstable and decomposable. This material transformation includes the effects of both sunlight and air.

Popular thermoplastics used in an automotive interior, such as polypropylene (PP) and LDPE, are highly photodegradable because of their weak tertiary carbon bonds. However, photodegradation of automotive thermoplastics can be inhibited with additives.

13.3. Transfer Molding, Compression Molding, and Compression Resin Transfer Molding (RTM)

In the process of compression RTM, the fiber preform is placed in the mold, but the mold is not closed entirely, creating a fiber-free channel on top of the preform. A properly measured amount of preheated resin is injected into the mold, which first fills this channel because of its high permeability. Then, the mold is closed and the preform is forcefully infused with resin and reshaped to fill the mold cavities and eventually into the part that has none of the following potential problems: wrinkles, warpage, edge effect (delamination), blisters (voids, cuts, nods), rough panel surface finish, irregular panel thickness, and flash.

13.3.1. Transfer Molding of Electronic Components

Transfer molding, just next to injection molding, is the second most widely used molding process in the semiconductor industry because of its capability to mold small parts with complex features. In the transfer molding process, the molding compound (e.g., encapsulant material) is first preheated prior to loading it into the molding chamber, as shown in Figure 13.3.1. After preheating, the mold compound is forced by a hydraulic plunger into the pot where it reaches the melting temperature and becomes fluid.

FIGURE 13.3.1   Overview of transfer molding of encapsulants on IC-chip array [Henderson 2012].

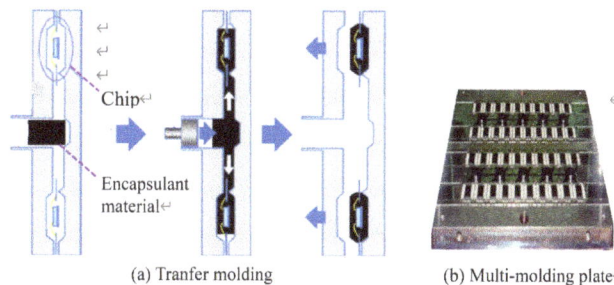

Image and photo courtesy of Semitracks, Inc.

(a) Tranfer molding   (b) Multi-molding plate

13.3.2. Compression RTM of Structural Composites

In the process of compression RTM, the dry fiber preform is first placed in the mold, but the mold is not closed entirely, creating a fiber-free channel on top of the preform. A properly measured amount of preheated resin is injected into the mold, which first fills this channel because of its high permeability. The mold is closed to a predefined gap, and the cavity is degassed. The degassing valve is shut, and thus, the tool gap is reduced. Then, the mold is fully closed and the preform is forcefully infused with resin, while the excess resin is removed from the preform until a prescribed fiber volume fraction is achieved. The impregnated preform is compacted and reshaped to fill the mold cavities as cured at the specified pressure, and eventually into its final part thickness, hopefully saturating the entire preform with resin. Lastly, the mold is opened, and the part and excess resin are ejected in one piece.

FIGURE 13.3.2 **Comparison of crushed steel and carbon fiber underbody assembly [Aitharaju 2020].**

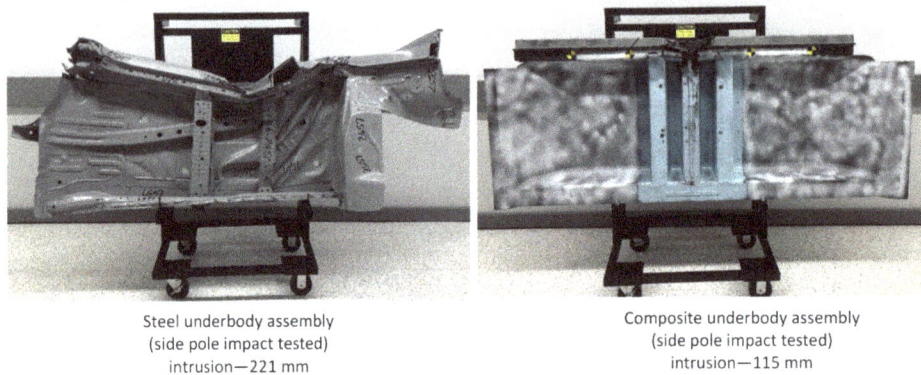

Steel underbody assembly
(side pole impact tested)
intrusion—221 mm

Composite underbody assembly
(side pole impact tested)
intrusion—115 mm

Composite materials can be tailored to have better mechanical properties than metals. For example, here is the automotive underbody assembly of a passenger vehicle as exhibited in Figure 13.3.2. Ten RTM process variables, including injection pressure, hydro-check pressure, mold temperature, post-cure temperature, valve position after filling, stroke length, vacuum level, type and number of layers of reinforcement, and resin-to-catalyst ratio, were studied by Jarugu [1996] using screening DOE.

13.3.3. Resin Delivery System for Large Composite Parts Molded Using the Resin Transfer Method

The resin delivery system for large composite parts, such as wind turbine blades, molded using the resin transfer method, has to be carefully controlled to meet demanding design specifications and product reliability. The differential pressure between the near-vacuum source and the resin reservoir pulls the resin into the preform, impregnating the empty space between the fibers of the preform. The vacuum compacts the preform during infusion into the mold to form the part. The resin delivery (injection) is allowed to continue until the entire preform is completely infused. A typical experimental setup and the related dimensions are given in Figure 13.3.3. The mold is a flat 1016 mm × 2032-mm clear polyvinyl chloride (PVC) sheet with a 6.35 mm in thickness that is supported by a wood frame and secured to a table frame. A detailed experimental setup for vacuum infusion can be found in Yuksel [2014]. As listed in Table 13.3.1, three process factors and their design levels based on full factorial design 2^3 are identified. The objective is to search for a robust design configuration that would reduce the loss of control distance in the resin delivery system. The design matrix and data of loss of control distance (column Y) are listed in Table 13.3.2.

FIGURE 13.3.3 Experimental setup and related dimensions for RTM of large composite parts [Yuksel 2014].

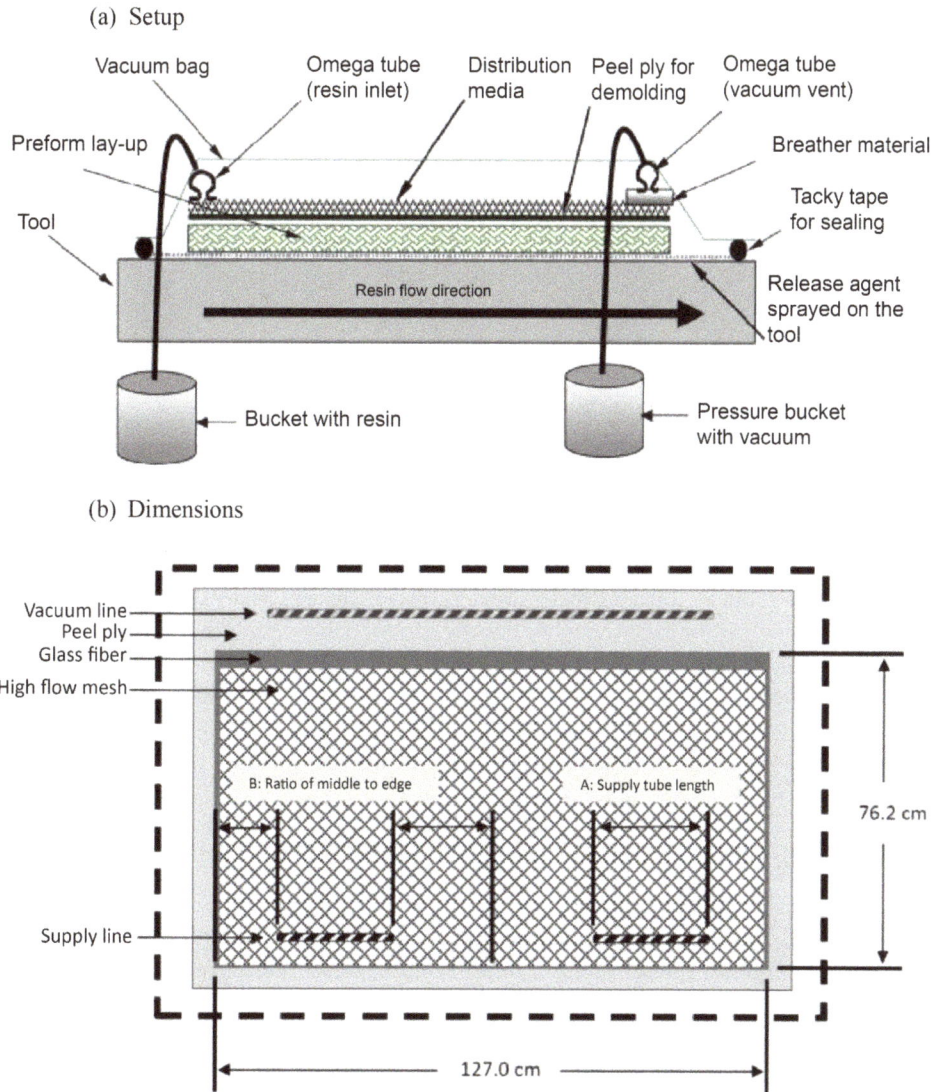

(a) Setup

(b) Dimensions

© ASME.

TABLE 13.3.1 Process factors and design levels of a resin delivery system for large long fiber-reinforced composites formed by RTM [Yuksel 2014].

Variables	(−1)	(+1)	Coded variables
A (mm): Supply tube length	152	356	$a = (A − 254)/102$
B: Ratio of middle to edge (Figure 13.3.1)	1/4	4	$b = (B − 2.125)/1.875$
C. Number of glass fiber sets	1	4	$c = (C − 2.5)/1.5$

© ASME.

TABLE 13.3.2 2^3 factorial design for resin delivery system for large resin transfer molded composite parts [Yuksel 2014].

Run	a	b	c	ab	ac	bc	abc	Y	Y_p
1	−1	−1	−1	1	1	1	−1	152	162
2	1	−1	−1	−1	−1	1	1	381	352
3	−1	1	−1	−1	1	−1	1	457	486
4	1	1	−1	1	−1	−1	−1	305	295
5	−1	−1	1	1	−1	−1	1	305	314
6	1	−1	1	−1	1	−1	−1	305	314
7	−1	1	1	−1	−1	1	−1	457	448
8	1	1	1	1	1	1	1	457	448
Contrast	19.25	133.25	57.25	−95.25	−19.25	18.75	95.25		
Effect	9.625	66.63	28.63	−47.63	−9.625	9.375	47.63		
Error	9.542	9.542	9.542	9.542	9.542	9.542	9.542		
t-Ratio	1.009	6.982	3.000	4.991	1.009	0.982	4.991		
$t_{3,1-\alpha} = t_{3,90\%}$	1.638	1.638	1.638	1.638	1.638	1.638	1.638		
Significant	No	Yes	No	Yes	No	No	Yes		

© ASME.

Y (mm): Loss of control distance

The effects of individual factors and their interactions are charted as a Pareto plot as shown in Figure 13.3.4. It is seen that factor B (ratio of the mold middle to supply line edges) has the largest impact on the crack initiation among all the factors in the given respective working ranges. The main effect of factor C (number of glass fiber sets) is also statistically significant. One two-factor interaction (AB) and the three-factor interaction (ABC) are also statistically effective, and their influences have to be taken into consideration in quest of product reliability.

FIGURE 13.3.4   Pareto plot of factorial impacts on loss of control distance (mm).

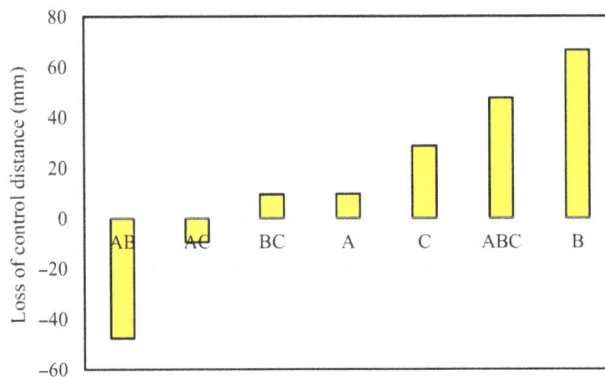

© SAE International.

Three small effects, i.e., main effect of factor A and interactive effect of AB and BC, are here defined as being less than 5% of the average (352.375) in magnitude. Based on t_3-statistic with the parting significance level of 10%, the calculated main and interactive effects and their statistical significances are demonstrated in Table 13.3.2. At a confidence level of 90% in light of t-distribution, the predictive equation for the loss of control distance can be expressed in terms of dimensionless coded variables as

$$Y_p = 352.375 + 66.625\,b + 28.625\,c - 47.625\,a\,b + 47.625\,a\,b\,c. \tag{13.3.1}$$

Let c = 1; i.e., there are four sets of glass fibers. Then, Equation (14.3.1) reduces to

$$Y_p = 381 + 66.625\,b. \tag{13.3.2}$$

It means that the loss of control distance is proportional to the ratio of the mold middle to supply line edges (factor B). The impact of factor B can be understood by observing the resin flow direction identified in Figure 13.3.1.

On the other hand, given that there is only one set of glass fibers (c = −1), Equation (14.3.1) becomes

$$Y_p = 323.75 + 66.625\,b - 95.25\,a\,b. \tag{13.3.3}$$

Substituting equations of coded variables, i.e., a = (A − 254)/102 and b = (B − 2.125)/1.875 given in Table 13.3.1, into the above equation leads to a response surface plot of Y_p versus factors A and B as shown in Figure 13.3.5.

FIGURE 13.3.5   Response surface of loss of control distance versus supply tube length (factor A) and ratio of middle to edge (factor B).

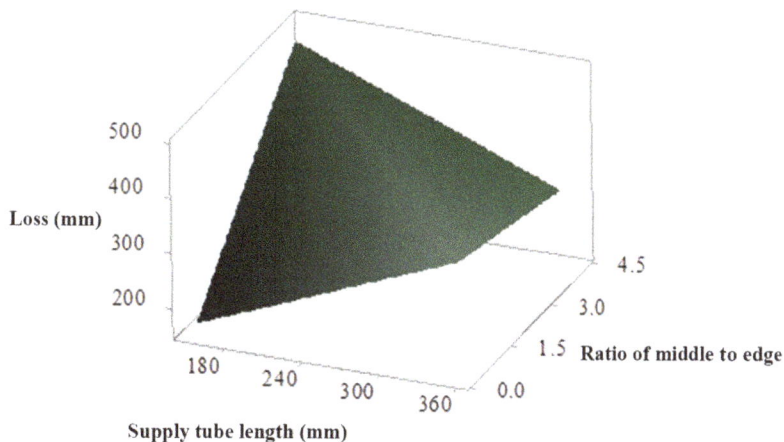

13.4. Metal Forming and Clinching

Metal-forming processes are characteristic in that the metal (sheet metal) is forcefully transformed into a desired shape with plastic deformations subjected to tension, compression, shearing, bending, and friction. Plastic deformation occurs when the applied stress level goes beyond the yield strength of the workpiece material. There is a considerable elastic recovery during unloading, and it may lead to springback after the tool is released. Friction in the contact surfaces between the workpiece and the tools has mostly negative effects, but it can be mitigated using lubricants or used to guide the material flow and form more uniformly strained components with no defects. Metal forming can be divided into two groups:

(a) Bulk Deformation: rolling, forging, extrusion, and drawing (bar or wire)

(b) Sheet Metal Forming: bending, stretching, shearing, and stamping/drawing (cup or deep)

It is well known that under a high blank holder pressure in the sheet metal-forming process wrinkling and tearing may occur, as shown in Figure 13.4.1. Finite element methods are usually used to predict these two failure modes [Arezodar and Eghbali 2000].

FIGURE 13.4.1   Wrinkling and tearing in sheet metal-drawing process [Arezodar and Eghbali 2000].

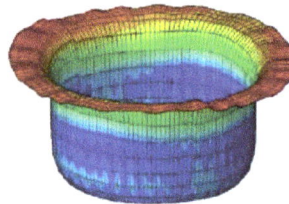

(a) Wrinkling   (b) Wrinkling by FEA

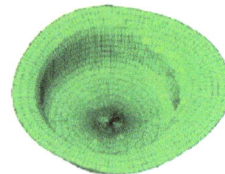

(c) Tearing   (d) Tearing by FEA

Reprinted with permission. © WSEAS.

13.4.1. Stamping

Stamping is one of the metal-forming processes, which involve tools such as punch, die, blank holder, draw beads, lubricants, and the blank that is to be formed into a desired shape. The example DOE given here is to draw a cylindrical cup in the stamping laboratory. Major influential parameters include die entry radius, punch nose radius, punch velocity, binder holder pressure, interfacial friction (as lubricated), blank thickness, and blank material.

Two key operating factors are taken into consideration with regard to the automotive body forming with circular blank (radius = 117.475 mm and thickness = 0.9906 mm) made of aluminum alloy 6111-T4 [Echempati and Sathya Dev 2002]:

(a) Coefficient of friction

(b) Punch velocity (mps)

Their influences on the binder pressure, which is the vital forming parameter for getting a successful draw, are to be sought in light of design matrix 3^2, as given in Table 13.4.1.

The predictive equation for the binder pressure is obtained using Minitab (Stat → DOE → Regression → Regression → Fit Regression Model). After the insignificant terms are dropped one by one, the predictive equation is derived with the statistical significance level of 5% (α = 5%) by means of the F-distribution as in Table 13.4.2 as follows:

$$Y_p = 203.8 - 1146\,A + 2246\,A^2 - 0.00288\,B^2$$

$$(R = 97.0\%,\ R_{adj} = 95.1\%,\ \text{and}\ R_{pred} = 89.1\%). \tag{13.4.1}$$

Effectiveness of the above equation is validated by the three different correlations given above. The predicted values of binder pressure using Equation (13.4.1) are listed in column Y_p of Table 13.4.1, which correlate very well with the data obtained from physical tests. A contour plot of the binder pressure versus the coefficient of friction and the punch speed is displayed in Figure 13.4.2.

TABLE 13.4.1 Automotive body stamping using the 3^2 design [Echempati and Sathya Dev 2002].

Run	a	b	A	B	Y (kPa)	Y_p
1	−1	−1	0.08	53	120.66	118.4
2	0	−1	0.08	80	120.66	108.1
3	1	−1	0.08	100	82.74	97.7
4	−1	0	0.20	53	48.27	56.4
5	0	0	0.20	80	44.82	46.0
6	1	0	0.20	100	44.82	35.6
7	−1	1	0.30	53	55.16	54.1
8	0	1	0.30	80	41.37	43.7
9	1	1	0.30	100	34.48	33.3

Notes:
Factor A: Coefficient of friction
Factor B: Punch velocity (mps)
Y (kPa): Binder pressure; the smaller the better

FIGURE 13.4.2 Binder pressure versus coefficient of friction and punch speed in a stamping process.

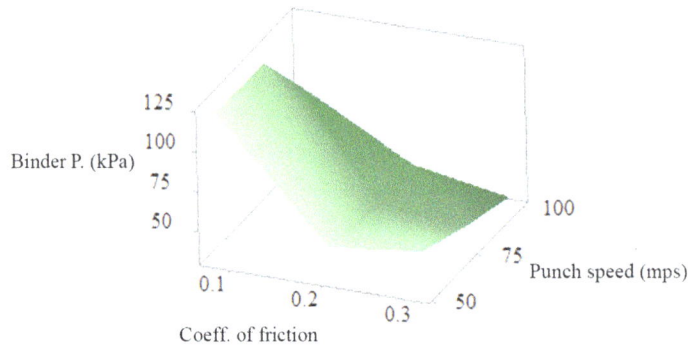

13.4.2. Clinching

Clinching creates a mechanical interlock that is formed using local plastic deformation without applying any additional joining elements [Varis 2006]. As the automotive industry strives to reduce vehicle weight by incorporating aluminum and magnesium alloys [Chen 2018], alternative joining techniques for sheet metals other than spot welding, such as clinching, are viable. Dissimilar or pre-painted materials cannot be welded together, such as steel to aluminum, steel to magnesium, and aluminum to magnesium. Clinching techniques have been used for on-ground vehicles, aircrafts, domestic appliances, and electronics joining applications [JMBS 2015]. It is a simple technique as schematically shown in Figure 13.4.3, for joining metal sheets that typically range from 0.2 to 4 mm in thickness. A typical final joint thickness is approximately 6 mm.

FIGURE 13.4.3 Schematic depiction of clinching process and viable parameters. [Lambiase and Di Ilio 2016].

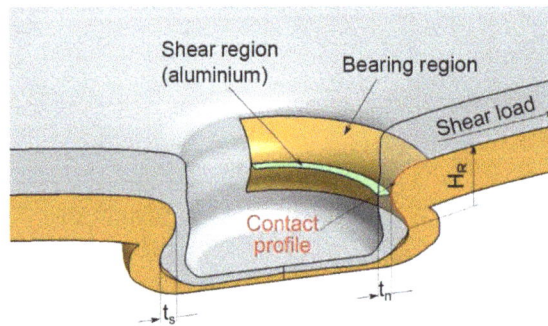

FIGURE 13.4.4 Important parameter in a clinching process [Oudjene et al. 2009].

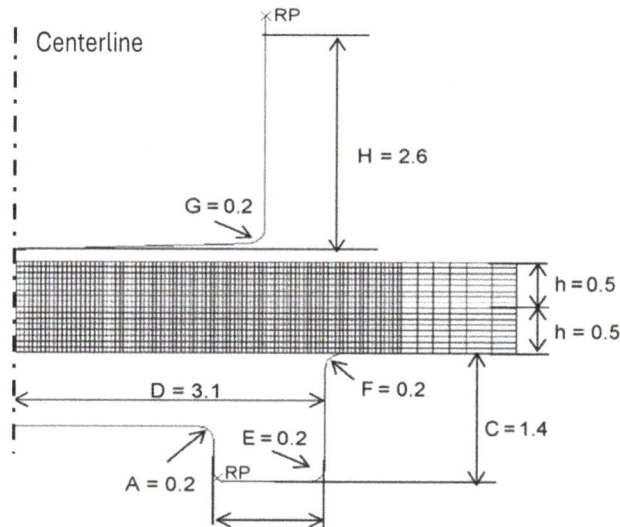

When geometric data and material properties are available, a sensitivity study using finite element methods in combination with DOE can be conducted for evaluating the structural integrity of a clinching joint (Figure 13.4.4). The geometric factors and their design levels are listed in Table 13.4.2. The sheet aluminum is Al 5754 of 0.5 mm (nominal value) in thickness, and it comes with the following material properties:

1. Young's modulus: E = 70 GPa

2. Poisson's ratio: $\nu = 0.3$

3. Isotropic hardening: $\sigma_{eq} = 627\,(0.023 + \varepsilon_p)^{0.31}$ MPa

4. Friction: $\mu = 0.35$ (between aluminum sheets) and $\mu = 0.1$ (between aluminum sheet and tools)

The objective is to find out the retaining load, Y (N), i.e., the maximum tensile force to separate the two clinched sheet metals.

TABLE 13.4.2 Dimensions and their design levels for clinching [Oudjene et al. 2009].

Dimensions (mm)	(−)	(0)	(+)
A	0.10	—	0.40
B	0.70	0.95	1.20
C	1.20	1.35	1.40
D	3.00	3.15	3.30
E	0.10	0.25	0.40
F	0.10	0.25	0.40
G	0.10	0.25	0.40
H	2.40	2.50	2.60

A DOE based on $2^1 \times 3^{7-5}$ was conducted by Oudjene et al. [2009]. Design matrix 3^{7-5} of factors B~H is nested within factor A that has two design levels, namely A– and A+ levels. The FEA is conducted using Abaqus/Explicit codes with remeshing algorithms [Oudjene et al. 2009]. The design matrix and separation forces obtained from FEAs for the 18 treatments are listed in Table 13.4.3.

TABLE 13.4.3 Shape optimization based on $2^1 \times 3^{7-5}$ for clinching of sheet aluminum [Oudjene et al. 2009].

Run	A	B	C	D	E	F	G	H	Y (N)	Y_p
1	1	1	1	1	1	1	1	1	645.9	618.7
2	1	1	2	2	2	2	2	2	775.5	718.2
3	1	1	3	3	3	3	3	3	26.0	24.9
4	1	2	1	1	2	2	3	3	615.6	604.3
5	1	2	2	2	3	3	1	1	752.2	774.7
6	1	2	3	3	1	1	2	2	697.3	651.3
7	1	3	1	2	1	3	2	3	615.3	718.2
8	1	3	2	3	2	1	3	1	12.1	24.9
9	1	3	3	1	3	2	1	2	615.0	618.7
10	2	1	1	3	3	2	2	1	194.7	—
11	2	1	2	1	1	3	3	2	668.9	—
12	2	1	3	2	2	1	1	3	676.2	—
13	2	2	1	2	3	1	3	2	605.7	—
14	2	2	2	3	1	2	1	3	745.9	—
15	2	2	3	1	2	3	2	1	635.2	—
16	2	3	1	3	2	3	1	2	341.0	—
17	2	3	2	1	3	1	2	3	525.7	—
18	2	3	3	2	1	2	3	1	308.0	—

The analysis is carried out using Minitab (Stat ➜ Regression ➜ Regression ➜ Fit Regression Model). Since factors B~H are nested within factor A, the analysis leads to two different equations for levels A– and A+. As it is a resolution III design, only individual factors and their quadratic terms are examined and the final ANOVA results are presented in Table 13.4.4 after weeding out the nonsignificant factors at a statistical significance level (p-value) of 10%.

The predictive equations by regression at A⁻ (A = 0.1 mm) are

$$Y_{A-} = -5778 + 2031 \, D + 33526 \, G - 9906 \, D \, G - 7712 \, G^2$$

$$(R = 98.7\%, R_{adj} = 97.4\%, \text{ and } R_{pred} = 83.6\%). \tag{13.4.2}$$

The effectiveness of the above equation is justified using the three corresponding correlative functions given above. The highly insignificant p-value for lack of fit means that the model is adequate. In search of the maximal retention force, one would redeem the following two equations:

$$\partial Y_{A-} / \partial D = 2031 - 9906 \, G = 0, \tag{13.4.3}$$

and $\quad \partial Y_{A-} / \partial G = 33526 - 9906 \, D - 15424 \, G = 0. \tag{13.4.4}$

These lead to D = 3.0652 and G = 0.20503, which is validated by the response surface plot of the retention force in terms of inner radius of die (factor D) and punch corner radius (factor G), as displayed in Figure 13.4.5.

Readers are encouraged to explore the predictive equation for the retention force of clinching joints in sheet aluminum of 0.4 mm in thickness (Problem P13.3).

TABLE 13.4.4 ANOVA based on $2^1 \times 3^{7-5}$ for clinching of 0.1 mm (A$^-$ level) sheet aluminum.

Source of Variance	SS_{adj}	DOF	MS_{adj}	$F_{u,v}$	p-Value
Regression	677,570	4	169,392	38.64	0.2%
D	41,044	1	41,044	9.36	3.8%
G	171,986	1	171,986	39.23	0.3%
DG	115,923	1	115,923	26.44	0.7%
GG	37,468	1	37,468	8.55	4.3%
Error	17,535	4	4384	—	—
Lack of fit	4129	1	4129	0.92	40.7%
Pure error	13,406	3	4469	—	—
Subtotal	695,105	8			
Grand average	—	1			
Total	—	9			

© SAE International.

FIGURE 13.4.5 Retention force (N) as a function of punch corner radius and inner radius of die.

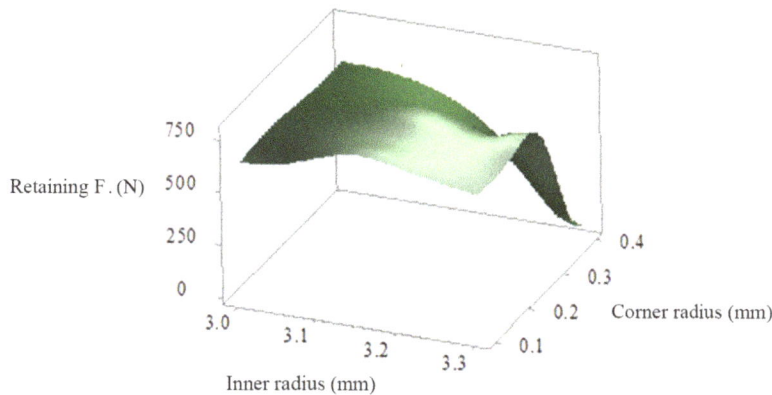

© SAE International.

13.4.3. Hemming

Hemming is one of the last metal-forming operations to be applied to automotive body panels, while flanging occurs right before the hemming operation. Both flanging and hemming are characterized by applying a bending effort as shown in Figure 13.4.5. The amount of roll-in (roll-out) of the hem edge is a measure of hemming quality at the workbench (Figures 13.4.6 and 13.4.7). Aluminum alloy sheet metal is more difficult to hem than its steel counterpart because of its susceptibility to in-process strain localization, which prompts edge cracking on the hemmed edge [Golovashchenko 2005]. A fractional factorial design was conducted by Lin et al. [2005] to investigate the hemming of AA 6111-T4 aluminum based on finite element methods. As listed in Table 13.4.5, six variables were identified to check against the interested output, i.e., the amount of roll-in that would retrain the dimension of enclosure.

FIGURE 13.4.6   Hemming operations of an automotive body part [Lin et al. 2005].

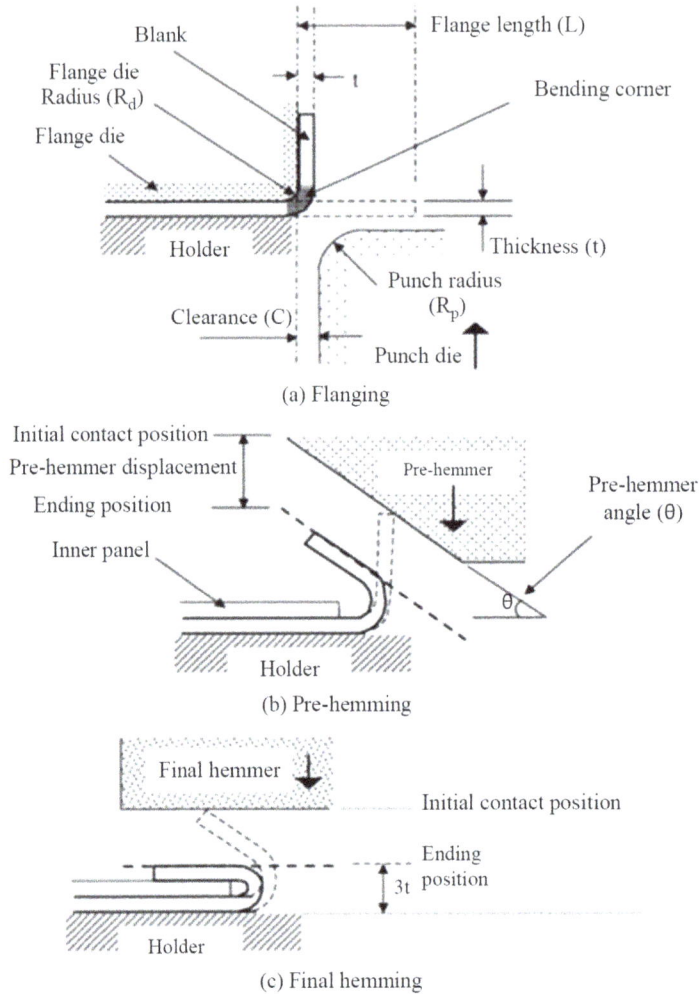

(a) Flanging

(b) Pre-hemming

(c) Final hemming

FIGURE 13.4.7   Hemming edges: roll-in, roll-out, recoil, and warp [Lin et al. 2005].

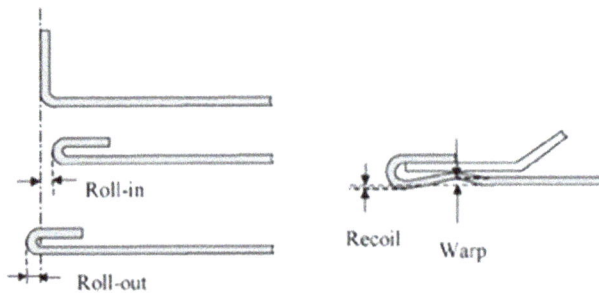

TABLE 13.4.5 Factors and their design levels for studying hemming operations [Lin et al. 2005].

Factors	Level (−)	Level (+)
A: Pre-hemming path (mm)	0.0663	0.078
B: Flange length (mm)	5.1	6.9
C: Flange die radius (mm)	1.19	1.61
D: Clearance (mm)	151.5	178.2
E: Pre-strain	10.625	14.375
F: Pre-hemming angle (deg)	212.5	250

Fractional factorial design 2^{6-2}_{IV}, as shown in Table 13.4.6, is deployed for studying their effects based on finite element methods. Finite element models for estimating the roll-ins corresponding to the 16 treatments (runs) developed for this study are built on the commercial codes Abaqus/Explicit and Abaqus/Standard (implicit) software. There are two alias equations enclosed in the design matrix, and they are E = ABD and F = BCD. In other words, AB = DE, AD = BE, and AE = BD; BC = DF, BD = CF, and BF = CD. Thus, there are merely nine independent interactive effects. The ten small effects, arbitrarily defined in terms of the size of effect that is less than 5% of the average (1.327 × 5% = 0.0663), are pooled together to estimate the random error (0.0372 in Table 13.4.6). Given that the significance level is 5%, the final predictive equation for roll-in (mm) can be written as

$$Y_p = 1.3267 + 0.2297\,c + 0.103\,d + 0.098\,e - 0.312\,f + 0.1025\,b\,f. \tag{13.4.5}$$

Note that interaction BF (flange length * pre-hemming angle) is confounded with interaction CD (flanging die radius × clearance). Further tests are required to resolve the issue.

Since the main effect of factor B is not insignificant and both factors C and D are significant, it tends to assume that interaction CD is effective. Thus, Equation (13.4.5) can be written as

$$Y_p = 1.3267 + 0.2297\,c + 0.1029\,d + 0.0981\,e - 0.3122\,f + 0.1024\,c\,d. \tag{13.4.6}$$

The predicted values of roll-in based on the above equation are listed in column Y_p of Table 13.4.6. The high correlation (R = 94.1%) between the predicted values and the results from FEA redeems the effectiveness of the predictive equation. Assume that e = 0 (pre-strain E is set to 12.5) and f = 0 (pre-hemming angle f is set to 231.5°) in the flanging stage, Equation (13.4.6) reduces to

$$Y_p = 1.3267 + 0.2297\,c + 0.103\,d + 0.1025\,c\,d. \tag{13.4.7}$$

A plot of response surface of roll-in in terms of factor C [flange die radius (mm)] and factor D [clearance (mm)] in hemming operations of automotive body sheet aluminum, as shown in Figure 13.4.8. It is shown that both factors and their interaction are effective to promote the length of roll-in positively.

TABLE 13.4.6 Factorial design 2^{6-2}_{IV} (E = ABD and F = BCD) for assessing roll-in due to hemming operations.

(a) Main effects and roll-in Y (mm) data							
Run	**A**	**B**	**C**	**D**	**E**	**F**	**Y**	**Y_p**
1	−1	−1	−1	−1	−1	−1	1.281	1.311
2	1	−1	−1	−1	1	−1	1.353	1.507
3	−1	1	−1	−1	1	1	1.050	0.882
4	1	1	−1	−1	−1	1	0.702	0.686
5	−1	−1	1	−1	−1	1	0.919	1.137
6	1	−1	1	−1	1	1	0.763	0.941
7	−1	1	1	−1	1	−1	1.827	1.565
8	1	1	1	−1	−1	−1	1.895	1.761
9	−1	−1	−1	1	−1	1	0.805	0.687
10	1	−1	−1	1	1	1	1.048	0.883
11	−1	1	−1	1	1	−1	1.475	1.508
12	1	1	−1	1	−1	−1	1.062	1.312
13	−1	−1	1	1	1	−1	2.082	2.172
14	1	−1	1	1	−1	−1	2.136	1.976
15	−1	17	1	1	−1	1	1.253	1.351
16	1	1	1	1	1	1	1.576	1.549
Contrast	−0.02	0.057	0.459	0.206	0.196	−0.624	—	
Effect	_−0.01_	_0.028_	0.2297	0.103	0.098	−0.312	1.327 (average)	
Error	0.0372	0.0372	0.0372	0.0372	0.0372	0.0372	0.0186	
t_{10}	0.264	0.762	6.181	2.770	2.639	8.401	71.41	
$t_{10,5\%}$	1.812	1.812	1.812	1.812	1.812	1.812	1.812	
Significant?	No	No	Yes	Yes	Yes	Yes	Yes	
(b) 2-Factor interactions								
Run	**AB**	**AC**	**AD**	**AE**	**AF**	**BC**	**BF**	**CE**	**EF**
1	1	1	1	1	1	1	1	1	1
2	−1	−1	−1	1	−1	1	1	−1	−1
3	−1	1	1	−1	−1	−1	1	−1	1
4	1	−1	−1	−1	1	−1	1	1	−1
5	1	−1	1	−1	−1	−1	−1	1	1
6	−1	1	−1	−1	1	−1	−1	−1	−1
7	−1	−1	1	1	1	1	−1	−1	1
8	1	1	−1	1	−1	1	−1	1	−1
9	1	1	−1	1	−1	1	−1	1	−1
10	−1	−1	1	1	1	1	−1	−1	1
11	−1	1	−1	−1	1	−1	−1	−1	−1
12	1	−1	1	−1	−1	−1	−1	1	1
13	1	−1	−1	−1	1	−1	1	1	−1
14	−1	1	1	−1	−1	−1	1	−1	1
15	−1	−1	−1	1	−1	1	1	−1	−1
16	1	1	1	1	1	1	1	1	1
Contrast	−0.073	0.092	0.071	0.106	0.035	0.106	0.205	−0.073	0.071
Effect	_−0.036_	_0.046_	_0.036_	_0.053_	_0.0176_	_0.053_	0.1025	_−0.036_	_0.036_
Error	0.0376	0.0376	0.0376	0.0376	0.0376	0.0376	0.0376	0.0376	0.038
T_{10}	0.969	1.221	0.949	1.411	0.467	1.411	2.723	0.969	0.949
$T_{10,5\%}$	1.812	1.812	1.812	1.812	1.812	1.812	1.812	1.812	1.812
Significant?	No	No	No	No	No	No	Yes	No	No

Y (mm): Roll-in; data obtained from Lin et al. [2005]

FIGURE 13.4.8 Plot of response surface of roll-in in terms of flange die radius (mm) and clearance (mm) in hemming operations of automotive body sheet aluminum.

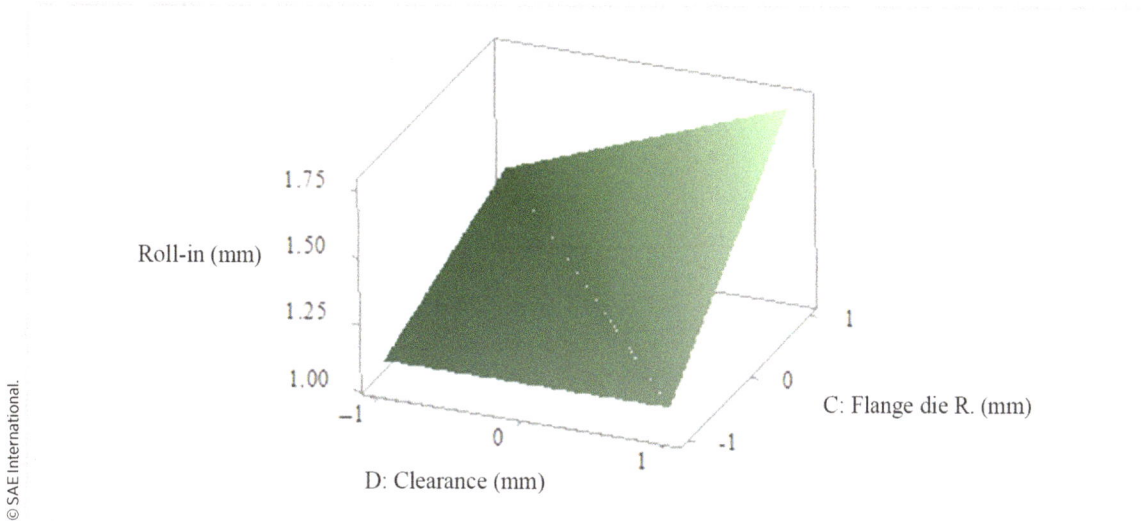

13.5. Casting

Casting is a hot metal-forming process. Molten metal is poured into a mold cavity configured in the desired shape, and the metal solidifies into the part. Casting is the most widely used process in bulky structural parts in the automotive industry. There are two major metal casting processes, i.e., sand casting and die casting.

13.5.1. Sand Casting

Sand casting is a metal casting process that utilizes the sand mold to contain the molten metal and solidify it into the desired shape. Part quality resulting from a sand casting is affected very much by the properties of molding sand. A face composite design is set up by Patel et al. [2019] to explore the mechanical properties of a molding material that is a mixture of resin, hardener, and silica sand. Three parameters were identified for this study, i.e., resin, hardener, and cure time, and their design levels are identified in Table 13.5.1.

TABLE 13.5.1 Factors and their design levels for examining silica sand mold [Patel et al. 2019].

Factor	Level (−1)	Level (0)	Level (+1)	Coded variable
A: Resin (%)	1.8	2.0	2.2	$a = (A - 2.0)/0.1$
B: Hardener (%)	0.2	0.3	0.4	$b = (B - 0.3)/0.1$
C: Cure time (h)	1	1.5	2	$c = (C - 1.5)/0.5$

TABLE 13.5.2 Face composite design based on factorial design 2^3 for examining silica sand mold [Patel et al. 2019].

Run (order)	a	b	c	A	B	C	σ_{ucs}	σ_{cal}	G_e	H_B	P_n
1 (8)	−1	−1	−1	1.8	0.2	1	168.6	107.8	8.54	60.7	203.5
2 (9)	1	−1	−1	2.2	0.2	1	368.7	280.5	8.86	74.2	109.3
3 (5)	−1	1	−1	1.8	0.4	1	252.8	181.7	6.63	67.1	159.5
4 (6)	1	1	−1	2.2	0.4	1	390.7	290.3	9.01	73.1	108.9
5 (16)	−1	−1	1	1.8	0.2	2	250.7	149.6	8.78	69.1	178.6
6 (12)	1	−1	1	2.2	0.2	2	360.6	272.0	8.11	74.7	138.4
7 (13)	−1	1	1	1.8	0.4	2	352.8	264.6	8.52	73.9	131.6
8 (2)	1	1	1	2.2	0.4	2	388.9	285.2	9.28	75.6	106.6
9 (17)	−1	0	0	1.8	0.3	1.5	268.8	182.6	8.32	70.5	171.2
10 (15)	1	0	0	2.2	0.3	1.5	372.2	281.2	8.97	75.6	121.3
11 (7)	0	−1	0	2.0	0.2	1.5	351.8	257.5	8.70	71.8	131.9
12 (10)	0	1	0	2.0	0.4	1.5	421.6	328.2	8.49	77.7	101.4
13 (4)	0	0	−1	2.0	0.3	1	350.7	268.2	7.34	72.8	134.6
14 (3)	0	0	1	2.0	0.3	2	378.6	280.2	7.75	76.8	125.8
15 (1)	0	0	0	2.0	0.3	1.5	371.2	270.7	7.60	75.1	118.5
16 (11)	0	0	0	2.0	0.3	1.5	372.8	274.5	8.27	75.2	114.2
17 (14)	0	0	0	2.0	0.3	1.5	370.7	288.7	8.43	75.5	116.7

where

σ_{ucs} (kPa): Ultimate compressive strength; high σ_{ucs} results in high-dimensional stability and surface finish, indicating the resistance of the mold to stresses during pouring and cooling of a casting and the ease of shakeout

σ_{cal} (kPa): Collapsibility strength, i.e., strength exposed to 650°C temperature in a muffle furnace for 2 min; low σ_{cal} results in hot tears, high residual stress, but easy shakeout

G_e (mL/g): Gas evolution; hot molten metal poured into the mold releases gas that results in surface defects of cast parts

H_B (0–100): Hardness, C-scale tester (Brinell type); high hardness results in high-dimensional stability and surface finish, but excessive hardness can cause cracks, scabs, blows, pinholes, and penetration

P_n (min^{-1}): Permeability number; low permeability results in undesired blows, pinholes, misrun, and expansion defects, but high permeability causes penetration-type defects and rough castings

The design matrix for the silica sand molding process and the related experimental data resulting from the 17 runs, with 15 distinct treatments, are also displayed in Table 13.5.2 [Patel et al. 2019]. It is a face composite design extended from full factorial design 2^3.

The data analysis is here done with "Minitab ➔ Stat ➔ Regression ➔ Regression ➔ Fit Regression Model," instead of "Minitab ➔ Stat ➔ DOE ➔ Response Surface." The "most insignificant" factor or interaction that comes with the largest p-value is weeded out one by one until all p-values are less than 10%, which is here defined as the parting significance level. Regression models, with three different correlations each, are given as follows:

(A) Permeability

$$P_n = 3262 - 2778\,A - 301.5\,C + 600.4\,A^2 - 1101\,B^2 + 31.9\,C^2$$

$$+ \ 323.6\,A\,B + 113.3\,A\,C - 46.1\,A\,B\,C$$

$$(R = 98.9\%, \ R_{adj} = 97.8\%, \ \text{and} \ R_{pred} = 91.5\%) \qquad (13.5.1)$$

(B) Gas Evolution

$$G_e = 68.9 - 54.6\,A - 113.3\,B + 5.63\,C + 12.45\,A^2 + 44.8\,B^2 - 2.407\,C^2$$

$$+\,37.67\,A\,B + 27.82\,B\,C - 10.57\,A\,B\,C$$

$$(R = 97.2\%,\ R_{adj} = 93.5\%,\ \text{and}\ R_{pred} = 94.7\%). \tag{13.5.2}$$

(C) Ultimate Compressive Strength

$$\sigma_{ucs} = -7186 + 6488\,A + 1382\,B + 659.8\,C - 1395\,A^2 + 1040\,B^2 - 46.6\,C^2$$

$$-\,850\,A\,B - 240\,A\,C$$

$$(R = 99.7\%,\ R_{adj} = 99.4\%,\ \text{and}\ R_{pred} = 98.3\%). \tag{13.5.3}$$

(D) Collapsibility Strength

$$\sigma_{col} = -6162 + 5839\,A - 1324\,A^2 + 265.6\,B^2 + 1334\,B\,C - 625.2\,A\,B\,C$$

$$(R = 99.2\%,\ R_{adj} = 98.9\%,\ \text{and}\ R_{pred} = 99.3\%). \tag{13.5.4}$$

(E) Hardness

$$H_B = -358.9 + 351.4\,A + 232.1\,B + 34.94\,C - 72.8\,A^2 - 121.2\,B^2 - 71.2\,A\,B - 15.25\,A\,C$$

$$(R = 98.3\%,\ R_{adj} = 97.0\%,\ \text{and}\ R_{pred} = 92.0\%). \tag{13.5.5}$$

The equations given above reveal that each performance index varies nonlinearly with respect to these three design factors. Both two-factor interactions (AB, AC, and BC) and the three-factor interaction (ABC) play significant roles in these four performance indices.

13.5.2. Die Casting

Die casting is a metal-forming process that utilizes dies and permanent metal molds to contain the molten metal and solidify it into the desired shape. Die casting molds are usually fabricated out of metal of high hardness and melting temperature, such as H13 steel that is able to hold out 250 MPa pressure. It allows a finer microstructure to arise and be reused for high-volume production. Passenger vehicles have many aluminum parts that have been manufactured through the die casting process. The surface finish and control of dimensional tolerances of die castings are thus superior to those of sand castings. Special attention to the draft angle has to be paid to make it possible for the part to be ejected easily.

Porosities are directly related to the mechanical strength of die-cast parts. Pressure may be applied to force molten material to get settle into mold details. The effects of injection parameters on the porosity and tensile properties of aluminum–silicon (Al-Si) alloy high-pressure die casting are of great concern [Adamane et al. 2015]. A lot of factors have to be considered to make sure that no air gets trapped inside the molten metal when it is injected. These include contamination of metal impurities, the way of pouring, potential melting loss and overheating, and volatilization and the way of spray of release lubricant (or agent).

Full factorial design 3^3 is utilized to investigate the potential influences of three process parameters (Table 13.5.3) on the mechanical strength of A413, i.e., a silicon-based aluminum alloy with a nominal specific gravity of 2.66, in die casting process under a squeeze pressure [Soundararajan et al. 2015]. The design matrix and test results are listed in Table 13.5.4. The three quality objective functions under study are Brinell hardness H_{BN}, tensile strength σ_{uts}, and yield strength σ_y in tension.

TABLE 13.5.3 Factors and their design levels for strength of die-cast parts [Soundararajan et al. 2015].

Factor	Level (−1)	Level (0)	Level (+1)
A: Squeeze pressure (MPa)	70	105	140
B: Die preheating temperature (°C)	150	225	300
C: Molten metal temperature (°C)	650	725	800

TABLE 13.5.4 Face composite design based on factorial design 2^3 for strength of die-cast parts [Soundararajan et al. 2015].

Run	A	B	C	H_{BN}	σ_{uts}	σ_y
1	70	150	650	72	252	149
2	70	225	650	76	266	147
3	70	300	650	65	236	136
4	70	150	725	71	249	146
5	70	225	725	78	269	152
6	70	300	725	66	237	135
7	70	150	800	72	239	135
8	70	225	800	75	248	141
9	70	300	800	62	232	131
10	105	150	650	84	285	158
11	105	225	650	87	289	167
12	105	300	650	79	258	142
13	105	150	725	89	284	166
14	105	225	725	88	293	167
15	105	300	725	81	274	158
16	105	150	800	78	265	156
17	105	225	800	87	291	161
18	105	300	800	79	258	149
19	140	150	650	94	298	171
20	140	225	650	92	299	174
21	140	300	650	87	286	162
22	140	150	725	94	301	167
23	140	225	725	95	305	176
24	140	300	725	87	287	169
25	140	150	800	91	285	166
26	140	225	800	92	299	171
27	140	300	800	83	284	162

The data analysis is carried out using Minitab (Stat → Regression → Regression → Fit Regression Model). The "most insignificant" factor or interaction that comes with the largest p-value is weeded out one by one until all p-values are less than 10%, which is here defined as the parting significance level based on the F-distribution. In light of the ANOVA given in Table 13.5.5, the regression model of the ultimate tensile strength, with three different correlations, is given as follows:

$$\sigma_{uts} = -504 + 1.822\ A + 0.861\ B + 1.649\ C$$

$$- 0.00553\ A^2 - 0.003042\ B^2 - 0.001264\ C^2 + 0.000593\ BC$$

$$(R = 98.4\%,\ R_{adj} = 97.8\%,\ \text{and}\ R_{pred} = 96.8\%). \tag{13.5.6}$$

High correlations redeem the effectiveness of the predictive equation of ultimate tensile strength of die-cast parts.

TABLE 13.5.5 ANOVA based on 3^3 for ultimate tensile strength of die-cast parts.

Source of variance	SS_{adj}	DOF	MS_{adj}	$F_{u,v}$	p-Value
Regression	12,964.4	7	1852.06	83.90	0.0%
A	671.7	1	671.71	30.43	0.0%
B	301.1	1	301.06	13.64	0.25%
C	242.5	1	242.51	10.99	0.4%
A^2	275.6	1	275.63	12.49	0.2%
B^2	1756.7	1	1756.74	79.58	0.0%
C^2	303.4	1	303.41	13.74	0.1%
BC	133.3	1	133.33	6.04	2.4%
Error	419.4	19	22.07		
Subtotal	13,383.9	26			
Grand average	—	1			
Total	—	27			

© SAE International.

When the molten temperature is set at 725°C, i.e., level 0 (Table 13.5.3), Equation (13.5.6) reduces to

$$\sigma_{uts} = 27.135 + 1.822\ A + 1.290925\ B$$

$$- 0.00553\ A^2 - 0.003042\ B^2 + 0.000593\ A\ B. \tag{13.5.7}$$

A plot of response surface of ultimate tensile strength in terms of die preheating temperature and squeeze pressure is demonstrated in Figure 13.5.1. In search of the maximal ultimate tensile strength of the die-cast parts, one can solve the following two equations:

$$\partial\sigma_{uts} / \partial A = 1.822 - 0.01106\ A + 0.000593\ B = 0, \tag{13.5.8}$$

and $\quad\partial\sigma_{uts} / \partial B = 1.290925 - 0006084\ B + 0.000593\ A = 0.$ $\tag{13.5.9}$

The above two equations lead to squeeze pressure A = 567 MPa and die heating temperature B = 195°C, which would maximize the ultimate tensile strength because $\partial^2\sigma_{uts}/\partial A^2 < 0$ and $\partial^2\sigma_{uts}/\partial B^2 < 0$. Although high squeeze pressure yields cast parts with a high ultimate tensile strength, the thermomechanical fatigue life of dies is of great concern.

© SAE International.

FIGURE 13.5.1 Plot of response surface of ultimate tensile strength in terms of die preheating temperature and squeeze pressure.

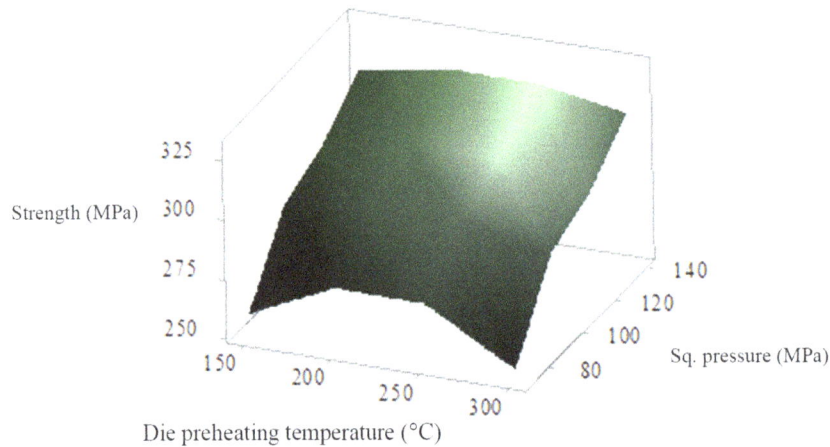

13.6. Shaping by Turning and Drilling

Machining is a very common and versatile manufacturing process to achieve the desired form by removing material from an object. Machining refers to combining various operations in a planned sequence to remove material from the workpiece for achieving the desired form [Nalbant et al. 2010]. Turning is the primary machining process for shaping round workpieces around the workpiece on a lathe, while drilling creates a round hole inside a workpiece. They are the two most applied machining actions. A machining operation consists of the following mechanical components:

(a) Workpiece: material properties, textures, and design features

(b) Cutting tool: geometry, material, coatings, and condition

(c) Machine tool: design, stiffness, and damping

(d) Fixture: workpiece holding devices

(e) Coolant: brand and application

(f) Working temperature

Shaping of automotive components, such as axles, crankshafts, connecting rods, studs, rams, pins, rolls, spindles, ratchets, torsion bars, sockets, worms, light gears, guide rods, and even bolts by machining, has been a traditional practice [Peters et al. 2014]. A P-diagram of a metal-cutting process is presented in Figure 13.6.1. The input parameters are described as follows:

(a) Cutting Speed (mm/s or m/min) and Spindle Speed (rpm): Cutting speed is the circumferential advancement of cutting per unit time. Spindle speed, an alias for the cutting speed, is the rotational frequency of the spindle of the machine. An excessive cutting speed will cause premature tool wear, breakages, and chatter, all of which may lead to potentially dangerous conditions.

(b) Feed Rate (mm/rev): Feed rate is the speed at which the cutter is fed in the radial direction, that is, advanced against the workpiece, typically expressed as millimeters per second (mm/s) or millimeters per revolution (mm/rev).

(c) Depth of Cut (mm): Depth of cut refers to how deep the tool is under the surface of material being cut. It gives the height of the chip produced, and it is typically expressed in millimeters (mm). The depth of cut should be less than or equal to the diameter of the cutting tool.

(d) Tool Size: Tool size is typically expressed in millimeters (mm). For example, drill bits that have been used to mill aluminum parts.

(e) Coolant: Coolants are applied for mitigating the heat build-up in the cutting zone and workpiece, reducing contact friction, flushing away chips, and debris.

(f) Working Temperature: Another primary parameter in machining is the cutting temperature that refers to the temperature distribution in the tool-workpiece contact zone. The cutting temperature is viable in tool life and finish surface quality.

FIGURE 13.6.1 P-Diagram of a machining process.

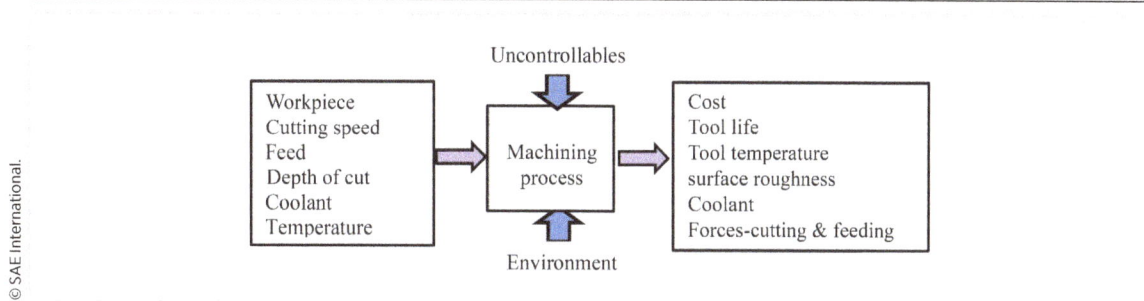

Primary motion of turning is the rotation of the workpiece driven by spindle rotation, and the feed is a translating motion of the cutting tool. Technical criteria used for assessing a turning process can be classified into the following:

1. R_a (μm): Surface roughness [Yang and Tarng 1998, Abbas et al. 2016]
2. R_R (cm³/min): Removal rate, workpiece material removed per unit time [Abbas et al. 2016]
3. F_T (N or kN): Tangential force [Cukor et al. 2011]
4. L_T (h or min): Tool life [Yang and Tarng 1998, Astakhov and Galitsky 2005]

13.6.1. Turning—Surface Roughness

Surface roughness has drawn great attention, as it is an important feature in many functional requirements such as fatigue strength, precision fits, tribological contacts, corrosion, and esthetics. It imposes a constraint for the selection of metal-cutting parameters. A considerable number of studies have been devoted to investigating the main and interactive effects in terms of cutting speed, feed rate of material, and depth of cut on the surface roughness and tool life.

The response data of surface roughness listed in Table 13.6.1 are taken from an extensive full factorial design 5^3 done by Abbas et al. [2016] but organized to meet full factorial design 4^3 as their five-level design is not statistically balanced. The uncoated cutters and workpieces used in the experiment are made of tungsten carbide steel (P-10) and heat-treated alloy steel material (J-Steel) with Vickers hardness in the range of HV 365-395. The applied process parameters and their design levels are given in Table 13.6.1. The ANOVA is employed to characterize the cutting process of J-steel bars using tungsten carbide cutting tools. The objective is to seek the proper surface roughness and machining time taken to remove a unit volume of workpiece material (Table 13.6.2).

TABLE 13.6.1 Factors and their design levels for studying cutting operations [Abbas et al. 2016].

Factors	Level 1	Level 2	Level 3	Level 4
A: Cutting speed (m/min)	150	200	250	300
B: Feed rate (mm/rev)	0.05	0.1	0.15	0.2
C: Depth of cut (mm)	0.1	0.2	0.3	0.4

TABLE 13.6.2 Full factorial design matrix 4^3 for evaluating surface roughness by turning [Abbas et al. 2016].

Treatment	a	b	c	A	B	C	R_a (μm)
1	1	1	1	150	0.05	0.1	0.370
2	1	1	2	150	0.05	0.2	0.339
3	1	1	3	150	0.05	0.3	0.328
4	1	1	4	150	0.05	0.4	0.348
5	1	2	1	150	0.1	0.1	0.868
6	1	2	2	150	0.1	0.2	0.856
7	1	2	3	150	0.1	0.3	0.907
8	1	2	4	150	0.1	0.4	1.261
9	1	3	1	150	0.15	0.1	1.873
10	1	3	2	150	0.15	0.2	2.162
11	1	3	3	150	0.15	0.3	2.389
12	1	3	4	150	0.15	0.4	1.929
13	1	4	1	150	0.2	0.1	3.200
14	1	4	2	150	0.2	0.2	3.229
15	1	4	3	150	0.2	0.3	3.152
16	1	4	4	150	0.2	0.4	2.798
17	2	1	1	200	0.05	0.1	0.357
18	2	1	2	200	0.05	0.2	0.389
19	2	1	3	200	0.05	0.3	0.309
20	2	1	4	200	0.05	0.4	0.434
21	2	2	1	200	0.1	0.1	0.821
22	2	2	2	200	0.1	0.2	0.770
23	2	2	3	200	0.1	0.3	0.908
24	2	2	4	200	0.1	0.4	0.956
25	2	3	1	200	0.15	0.1	1.859
26	2	3	2	200	0.15	0.2	1.843
27	2	3	3	200	0.15	0.3	1.837
28	2	3	4	200	0.15	0.4	2.046
29	2	4	1	200	0.2	0.1	3.501
30	2	4	2	200	0.2	0.2	3.417
31	2	4	3	200	0.2	0.3	3.593
32	2	4	4	200	0.2	0.4	3.048
33	3	1	1	250	0.05	0.1	0.371
34	3	1	2	250	0.05	0.2	0.329
35	3	1	3	250	0.05	0.3	0.462
36	3	1	4	250	0.05	0.4	0.557

(Continued)

TABLE 13.6.2 (Continued) Full factorial design matrix 4^3 for evaluating surface roughness by turning [Abbas et al. 2016].

Treatment	a	b	c	A	B	C	R_a (µm)
37	3	2	1	250	0.1	0.1	1.013
38	3	2	2	250	0.1	0.2	0.884
39	3	2	3	250	0.1	0.3	0.649
40	3	2	4	250	0.1	0.4	0.518
41	3	3	1	250	0.15	0.1	1.692
42	3	3	2	250	0.15	0.2	1.601
43	3	3	3	250	0.15	0.3	1.695
44	3	3	4	250	0.15	0.4	1.222
45	3	4	1	250	0.2	0.1	3.480
46	3	4	2	250	0.2	0.2	3.337
47	3	4	3	250	0.2	0.3	3.102
48	3	4	4	250	0.2	0.4	2.928
49	4	1	1	300	0.05	0.1	0.559
50	4	1	2	300	0.05	0.2	0.475
51	4	1	3	300	0.05	0.3	0.762
52	4	1	4	300	0.05	0.4	0.504
53	4	2	1	300	0.1	0.1	0.842
54	4	2	2	300	0.1	0.2	0.866
55	4	2	3	300	0.1	0.3	0.788
56	4	2	4	300	0.1	0.4	0.801
57	4	3	1	300	0.15	0.1	1.852
58	4	3	2	300	0.15	0.2	1.633
59	4	3	3	300	0.15	0.3	1.556
60	4	3	4	300	0.15	0.4	1.963
61	4	4	1	300	0.2	0.1	3.364
62	4	4	2	300	0.2	0.2	3.638
63	4	4	3	300	0.2	0.3	2.803
64	4	4	4	300	0.2	0.4	2.861

where R_a (µm): Surface roughness per arithmetic average deviation

The data analysis is carried out using Minitab (Stat → Regression → Regression → Fit Regression Model). The "most insignificant" factor or interaction that comes with the largest p-value is weeded out one by one until all p-values are less than 10%, which is here defined as the parting significance level based on the F-distribution. In light of the ANOVA presented in Table 13.6.3, the final predictive equation for the surface roughness (arithmetic averaged deviation) based on the factorial design is summarized as follows:

$$R_{ap} = -0.091 + 1.132\,C + 0.000003\,A^2 + 86.96\,B^2 - 0.0537\,A\,B\,C$$

$$(R = 98.6\%,\ R_{adj} = 98.5\%,\ \text{and}\ R_{pred} = 98.3\%), \tag{13.6.1}$$

of which high correlations confirm the effectiveness of the predictive equation for the surface roughness of machined parts.

In light of Equation (13.6.1), it can be seen that the surface roughness due to turning is proportional to the depth of cut (factor C), while it grows quadratically with increasing cutting speed (factor A) and increasing feed rate (factor B). Nevertheless, the three-factor interaction, i.e., ABC, is also statistically significant. Let C = 0.1 mm (low level), intended for low surface roughness. Then, Equation (13.6.1) reduces to

$$R_{ap} = 0.0222 + 0.000003 \ A^2 + 86.96 \ B^2 - 0.00537 \ A \ B. \tag{13.6.2}$$

A plot of response surface of surface roughness upon turning in terms of depth of cut (factor A) and feed rate (factor B) in a turning operation of heat-treated alloy steel material (J-Steel) using tungsten carbide (P-10) tools, as shown in Figure 13.6.2. It is shown that the feed rate (factor B) has a dominant effect on the surface roughness. The higher the feed rate, the higher the surface roughness.

TABLE 13.6.3 ANOVA for assessing surface roughness due to turning.

Variance	SS_{adj}	DOF	MS_{adj}	$F_{u,v}$	p-Value
Regression	73.8880	4	18.472	502	0.0%
C	0.2768	1	0.2768	7.52	0.8%
A^2	0.1453	1	0.1453	3.95	5.2%
B^2	26.950	1	26.950	732.4	0.0%
ABC	0.6759	1	0.6759	18.37	0.0%
Error	2.1711	59	0.0368		
Subtotal	76.059	63			
Grand average	—	1			
Total	—	64			

© SAE International.

FIGURE 13.6.2 Response surface plot of surface roughness as a function of cutting speed (factor A) and feed rate (factor B).

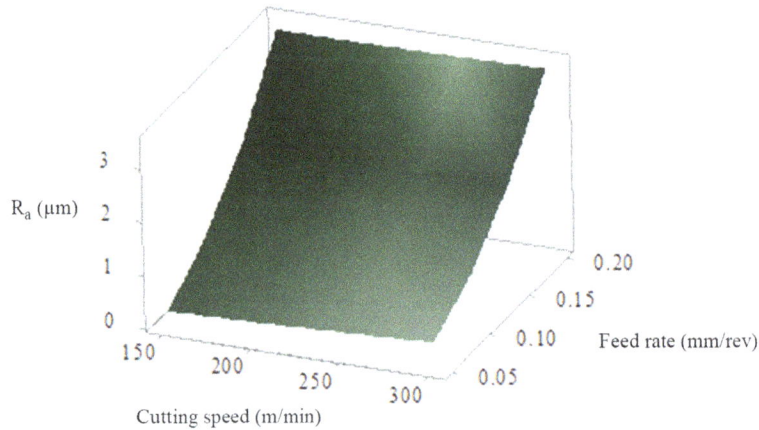

© SAE International.

13.6.2. Turning—Removal Rate

Removal rate is the volume of workpiece material removed per unit time, and it is an index for evaluating the machining efficiency that can be extended to disclose the operating cost. If the units for all three factors are consistent, one can expect that

$$\text{Removal Rate} \approx A\ B\ C,$$

of which factors A (cutting speed), B (feed rate), and C (depth of cut) are defined in Table 13.6.1. Each factor has a great impact on the removal rate. The units have to be unified when performing the calculation.

13.6.3. Turning—Tangential Force

An example taken from Cukor et al. [2011] is given here to demonstrate how to utilize the spherical composite design. The goal is to investigate how the tangential component of the cutting force is influenced by the three machining parameters given in Table 13.6.4 during a rough turning operation.

TABLE 13.6.4 Factorial levels for assessing the tangential force when turning a rough part [Cukor et al. 2011].

	Level				
Factor	**−1.682**	**−1**	**0**	**1**	**1.682**
A: Cutting speed (m/min)	265.9	300	350	400	434.1
B: Feed rate (mm/rev)	0.23	0.3	0.4	0.5	0.57
C: Depth of cut (mm)	0.99	1.5	2.25	3.0	3.51

The data analysis is carried out using Minitab (Stat → Regression → Regression → Fit Regression Model). The "most insignificant" factor or interaction that comes with the largest p-value is weeded out one by one until all p-values are less than 10%, which is here defined as the parting significance level based on the F-distribution. In light of the ANOVA presented in Table 13.6.5, the final predictive equation for the tangential force in the turning process based on the factorial design is summarized as follows:

$$Y_p = 824 - 3315\ B - 212.5\ C + 4008\ B^2 + 41.2\ C^2 + 2008\ B\ C. \tag{13.6.3}$$

A plot of response surface of tangential force in terms of feed rate (factor B) and depth of cut (factor C) in a turning operation is shown in Figure 13.6.3. It appears that the tangential force increases with increasing feed rate (factor B) and cutting speed (factor C) nonlinearly. The cutting speed (factor A) is statistically insignificant.

TABLE 13.6.5 Spherical composite design for assessing the tangential force when turning a rough part [Cukor et al. 2011].

Run	a	b	c	A	B	C	F (kN)	F_p (kN)
1	−1	−1	−1	300	0.3	1.5	879	868
2	1	−1	−1	400	0.3	1.5	894	868
3	−1	1	−1	300	0.5	1.5	1436	1448
4	1	1	−1	400	0.5	1.5	1408	1448
5	−1	−1	1	300	0.3	3	1754	1731
6	1	−1	1	400	0.3	3	1727	1731
7	−1	1	1	300	0.5	3	2896	2914
8	1	1	1	400	0.5	3	2861	2914
9	−1.682	0	0	265.9	0.4	2.25	1698	1677
10	1.682	0	0	434.1	0.4	2.25	1683	1677
11	0	−1.682	0	350	0.2318	2.25	1003	1043
12	0	1.682	0	350	0.5682	2.25	2609	2542
13	0	0	−1.682	350	0.4	0.9885	766	763
14	0	0	1.682	350	0.4	3.5115	2746	2722
15	0	0	0	350	0.4	2.25	1677.1	1677
16	0	0	0	350	0.4	2.25	1672.8	1677
17	0	0	0	350	0.4	2.25	1679.4	1677
18	0	0	0	350	0.4	2.25	1678.8	1677
19	0	0	0	350	0.4	2.25	1675.8	1677
20	0	0	0	350	0.4	2.25	1678.2	1677

Notes:
F (kN): Tangential cutting force obtained from experiment tests
F_p (kN): Tangential cutting force predicted using Equation (13.6.3)

© SAE International.

FIGURE 13.6.3 Response surface plot of tangential force (turning) as a function of feed rate (factor B) and depth of cut (factor C).

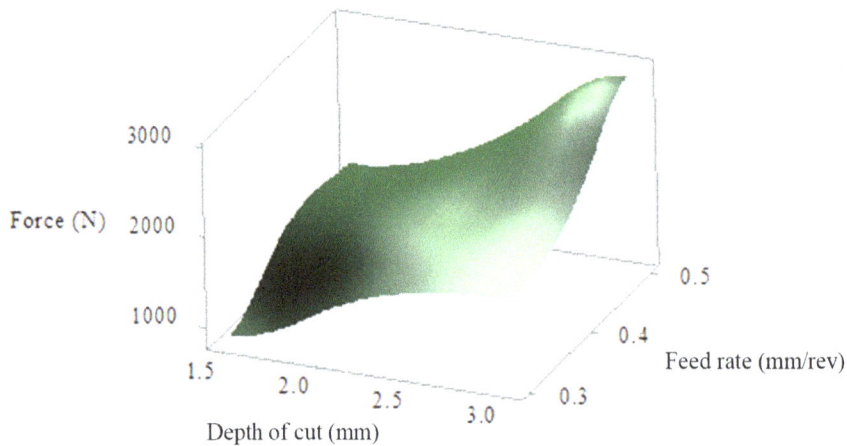

© SAE International.

13.6.4. Turning—Tool Life

Another major concern of turning operations is the tool life, which may require logarithmic transformation. A case study on the tool life in turning is presented in Volume II, Chapter 8, Section 8.3.

13.6.5. Drilling in Sheet Molding Compound (SMC) Composites

Drilling creates a round hole in a workpiece, and it is commonly employed to make holes for bolted and riveted joints. Defects like rough hole surface, delamination, or cracks may be observed, and severe tool wear is encountered frequently when performing drilling in SMC composites. As an example, tool flank wear of high-speed steel (HSS) tools in drilling SMC composites that consist of polyester (25% by weight)/calcium carbonate (35% by weight) reinforced with glass fibers (30% by weight) has drawn special attention [Uysal et al. 2012]. The operating factors and their design levels involved in drilling these SMC composites are given in Table 13.6.6.

TABLE 13.6.6 Operating factors and their design levels for drilling of SMC composites [Uysal et al. 2012].

Factors	Level (−1)	Level (0)	Level (1)
A: Drill point angle (deg)	80	100	120
B: Feed speed (mm/rev)	0.1	0.3	0.6
C: Cutting speed (m/min)	15	25	35

Reprinted with permission. © Springer.

Fractional factorial design 3_{III}^{3-1} and the related ANOVA were utilized to determine the optimal cutting parameters and to analyze the effects of them on the wear of tool flank. The wear data of HSS tool flanks subjected to these three drilling parameters are listed in Table 13.6.7 [Uysal et al. 2012].

TABLE 13.6.7 Factorial design 3_{III}^{3-1} (C = AB) for drilling of SMC composites [Uysal et al. 2012].

Run	a	b	c	A	B	C	W	W$_p$
1	−1	−1	−1	80	0.1	15	69.97	69.62
2	−1	0	0	80	0.3	25	61.80	61.31
3	−1	1	1	80	0.6	35	50.81	50.50
4	0	−1	0	100	0.1	25	104.12	104.92
5	0	0	1	100	0.3	35	80.31	81.02
6	0	1	−1	100	0.6	15	41.65	42.22
7	1	−1	1	120	0.1	35	128.95	128.48
8	1	0	−1	120	0.3	15	78.80	78.51
9	1	1	0	120	0.6	25	68.30	67.91

Notes:
W (μm): Data obtained from experimental tests
W$_p$ (μm): Predicted value of W based on Equation (13.6.4)

Multivariable linear regression analysis was also employed to determine the correlations between the factors and the tool wear. The analysis is conducted using the ANOVA based on the F-distribution on the wear data, as demonstrated in Table 13.6.8. After removing the least insignificant factor or interaction one by one via the comparison of their resulting p-values, with a cutoff significance level (p-value) of 10%, an experimenter obtains the following regression equation:

$$W_p = -47.27 + 1.0995 \, A - 96.8 \, B + 3.875 \, C$$

$$+ 140.1 \, B^2 - 0.05877 \, C^2 - 0.962 \, A \, B$$

$$(R = 99.98\%, R_{adj} = 99.91\%, \text{ and } R_{pred} = 99.14\%). \tag{13.6.4}$$

The predicted responses corresponding to the nine treatments are listed in column W_p, Table 13.6.7, of which the maximum error is 1.36%. All three main effects of factors A, B, and C are statistically significant. However, the influence of factor B (feed) is quadratically nonlinear, and so is the influence of factor C (cutting speed).

Given that the drill point angle is set at the low level (A = 80°),

$$W_p = 140.1 \, B^2 - 173.76 \, B - 0.05877 \, C^2 + 3.875 \, C + 40.69. \tag{13.6.5}$$

Thus, the influences of factor B and factor C on the wear of tool flank become quadratically nonlinear as shown in Figure 13.6.4. Setting the partial differentiations with respect to factors B and C to zeros, i.e., $\partial W_p / \partial B = 0$ and $\partial W_p / \partial C = 0$, leads to the minimum wear at B = 0.6 mm/rec (feed) and C = 15 m/min (cutting speed).

TABLE 13.6.8 ANOVA for HSS tool wear of drilling SMC composites.

Source	SS_{Adj}	DOF	MS_{Adj}	F-value	p-Value
Regression	5730.16	6	955.027	800.83	0.1%
A	463.23	1	463.231	388.44	0.3%
B	34.40	1	34.402	28.85	3.3%
C	94.83	1	94.827	79.52	1.2%
B^2	139.46	1	139.462	116.95	0.8%
C^2	50.97	1	50.967	42.74	2.3%
AB	46.86	1	46.861	39.30	2.5%
Error	2.386	2	1.193		
Subtotal	5732.55	8			
Grand average	—	1			
Total	—	9			

FIGURE 13.6.4 Quadratic variation of tool flank wear with respect to feed and cutting speed.

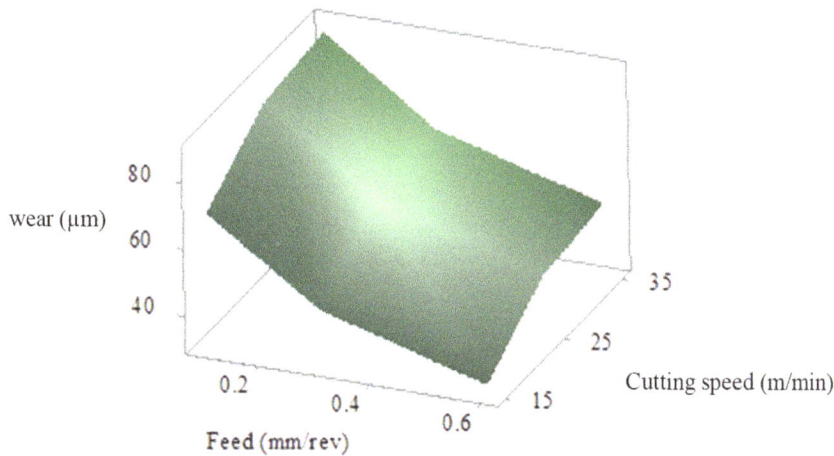

© SAE International.

13.7. Surface Finishing by Boring, Milling, and Grinding

Besides turning and drilling, three of the most common machining processes include, boring, milling, and grinding (abrasive processes), which are generally used for surface finishing. Surface finishing also includes some advanced material-removing procedures such as electrical, chemical, thermal, hydrodynamic, and laser processes.

13.7.1. Boring

Boring is a machining operation performed on the inside of a hollow workpiece, e.g., a drilled hole made previously by drilling, casting, or other processes. It increases the inner radius of the hole and provides an acceptable surface finish of the hole. Three force components are involved with the boring tools, i.e., tangential force, radial force, and feed force. Influences of four operating parameters, i.e., spindle speed (rpm), feed speed (mm/min), depth of cut (mm), and damper material [polytetrafluoroethylene (PTFE), polyvinyl ether (PVE), and butyl rubber (BR)], are examined against the surface roughness of drilled holes in EN8 steel, as proposed by Patil et al. [2020]. Design levels of these four operating parameters are given in Table 13.7.1. Factorial design $3^{3-1} \times 4$ matrix and test results (R_a) are listed in Table 13.7.2.

TABLE 13.7.1 Factors and levels used for examining surface roughness of bored holes in EN8 steel [Patil et al. 2020].

Factor	(1)	(2)	(3)	(4)
A: Spindle speed (rpm)	1000	1100	1200	—
B: Feed speed (mm/min)	0.02	0.04	0.06	—
C: Depth of cut (mm)	0.6	0.8	1.0	—
D: Damper	PTFE	PVC	BR	No damper
E: Length/diameter ratio of boring bar	Fixed (long bars inviting chattering)			

Note:
PTFE: polytetrafluoroethylene; PVC: polyvinyl chloride; BR: butyl rubber

© SAE International.

TABLE 13.7.2 Factorial design $3^{3-1} \times 4$ for examining surface roughness of bored holes in EN8 steel [Patil et al. 2020].

Run	A	B	C	D	R_a (µm)	R_{ap} (µm)
1	1000	0.02	0.6	PTFE	3.229	3.161
2	1000	0.04	0.8	PTFE	3.821	3.817
3	1000	0.06	1.0	PTFE	5.622	6.027
4	1100	0.02	0.8	PTFE	3.364	3.439
5	1100	0.04	1.0	PTFE	3.902	3.975
6	1100	0.06	0.6	PTFE	3.821	3.898
7	1200	0.02	1.0	PTFE	2.981	2.943
8	1200	0.04	0.6	PTFE	3.325	3.314
9	1200	0.06	0.8	PTFE	4.922	5.284
10	1000	0.02	0.6	PVC	3.651	3.611
11	1000	0.04	0.8	PVC	4.296	4.267
12	1000	0.06	1.0	PVC	6.345	6.477
13	1100	0.02	0.8	PVC	3.779	3.889
14	1100	0.04	1.0	PVC	4.410	4.425
15	1100	0.06	0.6	PVC	4.040	4.348
16	1200	0.02	1.0	PVC	3.352	3.393
17	1200	0.04	0.6	PVC	3.638	3.764
18	1200	0.06	0.8	PVC	5.549	5.734
19	1000	0.02	0.6	BR	3.881	3.871
20	1000	0.04	0.8	BR	3.903	4.527
21	1000	0.06	1.0	BR	6.747	6.737
22	1100	0.02	0.8	BR	4.036	4.149
23	1100	0.04	1.0	BR	4.684	4.685
24	1100	0.06	0.6	BR	4.651	4.608
25	1200	0.02	1.0	BR	3.563	3.653
26	1200	0.04	0.6	BR	4.005	4.024
27	1200	0.06	0.8	BR	5.905	5.994
28	1000	0.02	0.6	No	4.621	4.831
29	1000	0.04	0.8	No	5.441	5.487
30	1000	0.06	1.0	No	8.032	7.697
31	1100	0.02	0.8	No	4.805	5.109
32	1100	0.04	1.0	No	5.576	5.645
33	1100	0.06	0.6	No	5.538	5.568
34	1200	0.02	1.0	No	4.241	4.613
35	1200	0.04	0.6	No	4.769	4.984
36	1200	0.06	0.8	No	7.031	6.954

The data analysis is carried out using Minitab (Stat ➔ Regression ➔ Regression ➔ Fit Regression Model). The "most insignificant" factor or interaction that comes with the largest p-value is weeded out one by one until all p-values are less than 10%, which is here defined as the parting significance level based on the F-distribution. In light of the ANOVA presented in Table 13.7.3, the final predictive equations for the surface roughness (arithmetic averaged deviation) based on the factorial design are given as follows:

$$R_{ap,ptfe} = 25.50 - 0.0563\ A - 449.2\ B + 42.52\ C + 0.000032\ A^2 + 1943\ B^2$$
$$+ 0.2939\ AB - 0.03537\ AC, \tag{13.7.1}$$

$$R_{ap,pvc} = 25.95 - 0.0563\ A - 449.2\ B + 42.52\ C + 0.000032\ A^2 + 1943\ B^2$$
$$+ 0.2939\ AB - 0.03537\ AC, \tag{13.7.2}$$

$$R_{ap,br} = 26.21 - 0.0563\ A - 449.2\ B + 42.52\ C + 0.000032\ A^2 + 1943\ B^2$$
$$+ 0.2939\ AB - 0.03537\ AC, \tag{13.7.3}$$

and
$$R_{ap,no} = 27.17 - 0.0563\ A - 449.2\ B + 42.52\ C + 0.000032\ A^2 + 1943\ B^2$$
$$+ 0.2939\ AB - 0.03537\ AC$$
$$(R = 98.9\%, R_{adj} = 98.5\%, \text{ and } R_{pred} = 97.8\%), \tag{13.7.4}$$

of which high correlations confirm the effectiveness of these four predictive equations for the surface roughness of bored parts. The response data calculated using these four predictive equations for individual experimental treatments are listed in column R_{ap} in Table 13.7.2.

TABLE 13.7.3 ANOVA on examining surface roughness of bored holes in EN8 steel.

Source	SS$_{adj}$	DOF	MS$_{adj}$	F-value	p-Value
Regression	47.4187	10	4.74187	115.11	0.0%
A	0.5065	1	0.50654	12.30	0.2%
B	3.7401	1	3.74008	90.79	0.0%
C	3.6780	1	3.67798	89.28	0.0%
D	13.4990	3	34.49965	109.23	0.0%
A^2	0.8145	1	0.81451	19.77	0.0%
B^2	3.6830	1	3.68298	89.40	0.0%
AB	2.7650	1	2.76497	67.12	0.0%
AC	3.2035	1	3.20347	77.76	0.0%
Error	1.0299	25	0.04120		
Subtotal	48.4486	35			
Grand average	—	1			
Total	—	36			

It is obvious that the damping material is statistically significant and the PTFE damper yields the best surface finish. In order to find out the optimal design configuration, one can take partial differentiations of Equation (13.7.1) with respect to factors A, B, and C and set them to zeros, respectively, as

$$\partial R_{ap,ptfe} / \partial A = -0.0563\ A + 0.000064\ A + 0.2939\ B - 0.03537\ C = 0, \tag{13.7.5}$$

$$\partial R_{ap,ptfe} / \partial B = -449.2 + 3886\ B + 0.2939\ B = 0, \tag{13.7.6}$$

and
$$\partial R_{ap,ptfe} / \partial C = 42.52 - 0.03537\ A = 0. \tag{13.7.7}$$

The three equations given above can be solved for the optimal set of experimental parameters, i.e., A = 1202 rpm, B = 0.024675 mm/min, and C = 0.7882 mm, which would lead to the minimal surface roughness (2.881 μm).

13.7.2. Milling

Milling is a machining process with a rotating multi-tooth cutter to remove material and produce multiple chips in a single revolution. Two main types of milling operations are face milling and peripheral milling. Face milling shapes flat surfaces into the workpiece and flat-bottomed cavities (e.g., shallow slots), with either horizontal or vertical feeding. Peripheral milling shapes deep slots, threads, and gear teeth.

One objective of a milling operation is to select and control viable manufacturing factors to minimize surface roughness that is denoted using the center line average. Investigation into the effects of three manufacturing factors on the surface roughness was made by Bajic et al. [2010], and their design levels of concern are listed in Table 13.7.4. Data listed in Table 13.7.5 are collected from the experiments done using the rotatable composite design matrix based on factorial design 2^3. They are further used to determine the optimal operating conditions for a milling machine against the surface finish. The main objective of the study was to optimize a milling process utilizing the significant factors and interactions that affected the surface finish of milled parts.

TABLE 13.7.4 Factors and levels used for characterizing surface roughness of milled parts [Bajic et al. 2010].

Factor	(−1.682)	(−1)	(0)	(1)	(1.682)
A: Cutting speed (m/min)	140	150	160	180	190
B: Feed speed (mm/tooth)	0.20	0.24	0.30	0.36	0.40
C: Depth of cut (mm)	0.82	1.30	2.00	2.70	3.18
D: Coolant	Fixed	—	—	—	—
E: Direction of cut	Fixed	—	—	—	—
F: Number of cuts per work surface	1	1	1	1	1

TABLE 13.7.5 Composite design based on factorial design 2^3 for surface roughness of milled parts [Bajic et al. 2010].

Run	a	b	c	A	B	C	R_a	R_{ap}
1	0	0	0	165	0.3	2	0.57	0.58
2	1	1	1	180	0.36	2.7	0.88	0.88
3	1.682	0	0	190	0.3	2	0.45	0.50
4	1	−1	−1	180	0.24	1.3	0.28	0.17
5	0	0	0	165	0.3	2	0.48	0.58
6	−1	−1	1	150	0.24	2.7	0.47	0.54
7	0	0	−1.682	165	0.3	0.82	0.42	0.49
8	0	0	0	165	0.3	2	0.50	0.58
9	0	0	0	165	0.3	2	0.59	0.58
10	−1.682	0	0	140	0.3	2	0.92	0.86
11	0	0	1.682	165	0.3	3.18	0.81	0.75
12	1	1	−1	180	0.36	1.3	0.77	0.73
13	−1	1	1	150	0.36	2.7	1.10	1.09
14	1	−1	1	180	0.24	2.7	0.34	0.32
15	0	−1.682	0	165	0.2	2	0.23	0.12
16	0	0	0	165	0.3	2	0.60	0.58
17	0	1.682	0	165	0.4	2	1.15	1.05
18	−1	1	−1	150	0.36	1.3	0.97	0.94
19	0	0	0	165	0.3	2	0.62	0.58
20	−1	−1	−1	150	0.24	1.3	0.39	0.39

The data analysis is carried out using Minitab (Stat → Regression → Regression → Fit Regression Model). The "most insignificant" factor or interaction that comes with the largest p-value is weeded out one by one until all p-values are less than 10%, which is here defined as the parting significance level based on the F-distribution. In light of the ANOVA presented in Table 13.7.6, the final predictive equation for the surface roughness (arithmetic averaged deviation) based on the factorial design is summarized as follows:

$$R_{ap} = 4.71 - 0.0610\, A + 4.639\, B + 0.000163\, A^2 + 0.02693\, C^2$$

$$(R = 97.2\%, R_{adj} = 96.5\%, \text{ and } R_{pred} = 94.4\%), \tag{13.7.8}$$

of which high correlations confirm the effectiveness of the predictive equation for the surface roughness of machined parts. The response data predicted by Equation (13.7.8) corresponding to individual experimental treatments are listed in column R_{ap} in Table 13.7.5.

In light of Equation (13.7.8), it can be seen that the surface roughness of milled parts is proportional to the feed rate (factor B), while it grows quadratically with increasing cutting speed (factor A) and increasing depth of cut (factor C).

13.7.3. Grinding of Silicon Nitride Ceramics

An experiment was performed to study the effect of grinding on the strength of a high-performance silicon nitride ceramic material in the US—NIST Ceramics Division: Material Science and Engineering Laboratory. Grinding variables considered in the study and their design levels are given in Table 13.6.6. This is a very comfortable DOE since it has a resolution of V (five). It means that the experimenter can make sense of two-factor interactions by default on the assumption that three-factor interactions are generally insignificant and engineering senses. Five factors are a good size of independent variables to be covered in most applications. The design generator for the fractional factorial design matrix 2^{5-1} given in Table 13.7.7 is E = A BCD.

TABLE 13.7.6 Process factors and their design levels for grinding ceramics [NIST 2022].

Variable	Level (−1)	Level (+1)	Coded variable
A (m/s): Grinding table speed	0.025	0.125	a = (A − 0.075)/0.5
B (mm/min): Sample feed rate	0.05	0.125	b = (B − 0.075)/0.5
C: Grinding wheel grit coarseness	140–170	80–100	c = C
D: Grinding direction	Longitudinal	Transverse	d = D
E: Batch of material	1st	2nd	e = E

The data for design matrix 2^5 published by the original author are rephrased in P3.2 (Homework Problem in Volume I, Chapter 3). Nevertheless, 16 out of the published 32 data are picked for an alternative analysis using design matrix 2_V^{5-1}. Since there is no replication, the lowest three-factor interactions are here employed to formulate the sample random error, as underlined in Table 13.7.7. The predictive equation is

$$Y_p = 551.25 + 8\, a + 7.875\, b - 21.375\, c + 102.5\, d - 35.75\, e + 13.375\, a\, b - 9.875\, a\, c$$

$$+ 8\, a\, d + 8.875\, b\, d - 7.875\, c\, d + 12.63\, c\, e + 9.75\, d\, e, \tag{13.7.9}$$

where parameters a, b, c, d, and e are the five dimensionless coded variables derived from variables A, B, C, D, and E, respectively, using Equation (2.4.5).

It can be seen that factors D (grinding direction), E (batch), and C (grit coarseness) have strong individual main effects. Three-factor interactions "abc" and "abd" given in Table 13.7.7 are represented using their aliased terms "de" and "ce" in the above equation, respectively, because both factors A and B have weak main effects, and thus, their further reactions with factor C and factor D for form effective three-factorial interactions are assumed to be remote.

TABLE 13.7.7 The fractional factorial design matrix of 2_V^{5-1} [NIST 2022].

(I) Main effects						
Run	A	B	C	D	E = ABCD	Y (MPa)
1	−1	−1	−1	−1	1	607
2	1	−1	−1	−1	−1	722
3	−1	1	−1	−1	−1	702
4	1	1	−1	−1	1	638
5	−1	−1	1	−1	−1	704
6	1	−1	1	−1	1	586
7	−1	1	1	−1	1	602
8	1	1	1	−1	−1	669
9	−1	−1	−1	1	−1	492
10	1	−1	−1	1	1	434
11	−1	1	−1	1	1	418
12	1	1	−1	1	−1	568
13	−1	−1	1	1	1	392
14	1	−1	1	1	−1	410
15	−1	1	1	1	−1	429
16	1	1	1	1	1	447
Contrast	16	15.57	−42.75	−205	−71.5	
Effect	8	7.875	−21.38	102.5	−35.75	551.25 (average)
Error	2.368	2.368	2.368	2.368	2.368	
t_3	3.378	3.325	9.025	43.28	5.647	
$t_{3,95\%}$	2.353	2.353	2.353	2.353	2.353	
Significant	Yes	Yes	Yes	Yes	Yes	

(II) Two-factor and three-factor interactions										
Run	AB	AC	AD	BC	BD	CD	ABC	ABD	ACD	BCD
1	1	1	1	1	1	1	−1	−1	−1	−1
2	−1	−1	−1	1	1	1	1	1	1	−1
3	−1	1	1	−1	−1	1	1	1	−1	1
4	1	−1	−1	−1	−1	1	−1	−1	1	1
5	1	−1	1	−1	1	−1	1	−1	1	1
6	−1	1	−1	−1	1	−1	−1	1	−1	1
7	−1	−1	1	1	−1	−1	−1	1	1	−1
8	1	1	−1	1	−1	−1	1	−1	−1	−1
9	1	1	−1	1	−1	−1	−1	1	1	1
10	−1	−1	1	1	−1	−1	1	−1	−1	1
11	−1	1	−1	−1	1	−1	1	−1	1	−1
12	−1	−1	1	−1	1	−1	−1	1	−1	−1
13	1	−1	−1	−1	−1	1	1	1	−1	−1
14	−1	1	1	−1	−1	1	−1	−1	1	−1
15	−1	−1	−1	1	1	1	−1	−1	−1	1
16	1	1	1	1	1	1	1	1	1	1
Contrast	26.65	−19.75	16.0	−2.0	17.75	−15.75	19.5	25.25	5.75	5.5
Effect	13.375	−9.875	8.0	_−1.0_	8.875	−7.875	9.75	12.63	_2.875_	_2.75_
Error	2.368	2.368	2.368	2.368	2.368	2.368	2.368	2.368	2.368	2.368
t_3	5.65	4.17	3.38	0.42	3.75	3.33	4.12	5.33	1.21	1.16
$t_{3,95\%}$	2.353	2.353	2.353	2.353	2.353	2.353	2.353	2.353	2.353	2.353
Significant	Yes	Yes	Yes	No	Yes	Yes	Yes	Yes	No	No

13.8. Machining with Diamond Tools

As lightweight metals (e.g., aluminum alloys) and composite materials play a vital role in both on-ground vehicles and aircrafts, diamond tools are gaining more and more applications in automotive engineering.

Brazing is one of the fabrication processes for making diamond tools. It is a metallurgical bonding of diamond to its supporting metal part, which is filled with alloying materials (mainly metals) spread up in the supporting metal matrix that provides high-bonding strength. There are three ways for brazing diamond tools in industrial practice: single-layer brazing [e.g., chemical vapor deposition (CVD)], polycrystalline diamond (PCD) compact brazing, and diamond bit brazing.

13.8.1. Turning with Diamond Tools

Aluminum alloys and plastic composites are increasingly used by the automotive and aerospace industries because of their many advantageous mechanical and chemical properties. Aluminum alloys (e.g., Al–Si and Al–Mg–Si alloys) and plastic composites (e.g., glass-fiber-reinforced epoxy and carbon-fiber-reinforced epoxy) have been machined using diamond tools exclusively. Surface roughness, in terms of surface height (R_z), average of surface heights (R_a), or root-mean-square surface heights (R_m), is a quality indicator of machinability, especially in the case of aluminum fine turning.

Horváth and Drégelyi-Kiss [2013] used a 16-run central composite design based on 2^3, of which the process parameters and their design levels are displayed in Table 13.8.1, in an attempt to characterize the relationship between the surface roughness and three influential machining parameters, when CVD diamond tools are applied. Turning experiments were performed in dry conditions using carbon nano-coil (CNC) lathe type NCT Euroturn 12B, with 7 kW spindle power at a rotating speed of 6000 rpm. The workpiece material is a wear-resisting hyper-eutectic aluminum alloy, of which the chemical composition is Al-20.03Si-4.57Cu-1.06Fe (weight percentage) with a Brinell hardness (H_B) of 114.

Design levels of the three parameters of concern at high-speed machining for the purpose of better productivity are listed in Table 13.8.1. The response surface method (DOE-F) utilizes the central composite design based on design matrix 2^3. However, nontraditional "d = 1.28719" is set up for the central composite design as given in Table 13.8.1, instead of "d = 1.682" for spherical composite design or "d = 1" for face composite design (see Volume I, Chapter 3, Section 3.6). There are three runs for each treatment, and the test results are shown in Table 13.8.2.

TABLE 13.8.1 Factors and their design levels for experimental tests on turning with diamond tools [Horváth et al. 2014].

Process factors	Levels (d = 1.28719)				
	−d	−1	0	+1	+d
A: Cutting speed (m/min)	500	667	1250	1833	2000
B: Feed rate (mm/rev)	0.05	0.058	0.085	0.112	0.12
C: Depth of cut (mm)	0.2	0.267	0.5	0.733	0.8

TABLE 13.8.2 16 runs (treatments) for central composite design based on 2^3 [Horváth et al. 2014].

Run	A	B	C	R_a (µm)			$R_{a, ave}$	R_{ap}
1	667	0.058	0.267	0.47	0.48	0.48	0.477	0.481
2	667	0.058	0.733	0.62	0.51	0.35	0.493	0.440
3	667	0.112	0.267	1.06	1.00	1.03	1.030	1.168
4	667	0.112	0.733	1.55	1.44	1.54	1.510	1.561
5	1833	0.058	0.267	0.41	0.51	0.60	0.507	0.537
6	1833	0.058	0.733	0.7	0.58	0.67	0.650	0.593
7	1833	0.112	0.267	0.96	1.01	1.00	0.990	1.080
8	1833	0.112	0.733	1.34	1.23	1.31	1.293	1.302
9	500	0.085	0.5	0.98	1.05	1.10	1.043	0.966
10	2000	0.085	0.5	0.8	1.01	1.03	0.947	0.927
11	1250	0.05	0.5	0.55	0.48	0.49	0.507	0.506
12	1250	0.12	0.5	1.72	1.75	1.85	1.773	1.504
13	1250	0.085	0.2	0.82	0.82	0.91	0.850	0.703
14	1250	0.085	0.8	0.84	0.85	0.85	0.847	0.912
15	1250	0.085	0.5	0.76	0.90	0.86	0.840	0.946
16	1250	0.085	0.5	0.83	0.83	0.85	0.837	0.946

© SAE International.

The data analysis is carried out using Minitab (Stat → Regression → Regression → Fit Regression Model). The "most insignificant" factor or interaction that comes with the largest p-value is weeded out one by one until all p-values are less than 10%, which is here defined as the parting significance level based on the F-distribution. In light of the ANOVA presented in Table 13.8.3, the final predictive equation for the surface roughness (arithmetic averaged deviation) based on the factorial design is summarized as follows:

$$R_{ap} = 0.0424 + 47.7\ B^2 - 1.542\ C^2 + 0.000676\ A\ C + 22.99\ B\ C - 0.00856\ A\ B\ C$$

$$(R = 94.8\%,\ R_{adj} = 94.1\%,\ \text{and}\ R_{pred} = 93.3\%), \tag{13.8.1}$$

of which high correlations confirm the effectiveness of the predictive equation for the surface roughness of machined parts. The response data predicted by Equation (13.8.1) corresponding to individual experimental treatments are listed in column R_{ap} in Table 13.8.2. The significance of lack of fit (p-value = 0.0%) means that the replicated runs are not consistent.

TABLE 13.8.3 ANOVA for machining Al alloy AS17 with diamond tools.

Source	SS_{Adj}	DOF	MS_{Adj}	F-value	p-Value
Regression	5.6958	5	1.13915	74.15	0.0%
B^2	0.2865	1	0.28647	18.65	0.0%
C^2	0.2816	1	0.28158	18.33	0.0%
AC	0.1318	1	0.13179	8.58	0.5%
BC	0.4637	1	0.46365	30.18	0.0%
ABC	0.1613	1	0.16135	10.50	0.2%
Error	0.6452	42	0.01536	—	—
Lack of fit	0.4974	9	0.05526	12.33	0.0%
Pure error	0.1479	33	0.00448	—	—
Subtotal	6.3410	47			
Grand average	—	1			
Total	—	48			

© SAE International.

In light of Equation (13.8.1), it can be seen that the surface roughness of machined parts increases quadratically with increasing feed rate (factor B) but decreases quadratically with increasing depth of cut (factor C). Two two-factor interactions (AC and BC) and the three-factor interaction ABC are also statistically significant. Assume that the cutting speed (factor A) is held at 1250 m/min (middle level), and Equation (13.8.1) reduces to

$$R_{ap} = 0.0424 + 47.7\ B^2 - 1.542\ C^2 + 0.845\ C + 22.99\ B\ C - 10.7\ B\ C. \tag{13.8.2}$$

The response surface plot of surface roughness with respect to feed rate (factor B) and depth of cut (factor C) is exhibited in Figure 13.8.1. It can be seen that the surface roughness increases with increasing feed rate and increasing depth of cut.

FIGURE 13.8.1 Response surface plot of surface roughness of turning Al alloy AS17 turned with diamond tools.

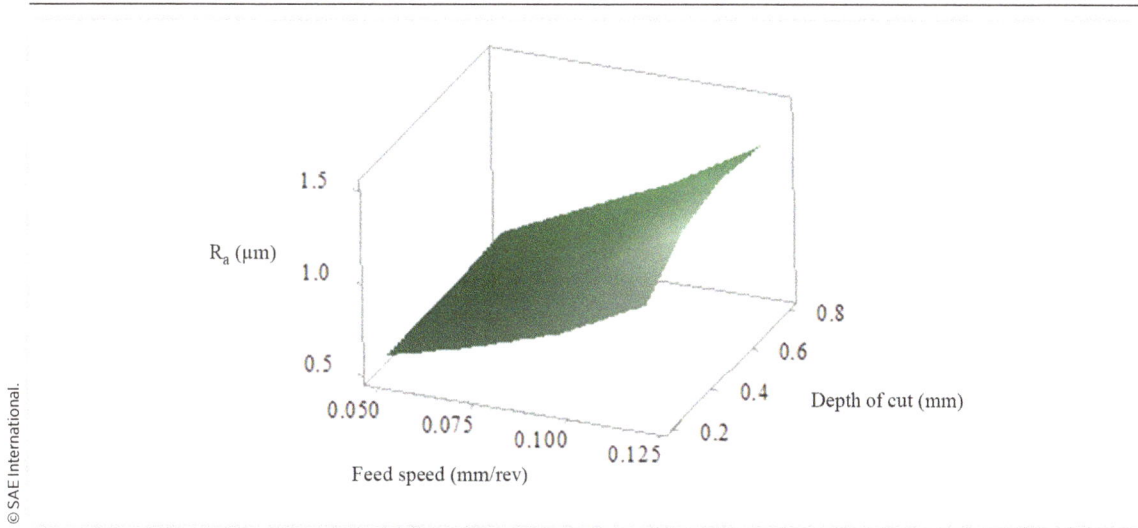

The study given above is extended to explore the influence of three different types of diamond tools on the machined surface roughness, including CVD, PCD, and monocrystalline diamond compact (MDC) [Horváth et al. 2014]. As cuttings by different diamond tools are applied individually and independently, PCD tools usually produce the best surface finish, followed by CVD tools.

13.8.2. Cutting with Diamond Wire Saws

Diamond wire saws are a conventional method for cutting Si ingots into wafers for making microelectronic devices and solar cells, because they have high cutting capacity and low material requirement and permit desired thickness with low surface roughness. A sawn surface usually exhibits both ductility mode and brittle fracture. The impact strength of ceramics is sensitive to brittle fracture.

Influence of diamond wire sawing parameters on surface roughness (μm) and substrate microcracks (μm) as formed in monocrystalline silicon wafers is presented in a comparative way using DOE by Costa et al. [2019]. Factors and their design levels are listed in Table 13.8.4, and the test results are given in Table 13.8.5.

TABLE 13.8.4 Factors and their design levels for examining surface roughness of Si ingot cut by diamond wire saws [Costa et al. 2019].

Factor	Level 1	Level 2	Level 3
A: Wire cutting speed (m/s)	10	15	20
B: Feed speed (mm/min)	20	30	40
C: Wire tension (N)	20	20	20

© SAE International.

TABLE 13.8.5 Design matrix and experimental test results for examining surface roughness and microcrack penetrations of wire-sawn monocrystalline silicon wafer [Costa et al. 2019].

Run	A	B	C_a	R_a	C_{ap}	R_{ap}
1	10	20	6.9	2.9	7.08	2.83
2	10	30	7.9	3.0	7.78	3.08
3	10	40	8.4	3.45	8.48	3.44
4	15	20	6.7	2.8	6.56	2.81
5	15	30	7.3	3.15	7.26	3.07
6	15	40	8.0	3.35	7.96	3.42
7	20	20	5.7	2.35	5.83	2.44
8	20	30	6.7	2.75	6.53	2.70
9	20	40	7.1	3.1	7.23	3.06

© SAE International.

where
R_a (μm): Average surface roughness across the surface profile of the sawn surface
C_a (μm): Average crack penetration into the substrate on the sawn surface
R_{ap} (μm): Predicted average surface roughness
C_{ap} (μm): Predicted average crack penetration

The data analysis is carried out using Minitab (Stat → Regression → Regression → Fit Regression Model). The "most insignificant" factor or interaction that comes with the largest p-value is weeded out one by one until all p-values are less than 10%, which is here defined as the parting significance level based on the F-distribution. In light of the ANOVA presented in Table 13.8.6, the final predictive equations for surface roughness and microcrack penetration based on the factorial design are summarized as follows:

$$R_{ap} = 1.606 + 0.1717 \, A - 0.00700 \, A^2 + 0.000511 \, B^2$$

$$(R = 97.8\%, R_{adj} = 96.5\%, \text{ and } R_{pred} = 92.9\%), \tag{13.8.3}$$

and

$$C_{ap} = 6.093 + 0.07000 \, B - 0.004153 \, A^2$$

$$(R = 98.7\%, R_{adj} = 98.3\%, \text{ and } R_{pred} = 96.6\%), \tag{13.8.4}$$

of which high correlations confirm the effectiveness of these two predictive equations. The values of both surface roughness and microcrack penetration predicted using the above two equations are listed in columns R_{ap} and C_{ap}, respectively.

TABLE 13.8.6 ANOVA for microcrack penetrations and surface roughness of wire-sawn monocrystalline silicon wafer.

(1) Surface roughness (R_a)					
Source	SS_{adj}	DOF	MS_{adj}	F-value	p-Value
Regression	0.85195	3	0.283984	37.32	0.001
A	0.04055	1	0.040554	5.33	0.069
A^2	0.06125	1	0.061250	8.05	0.036
B^2	0.57029	1	0.570287	74.95	0.000
Error	0.03805	5	0.007609		
Subtotal	0.89000	8			
Grand average	—	1			
Total		9			
(2) Microcrack penetration (C_a)					
Source	SS_{adj}	DOF	MS_{adj}	F-value	p-Value
Regression	5.2899	2	2.64493	114.14	0.0%
B	2.9400	1	2.94000	126.87	0.0%
A^2	2.3499	1	2.34985	101.41	0.0%
Error	0.1390	6	0.02317		
Subtotal	5.4289	8			
Grand average	—	1			
Total		9			

Response surface plots of surface roughness and microcrack penetration versus wire cutting speed (factor A) and feed speed (factor B) are demonstrated in Figure 13.8.4(a) and (b), respectively. They have a similar trending pattern.

FIGURE 13.8.4   Response surface plots of surface roughness and microcrack penetration in monocrystalline silicon wafer as cut by diamond wire saws.

(a) Surface roughness (R_a)

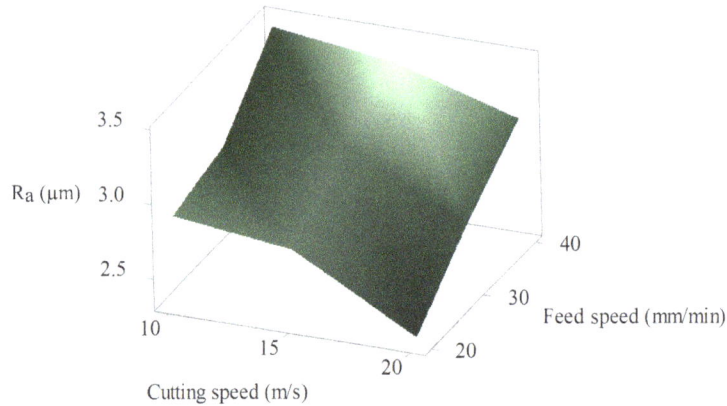

(b) Microcracks (C_a)

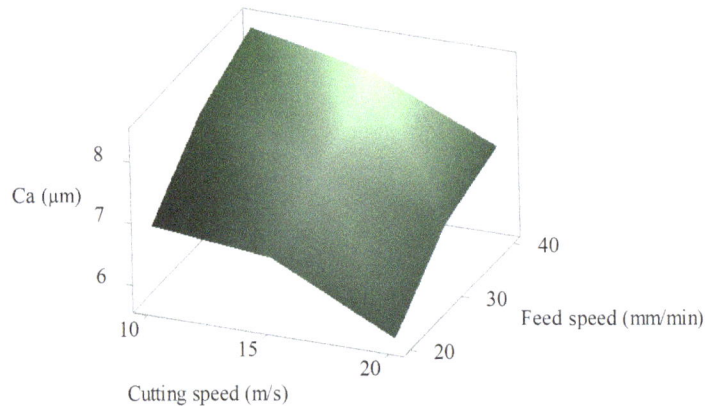

13.8.3. Cutting with Circular Diamond Blade Saws

Diamond blade saws and wire saws are very effective for cutting hard materials such as ceramics and stones. As shown in Table 13.8.7, four operating parameters and their design levels are employed to explore the effects on the cutting force (vector sum of the tangential and normal forces) of circular diamond blade saws [Karakurt 2014]. Data of the resultant force and its standard deviation obtained from experimental tests are listed in column F (N) and S_F (N), respectively, in Table 13.8.8. The efficacy of each measured resultant force is validated by its standard deviation.

TABLE 13.8.7 Factors and their design levels for studying cutting force with diamond sawblades [Karakurt 2014].

Factors	(1)	(2)	(3)	(4)
A (m/s): Cutting speed, peripheral	25	30	35	40
B (cm/min): Feed rate, transverse	40	50	60	70
C (cm): Depth of cut	0.5	1.0	1.5	2.0
D (10^{-3} L/s): Flow rate of coolant	50	100	150	200

TABLE 13.8.8 Factorial design 4^{4-2} for cutting rocks with diamond blade saws [Karakurt 2014].

Run	A	B	C	D	F (N)	S_F (N)	F_p
1	25	40	0.5	50	67.95	2.41	71.44
2	25	50	1.0	100	132.2	4.80	130.8
3	25	60	1.5	150	179.2	3.31	178.0
4	25	70	2.0	200	212.8	5.88	213.1
5	30	40	1.0	150	89.58	1.17	83.74
6	30	50	0.5	200	55.99	1.36	57.97
7	30	60	2.0	50	214.8	6.22	212.8
8	30	70	1.5	100	151.4	5.37	150.1
9	35	40	1.5	200	90.81	1.73	94.0
10	35	50	2.0	150	123.8	2.72	127.9
11	35	60	0.5	100	68.98	1.79	65.64
12	35	70	1.0	50	133.8	3.01	135.6
13	40	40	2.0	100	117.3	1.62	113.5
14	40	50	1.5	50	151.3	4.35	54.2
15	40	60	1.0	200	71.15	1.33	68.97
16	40	70	0.5	150	48.11	1.72	48.90

The data analysis is carried out using Minitab (Stat ➜ Regression ➜ Regression ➜ Fit Regression Model). The "most insignificant" factor or interaction that comes with the largest p-value is weeded out one by one until all p-values are less than 10%, which is here defined as the parting significance level based on the F-distribution. In light of the ANOVA presented in Table 13.8.9, the final predictive equation for the resultant cutting force based on the factorial design is summarized as follows:

$$F_p = 528.9 - 49.76\,A + 7.59\,B + 2.496\,D + 0.857\,A^2 - 0.0536\,B^2 + 15.43\,C^2$$

$$- 3.770\,A\,C - 0.02419\,A\,D + 2.736\,B\,C - 0.03647\,B\,D$$

$$(R = 99.8\%,\ R_{adj} = 99.5\%,\ \text{and}\ R_{pred} = 97.4\%), \tag{13.8.5}$$

of which high correlations confirm the effectiveness of the predictive equation for the cutting force of machined parts. The response data predicted by the above equation for individual experimental treatments are listed in column F_p in Table 13.8.8.

TABLE 13.8.9 ANOVA for cutting rocks with diamond blade saws.

Source	SS_{Adj}	DOF	MS_{Adj}	F-value	p-Value
Regression	42,036.8	10	4203.68	162.88	0.0%
A	1373.2	1	1373.23	53.21	0.1%
B	643.2	1	643.20	24.92	0.4%
D	1828.4	1	1828.36	70.84	0.0%
A^2	1521.1	1	1521.12	58.94	0.1%
B^2	459.1	1	459.14	17.79	0.8%
C^2	170.0	1	169.97	6.59	5.0%
AC	696.6	1	696.59	26.99	0.3%
AD	767.9	1	767.89	29.75	0.3%
BC	1939.3	1	1939.26	75.14	0.0%
BD	1390.8	1	1390.81	53.89	0.1%
Error	129.0	5	25.81		
Subtotal	42,165.8	15			
Grand average	—	1			
Total	—	16			

© SAE International.

13.8.4. Rotary Ultrasonic Face Milling of Ceramics with Diamond Bit Tools

Diamond bit brazing is a joining method that lets the matrix and diamond adhere to each other by diffusion brazing, which enhances the bonding force of the particular composite–diamond particles in pre-alloyed matrix [Fang et al. 2018]. Active pre-alloyed powder can improve the lifetime of diamond bit tools. The effect of Sn (zinc) content on the sintered diamond matrix under high compression via microstructural analysis in terms of porosity, hardness, and bending strength is of great interest [Liu et al. 2019]. Of course, a lot of attention has been directed to the resultant cutting force (F), material removal rate (M_{RR}), and surface roughness (R_a). Nevertheless, the workpiece material, i.e., magnesia-stabilized zirconia here, is expected to be removed in ductility mode (percentage of ductility), instead of creating brittle fracture.

A fractional factorial design 2_V^{5-1} (E = ABCD) is used to conduct an experiment [Pei and Ferreira 1999] to explore the performance of rotary ultrasonic face milling of magnesia-stabilized zirconia (ceramics) with diamond bit tools with a vibration at 20 Hz. The experimental setup is schematically depicted in Figure 13.8.5. Each workpiece is a 24 mm × 12 mm × 7-mm plate made of magnesia-stabilized zirconia. Potential effects of the process parameters, such as applied pressure, rotating speed, ultrasonic vibration amplitude, ultrasonic vibration frequency, diamond type, diamond concentration, bond type, and abrasive size, are of great concern. The purpose of the study is to exploit the main and interactive effects of five process parameters, which are listed in Table 13.8.10, on ductility mode. The test results are listed in Table 13.8.11.

FIGURE 13.8.5   Schematic drawing of an ultrasonic face-milling setup [Khanna et al. 1995].

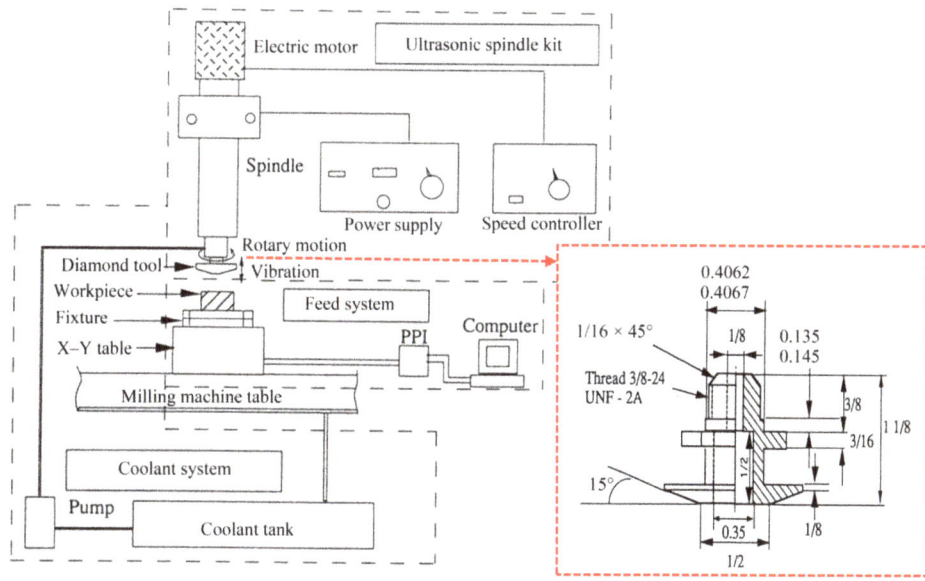

TABLE 13.8.10 Factors and their design levels for studying milling with diamond tools [Pei and Ferreira 1999].

Factor	Level (−1)	Level (+1)
A: Rotating speed (rpm)	1000	3000
B: Vibration amplitude (μm)	23	33
C: Feed rate (BLU/s)	40	80
D: Cutting tool, grain size	Coarse (170/200)	Fine (60/80)
E: Depth of cut (mm)	0.05	1.10

where
BLU: Basic length unit of the stepper motor feed system
D: Cutting tool, grain size-grit concentration: 100 (25% by volume)

TABLE 13.8.11 Factorial design 2_V^{5-1} (E = ABCD) for face milling by diamond tools.

Run	a	b	c	d	e	A	B	C	D	E	M
1	−1	−1	−1	−1	1	1000	23	40	−1	1.10	0.86
2	1	−1	−1	−1	−1	3000	23	40	−1	0.05	0.80
3	−1	1	−1	−1	−1	1000	33	40	−1	0.05	0.77
4	1	1	−1	−1	1	3000	33	40	−1	1.10	0.87
5	−1	−1	1	−1	−1	1000	23	80	−1	0.05	0.81
6	1	−1	1	−1	1	3000	23	80	−1	1.10	0.82
7	−1	1	1	−1	1	1000	33	80	−1	1.10	0.82
8	1	1	1	−1	−1	3000	33	80	1	0.05	0.79
9	−1	−1	−1	1	−1	1000	23	40	1	0.05	0.90
10	1	−1	−1	1	1	3000	23	40	1	1.10	0.99
11	−1	1	−1	1	1	1000	33	40	1	1.10	0.89
12	1	1	−1	1	−1	3000	33	40	1	0.05	0.94
13	−1	−1	1	1	1	1000	23	80	1	1.10	0.97
14	1	−1	1	1	−1	3000	23	80	1	0.05	0.94
15	−1	1	1	1	−1	1000	33	80	1	0.05	0.94
16	1	1	1	1	1	3000	33	80	1	1.10	0.91

M: Ductility mode, i.e., percentage of ductility of machine surface; the higher the better

Since factor D (grain size of cutting tool) is a categorical factor, the analysis is divided into two parts. There will be two predictive equations: one for coarse grains (170/200) and the other for fine grains (60/80). The data analysis is carried out using Minitab (Stat → Regression → Regression → Fit Regression Model). The "most insignificant" factor or interaction that comes with the largest p-value is weeded out one by one until all p-values are less than 10%, which is here defined as the parting significance level based on the F-distribution. In light of the ANOVA presented in Table 13.8.12, the final predictive equations for the ductility mode based on the factorial design are summarized as follows:

1. Coarse grains: 170/200 (d = −1)

$$M_{coarse} = 0.95606 - 0.000044\ A - 0.006102\ B + 0.000002\ A\ B + 0.003047\ B\ E$$

$$- 0.000631\ C\ E \quad (R = 99.98\%,\ R_{adj} = 99.94\%,\ \text{and}\ R_{pred} = 99.81\%)\ . \tag{13.8.6}$$

2. Fine grains: 60/80 (d = 1)

$$M_{fine} = 0.75000 + 0.000126\ A + 0.002655\ C - 0.000001\ A\ B - 0.000001\ A\ C$$

$$+ 0.000165\ C\ E \quad (R = 99.97\%,\ \text{and}\ R_{adj} = 99.91\%,\ \text{and}\ R_{pred} = 99.72\%). \tag{13.8.7}$$

The contrast in ductility mode between milling with coarse-grain (170/200) tools and milling with fine-grain (60/80) tools can be assessed, comparing the average for coarse grains of the eight runs (81.75%) to the average for fine grains (93.5%).

Next, consider the ANOVA table for milling with fine grains. Based on the MS_{adj} (adjusted mean square) values given in Table 13.8.6(2), it can be understood that factors A and C are the two dominant factors. Let B = (23 + 33)/2 = 28 and E = (0.05 + 1.0)/2 = 0.525, and then, Equation (13.8.7) reduces to

$$M_{fine} = 0.75000 + 0.000098\ A + 0.002736625\ C - 0.000001\ A\ C. \tag{13.8.8}$$

According to the above equation, a response surface plot of the ductility mode versus rotating speed (factor A) and feed speed (factor C) is demonstrated in Figure 13.8.6. It can be seen that 100% of the ductility mode is feasible.

TABLE 13.8.12 ANOVA on ductility mode of zirconia milled by diamond bit tools.

(1) Coarse grains: 170/200 (d = −1)					
Source	SS_{adj}	DOF	MS_{adj}	F-value	p-Value
Regression	0.007947	5	0.001589	1261.56	0.1%
A	0.000294	1	0.000294	233.57	0.4%
B	0.001159	1	0.001159	920.02	0.1%
AB	0.000335	1	0.000335	265.58	0.4%
BE	0.002399	1	0.002399	1904.07	0.1%
CE	0.000519	1	0.000519	411.86	0.2%
Error	0.000003	2	0.000001		
Subtotal	0.007950	7			
Grand average	−	1			
Total	−	8			
(2) Fine grains: 60/80 (d = 1)					
Source	SS_{adj}	DOF	MS_{adj}	F-value	p-Value
Regression	0.008196	5	0.001639	803.20	0.1%
A	0.007762	1	0.007762	3803.36	0.0%
C	0.004404	1	0.004404	2157.76	0.0%
AB	0.002116	1	0.002116	1036.80	0.1%
AC	0.005000	1	0.005000	2450.00	0.0%
CE	0.000236	1	0.000236	115.60	0.9%
Error	0.000004	2	0.000002		
Subtotal	0.008200	7			
Grand average	−	1			
Total	−	8			

© SAE International.

FIGURE 13.8.6 Response surface plot of ductility mode of zirconia milled by diamond bit tools.

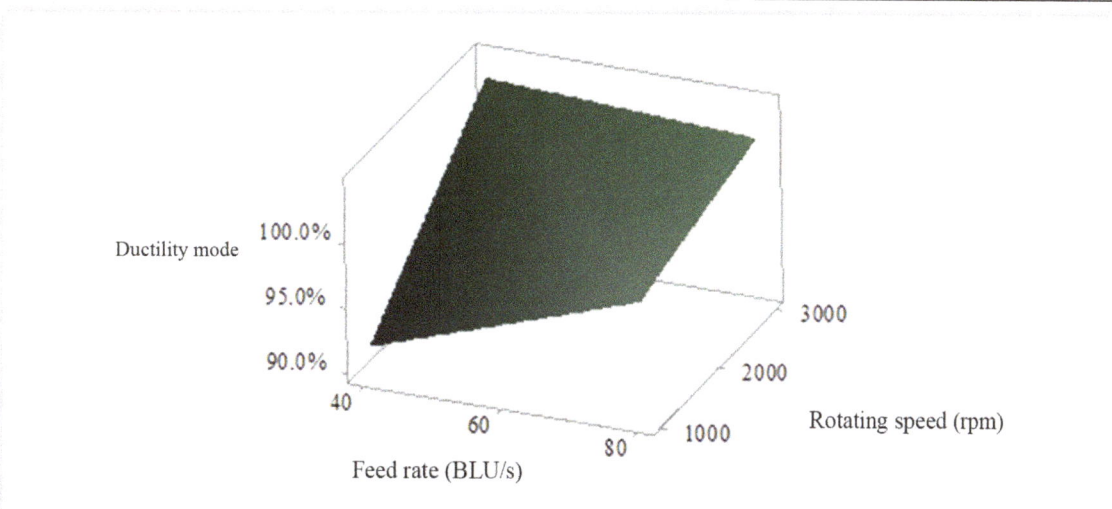

© SAE International.

13.9. Heat Treatments and Thermal Diffusions

A heat treatment process is to optimize the material microstructure in a specified environment for enhancing both physical and chemical characteristics in a time- and temperature-varying environment. How the temperature is heated up and gets cooled (e.g., in oil or air) can change the material properties dramatically. For example, heat treatment of bearing material AISI 52100 is to minimize the retained austenite content while maximizing hardness. Determination of proper time- and temperature-varying process parameters for a heat-treating procedure in pursuit of low retained austenite and high hardness can be achieved using DOE.

13.9.1. Steel Hardening by Carburizing and Nitriding

Hardening of steels can be achieved by heat treatment, in which diffusion occurs between a steel and its surrounding environment filled with carbon at a high temperature. The carbon slowly penetrates into workpiece surface and produces a case of up to 1.5 mm in thickness, while the inner core maintains its original mechanical properties. The procedure results in a hard case and a ductile substrate, and thus, it is called case hardening by carburizing. It provides a hard surface with enhanced properties, such as better scuff resistance, corrosion prevention, and fatigue strength, but still comes with a tough core to survive well under mechanical impacts.

Key process parameters of gas carburizing of SAE 8620 alloy steel carried out by Rehman et al. [2018] include delay quenching interval, hardening temperature, and soaking time. The design levels of interest are given in Table 13.9.1. The performance of case hardening can be assessed in terms of effective case depth and surface hardness. Treatment conditions (runs) and test results are listed in Table 13.9.2.

TABLE 13.9.1 Factors and their design levels of case hardening by carburizing [Rehman et al. 2018].

Factors	(−1)	(0)	(1)	Coded variables
A: Delay quenching interval (sec)	45	60	90	$a = (A − 60)/15$
B: Hardening temperature (°C)	780	800	820	$b = (B − 800)/20$
C: Soaking time in oil (sec)	180	300	420	$c = (C − 300)/120$

TABLE 13.9.2 The fractional factorial design matrix of 3_{III}^{3-1} with C = AB for carburizing.

Run	a	b	c	A	B	C	Depth	H_{RC}—case/core
1	−1	−1	−1	45	780	180	0.90	57/26
2	−1	0	0	45	800	300	1.00	59/29
3	−1	1	1	45	820	420	1.20	60/34
4	0	−1	0	60	780	300	0.65	58/22
5	0	0	1	60	800	420	0.80	57/25
6	0	1	−1	60	820	180	0.90	58/28
7	1	−1	1	90	780	420	0.45	56/20
8	1	0	−1	90	800	180	0.60	57/23
9	1	1	0	90	820	300	0.70	57/26

Depth (mm): Case depth by carburizing
H_{RC}: Rockwell hardness on C-scale of both the outer case and inner core; data obtained from Rehman et al. [2018]

The data analysis is carried out using Minitab (Stat → Regression → Regression → Fit Regression Model). The "most insignificant" factor or interaction that comes with the largest p-value is weeded out one by one until all p-values are less than 10%, which is here defined as the parting significance level based on the F-distribution. In light of the ANOVA presented in Table 13.9.3, the final predictive equation for the case thickness based on the factorial design is

$$\text{Depth} = -2.950 - 0.04000 \text{ A} + 0.006667 \text{ B} + 0.000222 \text{ A}^2$$

$$(R = 99.6\%, R_{adj} = 99.4\%, \text{ and } R_{pred} = 98.7\%),$$

of which high correlations confirm the effectiveness of the predictive equation for the cutting force of machined parts. On the influence on the depth of hardened case by these three process factors, the delay quenching interval (factor A) has a quadratic impact and the hardening temperature (factor B) has a linear impact, but the soaking time in oil within the parametric range given in Table 13.9.1 has no impact at all.

TABLE 13.9.3 ANOVA on experiment of case hardening by carburizing.

Source	SS_{adj}	DOF	MS_{adj}	F-value	p-Value
Regression	0.411667	3	0.137222	205.83	0.0%
A	0.033005	1	0.033005	49.51	0.1%
B	0.106667	1	0.106667	160.00	0.0%
A^2	0.019286	1	0.019286	28.93	0.3%
Error	0.003333	5	0.000667		
Subtotal	0.415000	8			
Total	—	9			

© SAE International.

Another popular way to achieve case hardening is nitriding, which is a process of diffusing nitrogen into the surface of steel. The nitrogen forms nitrides with elements such as aluminum, chromium, molybdenum, and vanadium in the outer surface of the steel workpiece. Nitriding is typically carried out at a temperature between 450 and 520°C, while nitrocarburizing at a temperature between 540 and 580°C. Since the operating temperature is below the tempering and stress-relieving temperatures of steel, i.e., 600°C approximately, the procedure causes little or no distortion [Rolinski and Woods 2018].

13.9.2. Effects of PWHT on Al 7075

Heat-treatable aluminum alloys 7000 series are often utilized in automotive applications, including marine vessels, on-ground vehicles, and aviation crafts, because of their high strength-to-density ratio. Factorial DOE was conducted to study the effect of PWHT on the mechanical properties of aluminum alloy Al 7075 [Peasura 2012]. Samples were solutionized at a temperature of 200 or 250°C in induction furnace and then air cooled. Heat treatment is applied at a prescribed temperature (°C) with a certain time duration (h) as given in Table 13.9.4. Data of the ultimate tensile strength (MPa) and Vickers hardness are then recorded as shown in Table 13.9.5. There are three replicated runs for each treatment condition.

TABLE 13.9.4 Factors and their design levels identified for PWHT of alloy Al 7075 [Peasura 2012].

Factor	Level (−)	Level (+)	Coded variables
A: Solution temperature (°C)	200	250	$a = (A - 225)/25$
B: Heat treatment temperature (deg)	80	110	$b = (B - 95)/15$
C: Heat treatment time duration (h)	20	24	$c = (C - 22)/2$

© SAE International.

TABLE 13.9.5 2^3 factorial design for PWHT of alloy Al 7075 [Peasura 2012].

(1) Ultimate tensile strength (σ_{uts})									
Treatment	A	B	C	AB	AC	BC	ABC	σ_{uts} (MPa)		
1	−1	−1	−1	1	1	1	−1	74.2	75.5	73.8
2	1	−1	−1	−1	−1	1	1	83.4	85.8	80.8
3	−1	1	−1	−1	1	−1	1	135.9	137.4	133.8
4	1	1	−1	1	−1	−1	−1	94.3	97.6	98.3
5	−1	−1	1	1	−1	−1	1	86.1	87.9	84.5
6	1	−1	1	−1	1	−1	−1	98.4	105.6	103.8
7	−1	1	1	−1	−1	1	−1	143.3	145.4	142.8
8	1	1	1	1	1	1	1	140.3	139.4	138.6
Contrast	−4.53	42.28	20.44	−17.16	10.54	4.97	6.74			
Effect	−2.26	21.14	10.22	−8.58	5.27	2.49	3.37			
Error	2.934	2.934	2.934	2.934	2.934	2.934	2.934			
t-Ratio	0.771	7.204	3.483	2.924	1.796	0.848	1.149			
$t_{16,\ 1\text{-}10\%}$	1.337	1.337	1.337	1.337	1.337	1.337	1.337			
Significant?	No	Yes	Yes	Yes	Yes	No	No			
(2) Hardness (H_v: Vickers hardness)									
Treatment	A	B	C	AB	AC	BC	ABC	H_v		
1	−1	−1	−1	1	1	1	−1	67.5	64.2	65.3
2	1	−1	−1	−1	−1	1	1	71.9	69.1	78.3
3	−1	1	−1	−1	1	−1	1	97.4	95.3	99.5
4	1	1	−1	1	−1	−1	−1	80.5	84.7	86.2
5	−1	−1	1	1	−1	−1	1	75.7	71.3	73.2
6	1	−1	1	−1	1	−1	−1	72.7	69.8	76.1
7	−1	1	1	−1	−1	1	−1	120.0	117.8	125.3
8	1	1	1	1	1	1	1	84.5	88.4	86.3
Contrast	−10.33	25.90	8.43	−13.78	−7.25	4.68	−3.27			
Effect	−5.17	12.95	4.22	−6.89	−3.63	2.34	−1.63			
Error S	4.233	4.233	4.233	4.233	4.233	4.233	4.233			
t-Ratio	1.220	3.059	0.996	1.628	0.856	0.553	0.386			
$t_{16,\ 1\text{-}15\%}$	1.071	1.071	1.071	1.071	1.071	1.071	1.071			
Significant?	Yes	Yes	No	Yes	No	No	No			

© SAE International.

Design contrasts and effects, both main and interactive, are calculated according to the procedure laid out in Chapter 2, while random error S is calculated using Equation (2.5.3) as

$$S = \left(\frac{2\,S_1^2 + 2\,S_2^2 + 2\,S_3^2 + 2\,S_4^2 + 2\,S_5^2 + 2\,S_6^2 + 2\,S_7^2 + 2\,S_8^2}{2 + 2 + 2 + 2 + 2 + 2 + 2 + 2} \right)^{1/2}, \quad (13.9.2)$$

where S_q (q = 1, 2, …, 8) is the sample standard deviation of the qth treatment, as obtained from the three replicated runs for each treatment. There are two, i.e., 3-1, degrees of freedom for each treatment and 16, i.e., 8 × 2, degrees of freedom inherent with random error S.

(a) Ultimate Tensile Strength: The following predictive equation for the ultimate tensile strength is derived with the statistical significance level of 10% (p-value = 10%):

$$\sigma_{uts} = 107.79 + 21.14\,b + 10.22\,c - 8.58\,a\,b + 5.27\,a\,c \quad (R_{pred} = 98.3\%).$$

It can be easily understood that both factor heat treatment temperature (factor B) and heat treatment time duration (factor C) have great impacts on the ultimate tensile strength. The higher the better for either factor B or factor C.

(b) Hardness: The following predictive equation is derived with the statistical significance level of 15% ($\alpha = 15\%$) for the material hardness:

$$H_v = 84.21 - 5.17\,a + 12.95\,b - 6.89\,a\,b \quad (R_{pred} = 92.8\%).$$

The reason why $\alpha = 15\%$ is taken for the hardness test in the above equation, because coefficient a (i.e., 5.17) is quite significant when compared with the average value (i.e., 84.21). In fact, the engineering significance (i.e., $5.17/84.21 = 6.2\%$) must be also taken into consideration in addition to the statistical significance.

13.9.3. Influence of Heat Treatment on Wear of Al6061/SiC Composites

Effect of heat treatment process parameters on the abrasive wear behavior of Al6061/SiC composites, i.e., alloy Al6061 reinforced with SiC particulates, has been studied by Shiri et al. [2015]. The experimental conditions of four process factors are given in Table 13.9.6. The test data based on factorial design 2^3, expressed in terms of volumetric loss (mm^3), are listed in Table 13.9.7.

TABLE 13.9.6 Factors and their design levels for wear of Al 6061/SiC composites [Shiri et al. 2015].

Factor	Level (−)	Level (0)	Level (+)
A (%): Volumetric fraction of SiC	0	7.5	15
B (h): Time of solution	1	2	4
C (°C): Aging temperature	150	180	210
D (h): Aging time	2	4	6

TABLE 13.9.7 2^3 Factorial design for wear of Al 6061/SiC composites [Shiri et al. 2015].

Run	A (%)	B (h)	C (deg)	D (h)	Wear (mm³)
1	0	1	150	2	248.5
2	0	1	180	4	178.3
3	0	1	210	6	153.1
4	0	2	150	4	224.2
5	0	2	180	6	149.3
6	0	2	210	2	167.8
7	0	4	150	6	251.6
8	0	4	180	2	282.4
9	0	4	210	4	145.6
10	7.5	4	150	2	216.0
11	7.5	4	180	4	160.9
12	7.5	4	210	6	135.9
13	7.5	1	150	4	186.4
14	7.5	1	180	6	143.5
15	7.5	1	210	2	138.6
16	7.5	2	150	6	199.3
17	7.5	2	180	2	187.8
18	7.5	2	210	4	131.4
19	15	2	150	2	15.63
20	15	2	180	4	15.75
21	15	2	210	6	11.96
22	15	4	150	4	8.308
23	15	4	180	6	14.14
24	15	4	210	2	9.022
25	15	1	150	6	16.55
26	15	1	180	2	16.67
27	15	1	210	4	15.46

The data analysis is carried out using Minitab (Stat → Regression → Regression → Fit Regression Model). The "most insignificant" factor or interaction that comes with the largest p-value is weeded out one by one until all p-values are less than 10%, which is here defined as the parting significance level based on the F-distribution. In light of the ANOVA presented in Table 13.9.8, the final predictive equation for wear based on the factorial design is

$$\text{Wear} = 539.8 - 1437\,A + 12.34\,B - 1.552\,C - 36.9\,D$$

$$- 10620\,A^2 + 3.13\,D^2 - 98.1\,A\,B + 9.40\,A\,C + 81.1\,A\,D$$

$$(R = 98.8\%, R_{adj} = 98.6\%, \text{ and } R_{pred} = 96.3\%), \tag{13.9.3}$$

of which high correlations confirm the effectiveness of the predictive equation for the cutting force of machined parts. Every factor manifests its own individual main effect on the volumetric wear of Al6061/SiC composites. Factor A (volumetric fraction of SiC particulates) has the greatest impact on the material wear as a quadratic function, and its interactions with the other three factors are also significant. The wear can be reduced dramatically with a high volumetric fraction of SiC particulates. Nevertheless, this is only true within the specified range of volumetric fraction in reinforcement, while an extrapolation beyond the test conditions specified in Table 13.9.6 needs further experiments to verify. Factor C (aging temperature) has also a good size of impact on the wear. The higher the aging temperature, the less the wear.

TABLE 13.9.8 ANOVA on wear experiment of Al 6061/SiC composites.

Source	SS$_{adj}$	DOF	MS$_{adj}$	F-value	p-Value
Regression	202,368	9	22,485	77.45	0.0%
A	2732	1	2732	9.41	0.7%
B	2557	1	2557	8.81	0.9%
C	15,616	1	15,616	53.79	0.0%
D	1936	1	1936	6.67	1.9%
A^2	21,413	1	21,413	73.76	0.0%
D^2	939	1	939	3.23	9.0%
AB	1515	1	1515	5.22	3.5%
AC	5366	1	5366	18.48	0.0%
AD	1777	1	1777	6.12	2.4%
Error	4935	17	290.3		
Subtotal	207,303	26			
Grand average	—	1			
Total	—	27			

© SAE International.

13.9.4. Hydrogen Embrittlement (HE)

HE results from the diffusion of hydrogen into metallic alloys, such as steels for vessels, pipelines, and structural bolts, and subsequent embrittlement at various trapping sites, especially grain boundaries and interfaces of heteroatoms. It works just like a heat treatment process but either at its manufacturing temperature or service temperature. Hydrogen migrates into the substrate via either electrochemical reaction (e.g., with 3.5% NaCl salt concentration) or a high-pressure gaseous hydrogen environment because of its small size [Barrera et al. 2018]. Brittleness would cause tremendous strength degradation, fracture initiation, subcritical crack growth, and even hydrogen blistering.

13.9.4.1. High-Strength Fasteners

There are two opportunities to introduce hydrogen into steel fasteners: One occurs during manufacturing, and the other is at service. For example, in the fabrication of steel fasteners, it is common to heat them mostly by baking, right after the coating process. It tends to extract any diffusible hydrogen that was introduced in the course of such a baking process. Environmental hydrogen may also be introduced as a result of steel corrosion when fasteners are put into service. A comprehensive DOE was detailed in Grendahl et al. [2015] on how to mitigate HE of two special types of AISI 4340 steel.

13.9.4.2. Austenitic Stainless Steel

Austenitic stainless steel is widely applied in automobiles and household appliances, because of its low cost and good resistance to corrosion. However, austenitic stainless steel is not immune to chloride-induced stress-corrosion cracking. An experiment is conducted to understand the impact of HE on the yield strength of austenitic stainless steel in the chlorine-contained environment (e.g., being exposed to the seawater) via electrochemical charging [Surendran et al. 2015]. The specimen of interest is connected to the cathodic terminal, while the base material acts as the reference electrode and is connected to the anode. Effects of solution concentration of electrolyte, charging current, and pre-strain on the HE of austenitic stainless steel on the strength degradation are revealed.

13.10. Coating

Coating refers to a work task, via which a layer of substance is sprayed or painted over a part surface. It is capable of producing a layer of durable outer case that would protect the part from invasion of foreign objects (e.g., impact of debris, material corrosion, and HE). Because of the complexity of a coating process, DOE is suggested to characterize the impact of process parameters on the coat quality [Rössler 2014].

Vehicle body and trim esthetics are a critical factor in consumers' purchasing decisions. Surface finish consistency, quality, and variety play the foremost role during a purchase. Automotive coating and finishing methods are mainly divided into two categories: body spraying and trim painting. So far, both solvent-based and waterborne paints have been widely used for coating automotive parts.

13.10.1. Automotive Body Spraying

Spraying is so far the most feasible way to apply metallic body coatings for exemplifying gloss, color effects, and esthetic patterns that meet customers' expectations at a high operating efficiency. Nevertheless, it also has to meet environmental regulations. Suggested layered structures of automotive body paint and trim paint are depicted in Figure 13.10.1(a) and (b), respectively.

Multi-performance requirements driving the research, development, and engineering of automotive body painting (coating) are (a) esthetic appearance, (b) corrosion prevention, (c) durability (e.g., no chipping over time), (d) cost, (e) mass production, (f) rustproof, (g) smear proof, and (g) weather resistance (Figure 13.10.2). A coat system for automotive body mainly consists of six essential layers described as follows [Akafuah et al. 2016]:

1. Pretreatment: This removes and cleans excess contaminants, degreases metal and plastics, surface conditions, and phosphates. It forms an appropriate surface structure enabling bonding of a corrosion protection layer.

2. Electrocoating: The electrodeposition (ED) of an anti-corrosion (anti-rust) prevention layer provides the automobile protection against diverse climatic conditions.

3. Sealing: A sealer like PVC is applied for anti-corrosion, elimination of water leaks, and minimization of chipping and vibrational noise.

4. Primer: A primer is applied to promote adhesion between the surface and the basecoat, having anti-chipping properties. This layer enhances the automobile's finish by bringing out the smoothness, brightness, and gloss of its colors. It also imparts a smoother surface for subsequent layers.

5. Basecoat: This pigmented coat provides surface properties that are sought after, including color, appearance, gloss, smoothness, and weather resistance, and transforms the car body into a thing of true beauty.

6. Clear Coat: This is the top layer that protects other layers under it and helps to resist fading of the color. The glass transition temperature of the clear coat for car body is around 80°C, while cured at 140°C [Nichols 2016].

Hoods and roofs are usually painted by the automotive manufacturers, while others (e.g., cowls and fenders) are painted by the suppliers.

The applied coating is expected to withstand a number of tests including crosshatch adhesion, reverse impact, pencil hardness, methyl ethyl ketone (MEK) double rubs to substrate, and 504-h neutral-salt-spray (NSS) testing.

FIGURE 13.10.1 Layers of generic automotive coating [Nichols 2016].

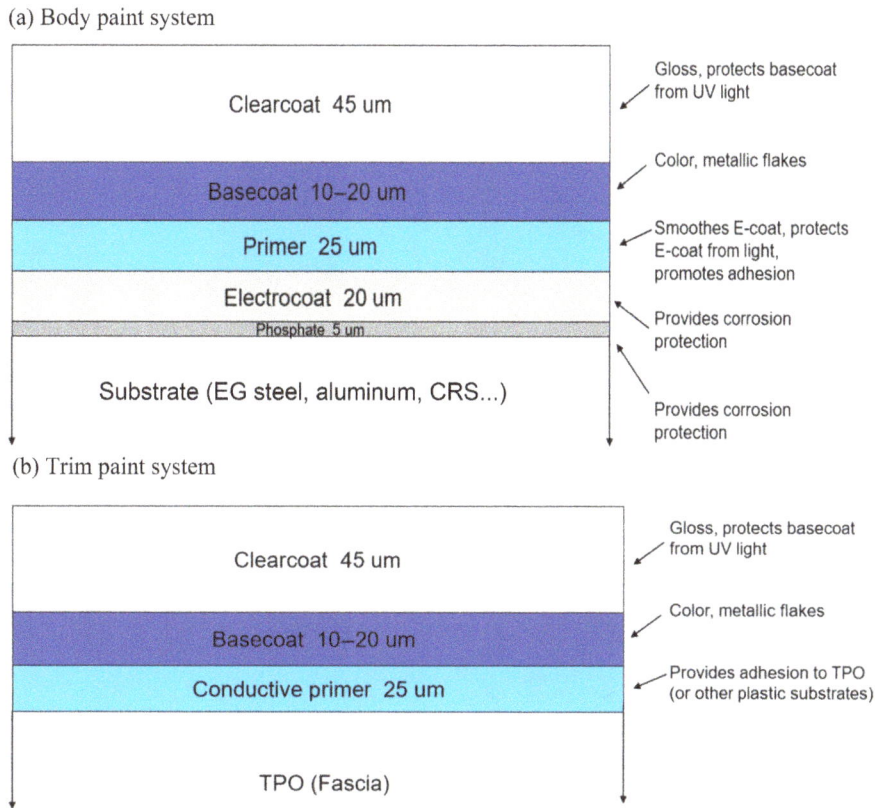

(a) Body paint system

Clearcoat 45 um — Gloss, protects basecoat from UV light

Basecoat 10–20 um — Color, metallic flakes

Primer 25 um — Smoothes E-coat, protects E-coat from light, promotes adhesion

Electrocoat 20 um — Provides corrosion protection

Phosphate 5 um

Substrate (EG steel, aluminum, CRS...) — Provides corrosion protection

(b) Trim paint system

Clearcoat 45 um — Gloss, protects basecoat from UV light

Basecoat 10–20 um — Color, metallic flakes

Conductive primer 25 um — Provides adhesion to TPO (or other plastic substrates)

TPO (Fascia)

FIGURE 13.10.2 Functional requirements of automotive body coatings [Akafuah et al. 2016].

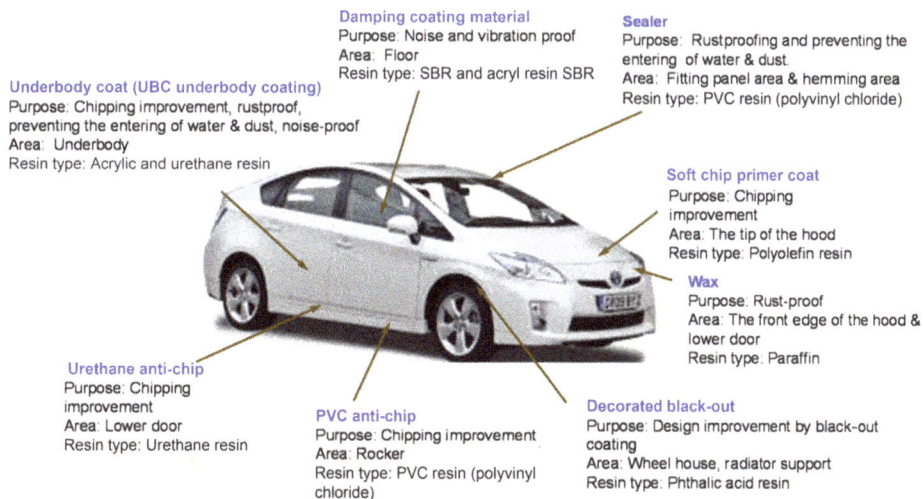

Damping coating material
Purpose: Noise and vibration proof
Area: Floor
Resin type: SBR and acryl resin SBR

Sealer
Purpose: Rustproofing and preventing the entering of water & dust.
Area: Fitting panel area & hemming area
Resin type: PVC resin (polyvinyl chloride)

Underbody coat (UBC underbody coating)
Purpose: Chipping improvement, rustproof, preventing the entering of water & dust, noise-proof
Area: Underbody
Resin type: Acrylic and urethane resin

Soft chip primer coat
Purpose: Chipping improvement
Area: The tip of the hood
Resin type: Polyolefin resin

Wax
Purpose: Rust-proof
Area: The front edge of the hood & lower door
Resin type: Paraffin

Urethane anti-chip
Purpose: Chipping improvement
Area: Lower door
Resin type: Urethane resin

PVC anti-chip
Purpose: Chipping improvement
Area: Rocker
Resin type: PVC resin (polyvinyl chloride)

Decorated black-out
Purpose: Design improvement by black-out coating
Area: Wheel house, radiator support
Resin type: Phthalic acid resin

13.10.2. Automated Spray Painting Process

The objective is to determine how the operating parameters of an automated spray painting process affect the paint distribution, paint thickness, and the color of the top coat of the automotive body. Liquid paint is atomized and deposited on the target surface. Proper paint viscosity is viable to have good surface finish quality including the thickness variation, surface roughness, and adherence of the coating. Automotive body paints are typically shear thinning, having lower viscosity at higher shear rates at the controlled temperature. Performance parameters such as paint flow, pressure of shape air, spray rotation, applied electric voltage, and flash between two layers are all linked directly to the viscosity of the liquid coating at the point of application.

Three primary factors and five design levels per factor, as given in Table 13.10.1, are selected to study the output quality of automated spray painting as measured in terms of dry film thickness variation [Chidhambara et al. 2011]. The dry film thickness variation is here expressed in terms of signal-to-noise ratio (S/N ratio) in light of Taguchi's method, as described in Volume II, Chapter 8, Section 8.8. Other concerned factors are fixed at their respective levels (Table 13.10.2).

TABLE 13.10.1 Factors and their five design levels for robotic spray painting [Chidhambara et al. 2011].

Factor	(−2)	(−1)	(0)	(1)	(2)
A (cc/min): Paint flow	200	250	300	350	400
B (bar): Pressure, shape air; 1 bar = 0.1 MPa	2	4	6	8	10
C (mPa·s, 10^{-3} Pa·s): Dynamic viscosity of paint	16	16.5	17	17.5	18
D (rpm): Rotation, spray	Fixed				
E (kV): Voltage	Fixed				
F (sec): Flash off between two layers	Fixed				

TABLE 13.10.2 Fraction factorial design 5^{3-1} for detecting thickness variation of automated spray painting [Chidhambara et al. 2011].

Run	A	B	C	S/N ratio	$(S/N)_p$
1	200	2	16.0	35.2686	35.3464
2	200	4	16.5	34.9638	34.8812
3	200	6	17.0	34.4855	34.4604
4	200	8	17.5	33.9794	34.084
5	200	10	18.0	33.6248	33.752
6	250	2	16.5	34.8073	35.12225
7	250	4	17.0	34.4855	34.573
8	250	6	17.5	34.1514	34.11625
9	250	8	18.0	33.9794	33.752
10	250	10	16.0	33.6248	33.154
11	300	2	17.0	34.8073	34.7982
12	300	4	17.5	34.4855	34.213
13	300	6	18.0	33.4420	33.7684
14	300	8	16.0	32.8691	33.2584
15	300	10	16.5	32.6694	33.1135
16	350	2	17.5	34.1514	34.37425
17	350	4	18.0	33.8039	33.8012
18	350	6	16.0	33.4420	33.5716
19	350	8	16.5	33.2552	33.2742
20	350	10	17.0	33.0643	33.1655
21	400	2	18.0	33.6248	33.8504
22	400	4	16.0	33.9794	34.0936
23	400	6	16.5	33.8039	33.5956
24	400	8	17.0	33.0643	33.3344
25	400	10	17.5	33.2552	33.31

The data analysis is carried out using Minitab (Stat → Regression → Regression → Fit Regression Model). The "most insignificant" factor or interaction that comes with the largest p-value is weeded out one by one until all p-values are less than 10%, which is here defined as the parting significance level based on the F-distribution. In light of the ANOVA presented in Table 13.10.3, the final predictive equation for the S/N ratio, referring to thickness variation, based on the factorial design is summarized as follows:

$$S/N = -34.7 + 9.14\ C - 0.00874\ A\ B - 0.296\ C^2 + 0.000481\ A\ B\ C$$

$$(R = 95.0\%,\ R_{adj} = 94.0\%,\ \text{and}\ R_{pred} = 92.0\%), \tag{13.10.1}$$

of which high correlations confirm the effectiveness of the predictive equation for the thickness variation of the automated spray painting. The high p-value of lack of fit (70.4%) means that Equation (13.10.1) is a good fit as measured using the pure error from replicated runs. The response data predicted by Equation (13.10.1) corresponding to individual experimental treatments are listed in column $(S/N)_p$ in Table 13.10.2.

TABLE 13.10.3 ANOVA resulting from fraction factorial design 5^{3-1} for detecting thickness variation of automated spray painting.

Source	SS_{adj}	DOF	MS_{adj}	F-value	p-Value
Regression	10.1907	4	2.54768	46.40	0.0%
C	0.3085	1	0.30848	5.62	2.8%
AB	1.8457	1	1.84566	33.62	0.0%
CC	0.3773	1	0.37730	6.87	1.6%
ABC	1.6028	1	1.60279	29.19	0.0%
Error	1.0980	20	0.05490	—	—
Lack of fit	1.0352	19	0.05448	0.87	70.4%
Pure error	0.0629	1	0.06287	—	—
Subtotal	11.2887	24			
Grand average	—	1			
Total	—	25			

In light of Equation (13.10.1), it can be seen that the thickness variation of automated spray painting decreases quadratically with increasing viscosity (factor C). Effects of two-factor interaction AB and three-factor interaction ABC are also statistically significant. In search of the extreme value of S/N ratio, one can solve the following three equations for the three unknowns (i.e., A, B, and C): d(S/N)/dA = 0, d(S/N)/dB = 0, and d(S/N)/dC = 0. They lead to

$$d(S/N) / dA = - 0.00874\ B + 0.000481\ B\ C = 0 \quad \rightarrow \quad C = 18.17$$

$$d(S/N) / dB = - 0.00874\ A + 0.000481\ A\ C = 0 \quad \rightarrow \quad C = 18.17$$

and $\quad d(S/N) / dC = 9.14 - 0.592\ C + 0.000481\ A\ B = 0 \qquad A\ B = 3361.$

It is interesting to notice that the applied levels of the paint flow (factor A) and the pressure of air shape (factor B) are supposed to compromise with each other in order to reduce the paint thickness variation. In fact, the unit for interaction AB is Watt (input energy per second), as derived below: 1 cc/min * bar = 1000 mm^3/min * 10^{-1} MPa = 100 mm^3/min * N/mm^2 = 100 Nmm/min = 10^{-1} Nm/min = 10^{-1} J/min = 600^{-1} W. Namely, interaction AB can be called paint delivery power in Watt (Figure 13.10.3).

FIGURE 13.10.3 S/N ratio as a function of interaction AB (paint delivery power) and factor C (paint viscosity).

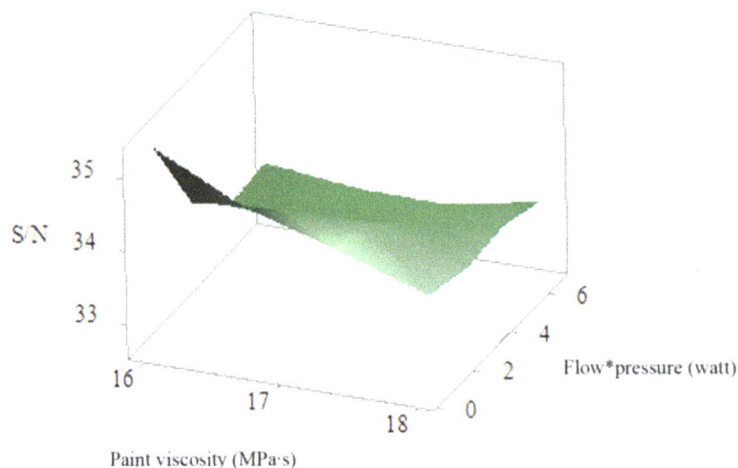

Assume that the paint flow (factor A) and the pressure of air shape (factor B) are fixed at 300 cc/min and 6 bar, respectively. It means the delivery power of paint is set at 3 W. Then, Equation (13.10.1) reduces to

$$S/N = -45.188 + 9.7172\ C - 0.296\ C^2. \tag{13.10.2}$$

The influence of paint viscosity (factor C) on the S/N ratio, according to the above equation, is depicted in Figure 13.10.4.

FIGURE 13.10.4 S/N ratio as a function of paint viscosity (sec) at paint delivery power = 3 W.

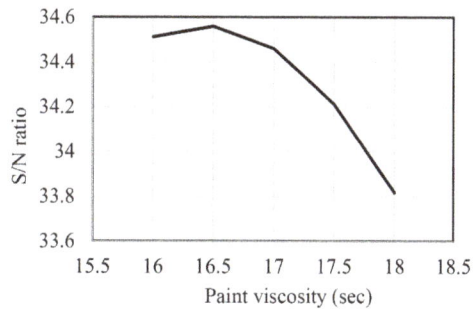

13.10.3. Scratch Creep of Body Paint

A DOE, as shown in Table 13.10.4, was conducted by Henkel Corporation [2018] to help improve the robustness of off-highway vehicle body coating in terms of scratch creep. It utilizes full factorial design 2^2 augmented by the center point and two sets of axial points, such that the design matrix "almost" fits the spherical composite design. Coded variables of the two process parameters, i.e., applied temperature (factor A) and time duration (factor B), are given as follows:

$$a = (A - 325) / 25$$

$$\text{and} \quad b = (B - 25) / 15,$$

The paint is applied to the vehicle panel, and then, a line is scribed all the way through to the bare metal. The panel is exposed to a saltwater fog for 21 days, and the total amount of coating which has been lost on both sides of the scribe, i.e., a value known as creep, is measured as shown in Figure 13.10.5. The process is described in detail in ISO 4628-8 [2012].

FIGURE 13.10.5 Scratched panel after exposure to 504-h NSS testing.

The observed gap size, i.e., Y value, is listed in Table 13.10.1. The objective is to reduce the "creep gap (mm)" via the response surface method. Hence, one can have the body coating to ensure the robustness in scratch creep despite variations in process conditions [Henkel Corporation 2018]. The ANOVA is summarized in Table 13.10.2. Accordingly, the regression model is

$$Y_p = 38.78 - 0.1791\ A - 0.978\ B + 0.00396\ A\ B + 0.00447\ B^2$$

$$(R = 92.3\%,\ R_{adjust} = 84.6\%,\ \text{and}\ R_{pred} = 76.8\%). \tag{13.10.3}$$

Note that the parting line for statistical significance to justify the above equation is taken as one-sided p-value = 13.5%, as shown in Table 13.10.5. The predicted values (Y_p) using the above equation are also listed in Table 13.10.4. The residual plot is given in Figure 13.10.6.

TABLE 13.10.4 DOE for reducing scratch creep of automotive body coating [Henkel Corporation 2018].

Run	a	b	A	B	Y	Y_p (mm)
1	−1	−1	148.9	10	7.9	8.68
2	1	−1	176.7	10	4.15	4.80
3	−1	1	148.9	40	3.75	3.73
4	1	1	176.7	40	3.3	3.15
5	0	0	162.8	25	3.4	4.08
6	−1.4	0	143.3	25	6.3	5.65
7	1.4	0	182.2	25	3.0	2.53
8	0	−1.333	162.8	5	9.0	8.07
9	0	1.333	162.8	45	3.35	3.68

Notes:
Factor A: Temperature (°C)
Factor B: Time elapsed after scratch (min)
(a = 0, b = 0): Original production setting
Y (mm): Measured scratch width
Y_p (mm): Predicted value

TABLE 13.10.5 ANOVA of scratch creep of automotive body coating.

Variance	SS_{adj}	DOF	MS_{adj}	$F_{u,v}$	p-Value
Regression	37.349	4	9.3373	11.94	1.7%
A	SS_A = 7.551	1	MS_A = 9.34	$F_{1,4}$ = 9.66	3.6%
B	SS_B = 5.784	1	MS_B = 7.55	$F_{1,4}$ = 7.40	5.3%
AB	SS_{AB} = 2.723	1	MS_{AB} = 5.78	$F_{1,4}$ = 3.48	13.5%
B^2	SS_{BC} = 4.018	1	MS_{BC} = 4.02	$F_{1,4}$ = 5.14	8.6%
Error	SS_E = 3.127	4	MS_E = 0.7818	—	—
Subtotal	SS_T = 40.48	8	—	—	—
Grand average	—	1	—	—	—
Total	—	9	—	—	—

FIGURE 13.10.6 Residual plot of scratch creep width of automotive body coating.

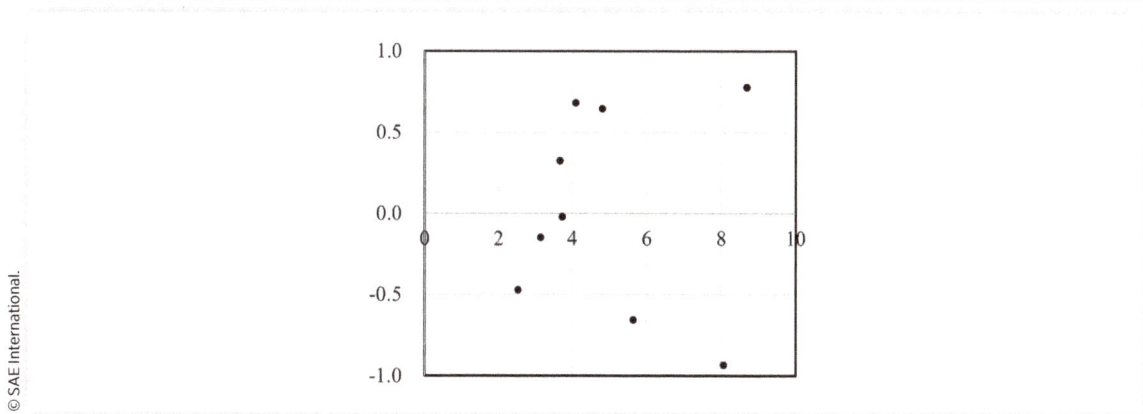

13.10.4. Automotive Trim Paint

Automotive plastic substrates such as bumpers, trim, door handles, and mirror housing require a different painting process from that used for metallic body parts. Two-coat systems are available for these applications, which are comprised of a waterborne base coat and a UV curing clear coat [Schulz 2013]. Plastic parts are provided with protection against mechanical, chemical, and physical influences such as scratches, stone chipping, weathering, and UV radiation, as well as ingredients used in cosmetics and cleaning agents. The glass transition temperature of the clear coat for plastic parts is around 20°C, while cured at 80°C [Nichols 2016]. A coat system for automotive trims mainly consists of three essential layers described as follows:

1. Primer: A primer is applied to promote adhesion between the surface and the basecoat, having anti-chipping properties. This layer enhances the automobile's finish by bringing out the smoothness, brightness, and gloss of its colors. It also imparts a smoother surface for subsequent layers.

2. Basecoat: The pigmented coat provides surface properties that are sought after, including color, appearance, gloss, smoothness, and weather resistance.

3. Clear Coat: This is the top layer that protects other layers under it and helps to resist fading of the color. The glass transition temperature of the clear coat for trims is around 20°C, while cured at 80°C [Nichols 2016].

Of course, cleaning and pretreatment of the plastic components are decisive for good painting results. Good adhesion can be achieved without pretreatment on many common plastic substrates, such as acrylonitrile butadiene styrene (ABS), ABS/PC, acrylonitrile styrene acrylate (ASA)/PC, polyamide (PA) (nylon), PS, styrene acrylonitrile (SAN), SMC (e.g., glass fiber-reinforced epoxy), and thermoplastic polyolefin (TPO) (rigid or flexible). Otherwise, adhesion characteristics can be optimized as a rule by activating the surface with the help of a flaming process. A typical automotive trim paint system is depicted in Figure 13.10.1(b).

Some plastic-based composite parts because of either injection molding or bulk molding compound (BMC) are subsequently coated to achieve esthetic appearances. The compatibility of substrate compositions with paint affects the final coating performance significantly. A detailed strategy of DOE for various coats can be found in Rössler [2014].

13.10.5. CVD Coating for Thermal Stability Protection

The CVD is a process which allows the creation of a deposit on the part surface to protect it from oxidation and corrosion. This protective coating is frequently used in the aeronautics industry, especially on the turbine blades of the reactors that work at very high temperatures.

13.10.6. CVD for Fabricating Nanocomposites

It has been also used for fabricating composite materials via infiltration techniques. Helical carbon nanofibers (HCNFs), including CNCs, carbon nano-twists, and multi-walled CNCs, can be synthesized by CVD [Suda et al. 2015].

References

Abbas, A. T. et al. (2016), "Multi-Objective Optimization of Turning Cutting Parameters for J-Steel Material," *Advances in Materials Science and Engineering*, 2016, Article ID 6429160.

Adamane, A. R. et al. (2015), "Influence of Injection Parameters on the Porosity and Tensile Properties of High-Pressure Die Cast Al-Si Alloys," *Int'l Journal of Metalcasting*, 9(1), pp. 43-53.

Aitharaju, V. (2020), "Development and Integration of Predictive Models for Manufacturing and Structural Performance of Carbon Fiber Composites in Automotive Applications," General Motors LLC, Warren, MI, March 30, 2020; Funded in part by the US-DOE (DE-EE0006826).

Akafuah, N. K. et al. (2016), "Evolution of the Automotive Body Coating Process - A Review," *Coatings*, 6(2), pp. 1-22.

Akbarzadeh, A. and Sadeghi, M. (2011), "Parameter Study in Plastic Injection Molding Process Using Statistical Methods and IWO Algorithm," *Int'l Journal of Modeling and Optimization*, 1(2), pp. 141-145.

Alton, M. (2010), "Reducing Shrinkage in Injection Moldings via the Taguchi, ANOVA and Neural Network Methods," *Materials & Design*, 31, pp. 599-604.

Arezodar, F. and Eghbali, A. (2000), "Evaluating the Parameters Affecting the Distribution of Thickness in Cup Deep Drawing of ST14 Sheet," in *Advances in Systems Theory, Signal Processing and Computational Science*, pp. 193-197; ISBN: 978-1-61804-115-9.

Astakhov, V. P. and Galitsky, V. (2005), "Tool Life Testing in Gundrilling: An Application of the Group Method of Data Handling (GMDH)," *Int'l Journal of Machine Tools and Manufacture*, 45, pp. 509-517.

Bajic, D. et al. (2010), "Design of Experiment's Application in the Optimization of Milling Process," *Metalurgija -Sisak then Zagreb*, 49(2).

Barrera, O. et al. (2018), "Understanding and Mitigating Hydrogen Embrittlement of Steels: A Review of Experimental, Modelling and Design Progress from Atomistic to Continuum," *Journal of Material Science*, 53, pp. 6251-6290.

Barringer, H. P. (2000), "Process Reliability and Six-Sigma," *National Manufacturing Week Conference Track for Manufacturing Process and Quality*, Chicago, IL, March 13, 2000.

Chen, C. (2018), "Investigation of the Two-steps Clinching Process for Joining Aluminum Alloy Sheets," PhD Dissertation, Tokushima University, Tokushima, Japan.

Chiang, Y. J. and Barber, G. C. (2002), "Operating Effort and Related Statistical Tolerance Analyses of Automotive Latching Mechanisms," *International Journal of Materials and Product Technology*, 17(5/6), p. 353-367.

Chidhambara, K. V. et al. (2011), "Optimization of Robotic Spray-Painting Process Parameters Using Taguchi Method," *IOP Conference Series: Materials Science and Engineering*, 310, p. 012108.

Costa, E. C. et al. (2019), "Influence of Diamond Wire Sawing Parameters on Substrate Microcracks Formation in Monocrystalline Silicon Wafer," *The 25th ABCM Int'l Congress of Mechanical Engineering*, Uberlândia, MG, Brazil, October 20–25, 2019.

Cukor, G., Jurkovi, Z., and Sekuli, M. (2011), "Rotatable Central Composite Design of Experiments versus Taguchi Method in the Optimization of Turning," *Metalurgija*, 50(1), pp. 17-20.

Echempati, R. and Sathya Dev, V. M. S. (2002), "Experimental Design Study of Metal Forming," *Proceedings of American Society for Engineering Education Annual Conference & Exposition*, Montreal, Canada, June 16–19, 2002.

Fang, J. et al. (2018), "Reliability Improvement of Diamond Drill Bits Using Design of Experiments," *Quality Engineering*, 30(2), pp. 339-350.

Gadiyar, A. U. (2001), "An Experimental Design Approach for Predicting Shrinkage of Plastics Injection Molded Parts," Retrospective Theses and Dissertations: 21214, Iowa State University, Ames, IA.

Golovashchenko, S. (2005), "Sharp Flanging and Flat Hemming of Aluminum Exterior Body Panels," *Journal of Materials Engineering and Performance*, 14, pp. 508-515.

Gou, J. et al. (2003), "Resin Transfer Molding Process Optimization Using Numerical Simulation and Design of Experiments Approach," *Polymer Composites*, 24(1), pp. 1-12.

Grendahl, S. M. et al. (2014), "Effect of Cleanliness on Hydrogen Tolerance in High-Strength Steel," ARL-TR-6885, Aberdeen Proving Ground, Aberdeen, MD, April 2014.

Grendahl, S. M. et al. (2015), "Design of Experiment Approach to Hydrogen Re-embrittlement Evaluation WP-2152," ARL-TR-7275, US Army Research Laboratory, Adelphi, MD, April 2015.

Henderson, C. (2012), "Transfer Molding," Infotracks, Issue 39, Semitracks Monthly Newsletter, September 2012.

Henkel Corporation (2018), "Design of Experiments Helps Improve the Robustness of Off-Highway Vehicle Coating," Madison Heights, MI, Retrieved December 2018.

Horváth, R. and Drégelyi-Kiss, Á. (2013), "Analysis of Surface Roughness Parameters in Aluminum Fine Turning with Diamond Tool," *Measurement 2013, Proceedings of the 9th Int'l Conference*, Smolenice, Slovakia, 2013.

Horváth, R. et al. (2014), "Application of RSM Method for the Examination of Diamond Tools," *Acta Polytechnica Hungarica*, 11(2), pp. 137-147.

ISO 4628-8 (2012), "Paints and Varnishes - Evaluation of Degradation of Coatings - Designation of Quantity and Size of Defects, and of Intensity of Uniform Changes in Appearance, Part 8: Assessment of Degree of Delamination and Corrosion around a Scribe or Other Artificial Defect."

Jarugu, S. R. (1996), "Effect of Various Process Variables on the Mechanical Properties of Laminates Produced through Resin Transfer Molding Process," Master Thesis, Concordia University, Canada.

JMBS (2015), "Our Guide to Batteries," Johnson Matthey Battery Systems, Milton Keynes, UK.

Karakurt, I. (2014), "Application of Taguchi Method for Cutting Force Optimization in Rock Sawing by Circular Diamond Sawblades," *Sadhana*, 39(5), pp. 1055-1070.

Kazmer, D. O., Westerdale, S., and Hazen, D. (2008), "A Comparison of Statistical Process Control (SPC) and On-Line Multivariate Analyses (MVA) for Injection Molding," *Int'l Polymer Processing*, 23, p. 447.

Khanna, N., Pei, Z., and Ferreira, P. M. (1995), "An Experimental Investigation of Rotary Ultrasonic Grinding of Ceramic Disks," Technical Papers of The North American Manufacturing Research Institution of SME, pp. 67-72.

Lambiase, F. and Di Ilio, A. (2016), "Damage Analysis in Mechanical Clinching: Experimental and Numerical Study," *Journal of Materials Processing Technology*, 230, pp. 109-120.

Lin, T. and Chananda, B. (2003), "Quality Improvement of an Injection-Molded Product Using Design of Experiments: A Case Study," *Quality Engineering*, 16(1), pp. 99-104.

Lin, G. et al. (2005), "A Computational Design-of-Experiments Study of Hemming Processes for Automotive Aluminum Alloys," *Journal of Engineering Manufacture*, 219(10), pp. 711-722.

Liu, D. et al. (2019), "Microstructure Evolution and Lifetime Extension Mechanism of Sn-Added Fe-Based Pre-Alloy Brazing Coating in Diamond Tools," *Coatings*, 9, p. 364.

MKS (2012), "The Optimization of Injection Molding Processes Using Design of Experiments," MKS Application Notes, MKS Instruments, Inc., Andover, MA.

Mukras, S. D. (2020), "Experimental-Based Optimization of Injection Molding Process Parameters for Short Product Cycle Time," *Advances in Polymer Technology*, 2020, Article ID 1309209, 15 pages.

Murguía, R. (2016), "Quality Improvement of a Casting Process Using Design of Experiments," *Prospect*, 14(1), p. 47.

Nalbant, M. et al. (2010), "Application of Taguchi Method in the Optimization of Cutting Parameters for Surface Roughness in Turning," *Materials and Design*, 28, pp. 1379-1385.

Nichols, M. (2016), "Paint Materials and Processes from an Automotive OEM Perspective," Ford Research and Advanced Engineering, Ford Motor Company, Dearborn, MI, July 20, 2016; https://www.nist.gov/system/files/documents/2016/11/04/nichols_-_paint_materials_and_processes_from_an_automotive_oem_persp.pdf.

NIST (2022), "5.4.7.1. Full Factorial Example," Engineering Statistics Handbook, Retrieved January 26, 2022, itl.nist.gov.

Oudjene, M. et al. (2009), "Shape Optimization of Clinching Tools Using the Response Surface Methodology with Moving Least-Square Approximation," *Journal of Materials Processing Technology*, 209(1), pp. 289-296.

Patel, M. G. C. et al. (2019), "Modelling and Optimization of Alpha-Set Sand Molding System Using Statistical Design of Experiments and Evolutionary Algorithms," in *Optimization of Manufacturing Process*, Edited by Gupta, K. and Gupta, M., Springer, Cham, pp. 1-28.

Patil, R. S. et al. (2020), "Optimization of Effect of Boring Parameters for Minimum Surface Roughness Using CNC Boring Machine," *Int'l Journal of Innovative Research in Technology*, 4(7), pp. 169-172.

Peasura, P. (2012), "Effects of Post Weld Heat Treatment on Aluminum Alloy 7075 in Gas Metal Arc Welding," *NANOCON*, Brno, Czech Republic, October 23–25, 2012.

Pei, Z. and Ferreira, P. M. (1999), "An Experimental Investigation of Rotary Ultrasonic Face Milling," *Int'l Journal of Machine Tools and Manufacture*, 39, pp. 1327-1344.

Pervez, H. et al. (2016), "Optimization of Injection Molding Parameters for HDPE/TiO$_2$ Nanocomposites Fabrication with Multiple Performance Characteristics Using the Taguchi Method and Grey Relational Analysis," *Materials*, 9, p. 710.

Peters, S., Lanza, G., Ni, J., Jin, X., Yi, P., and Colledani, M. (2014), "Automotive Manufacturing Technologies - An International Viewpoint," *Manufacturing Review*, 1(10), pp. 1-12.

Philip, M. (1996), *Injection Molding. Tool and Manufacturing Engineers Handbook*, Vol. 8, 4th Edition, Society of Manufacturing Engineering, Dearborn, MI.

Pimenta, C. D. et al. (2018), "Desirability and Design of Experiments Applied to the Optimization of the Reduction of Decarburization of the Process Heat Treatment for Steel Wire SAE 51B35," *American Journal of Theoretical and Applied Statistics*, 7(1), pp. 35-44.

Rehman, M. et al. (2018), "Design of Experiment Approach in the Industrial Gas Carburizing Process," IMTECH.

Rolinski, E. and Woods, M. (2018), "The Benefits of Nitriding and Nitrocarburizing," Machine Design, June 1, 2018.

Rössler, A. (2014), *Design of Experiment for Coatings*, Vincentz Network, Hanover, Germany; ISBN: 978-3-86630-885-5.

Schulz, D. (2013), "White Paper: Painting Trends for Automotive Plastic Parts; More than Just Perfect Protection," Materials, Business, Automotive and Mobility, Plastics Today, August 5, 2013.

Shiri, N. D. et al. (2015), "Effect of Heat Treatment Variables on the Abrasive Wear Behavior of Al-6061 Alloy and Al 6061-SiCP Composites Using Taguchi Technique," *American Journal of Materials Science*, 5(3C), pp. 151-156.

Soundararajan, R. et al. (2015), "Modeling and Analysis of Mechanical Properties of Aluminum Alloy (A413) Processed through Squeeze Casting Route Using Artificial Neural Network Model and Statistical Technique," *Advances in Materials Science and Engineering*, 2015, Article ID 714762, 16 pages.

Suda, Y. et al. (2015), "High-Yield Synthesis of Helical Carbon Nanofibers Using Iron Oxide Fine Powder as a Catalyst," *Crystals*, 5, pp. 47-60.

Surendran, K. R. et al. (2015), "Study on Effect of Failure Modes in Hydrogen Embrittlement of Austenitic Stainless Steel," *Int'l Journal of Scientific & Engineering Research*, 6(4).

Svec, P. et al. (1990), *Polystyrene and Its Modification*, 885 pages; ISBN: 978-0136824855.

Tariq, S. et al. (2022), "Minimizing the Casting Defects in High-Pressure Die Casting Using Taguchi Analysis," *Scientia Iranica, B*, 29(1), pp. 53-69.

Uysal, A., Altan, M., and Altan, E. (2012), "Effects of Cutting Parameters on Tool Wear in Drilling of Polymer Composite by Taguchi Method," *Int'l Journal of Advanced Manufacturing Technology*, 8, pp. 915-921.

Varis, J. (2006), "Ensuring the Integrity in Clinching Process," *Journal of Materials Processing Technology*, 174, pp. 277-285.

Yang, W. H. and Tarng, Y. S. (1998), "Design Optimization of Cutting Parameters for Turning Operations Based on the Taguchi Method," *Journal of Materials Processing Technology*, 84, pp. 122-129.

Yuksel, T. (2014), "Intelligent Resin Delivery System for Manufacturing Large Composite Structures," IMECE2014-36442, Montreal, Quebec, Canada, November 14–20, 2014.

Problems

P13.1: After computer simulations were validated using physical tests, an experimental design was set up to explore the influences on the maximum pressure (P_{max}) and mold filling time (F_{time}) in a RTM process by some eligible process factors, including factor A: flow rate of resin (CC/sec), factor B: volume fraction of fibers (%), factor C: number of gates, factor D: gate location (top or bottom), and factor E: number of vents. The corresponding 16 responses resulting from computer simulations are given as follows [Gou et al. 2003]:

Run	A	B (%)	C	E	D	P_{max} (kPa)	F_{time} (sec)
1	1.0	45	4	8	Bottom 92.3	759	
2	2.0	45	4	4	Bottom 212	379	
3	1.0	50	4	4	Bottom 234	689	
4	2.0	50	4	8	Bottom 408	345	
5	1.0	45	8	4	Bottom 89.3	754	
6	2.0	45	8	8	Bottom 151	377	
7	1.0	50	8	8	Bottom 167	686	
8	2.0	50	8	4	Bottom 395	343	
9	1.0	45	4	4	Top	104	779
10	2.0	45	4	8	Top	182	391
11	1.0	50	4	8	Top	201	711
12	2.0	50	4	4	Top	462	354
13	1.0	45	8	8	Top	71.7	774
14	2.0	45	8	4	Top	171	385
15	1.0	50	8	4	Top	189	701
16	2.0	50	8	8	Top	317	352

Please

1. Identify the alias equation.
2. Based on the t-distribution, formulate the predictive equations for both P_{max} and F_{time}.

P13.2: Reliability improvement of a sand casting process based on the Box–Behnken design takes aim at enhancing the production process of pump impellers (gray iron castings). The sand mold is made of A% of clay and B% of water and 100%-A%-B% of molding sand. Given that the mold hardness is the third controllable factor, the design matrix and test results are listed as follows [Murguía 2016]:

Run	A	B	C	Y (%)
1	3	5	6	50
2	3	5	5.5	10
3	3	3	5	20
4	3	4	5.5	20
5	3	5	5	30
6	2	5	6	20
7	2	4	5	0
8	3	4	5.5	25
9	4	4	6	30

Run	A	B	C	Y (%)
10	3	4	5.5	20
11	4	5	5.5	20
12	2	3	5.5	10
13	4	3	5.5	20
14	3	3	6	30
15	4	4	5	20

where
A (%): Clay content
B (%): Moisture content
C (kg/cm^2): Mold hardness
Y (%): Defect in percentage

Please identify the effects that have influences on the defect ratio, i.e., Y.

P13.3: Please use the second half of the data in Table 13.4.3 (runs 10–18) to unveil the predictive equation for the retention force of clinching joints in sheet aluminum of 0.4 mm in thickness (at A$^+$).

P13.4: SAE 51B35 drawn steel wires of the same manufacturing batch are selected to study the parametric effects in a heat treatment process on the decarburization and hardness change [Pimenta et al. 2018]. The process parameters of concern, design matrix, and test results (i.e., decarburization and hardness) are given as follows:

(1)	Process parameters	Level (−1)	Level (1)			
	Oxidation	Yes	No			
	Heat treatment cycle	Plan 1	Plan 2			
	Furnace pressure	−35	−25			
	Moisture	High	Low			
(2)	**Design matrix and test result**					
	Run	a	b	c	d	Decarburization	Hardness
	1	−1	−1	−1	−1	0.18	260
	2	1	−1	−1	−1	0.08	290
	3	−1	1	−1	−1	0.17	256
	4	1	1	−1	−1	0.07	281
	5	−1	−1	1	−1	0.22	270
	6	1	−1	1	−1	0.12	294
	7	−1	1	1	1	0.23	264
	8	1	1	1	−1	0.10	290
	9	−1	−1	−1	1	0.18	250
	10	1	−1	−1	1	0.09	280
	11	−1	1	−1	1	0.17	258
	12	1	1	−1	1	0.08	281
	13	−1	−1	1	1	0.213	254
	14	1	−1	1	1	0.09	293
	15	−1	1	1	1	0.23	258
	16	1	1	1	1	0.10	290

P13.5: A DOE was carried out to maximize the density of die-cast parts by Tariq et al. [2022] using Taguchi's method (S/N ratio). Apply the traditional ANOVA to the raw data (density) and make comments on these two methodologies (traditional ANOVA and Taguchi's method). The process parameters identified are given as follows:

A (bar): Injection pressure
B (°C): Molten temperature
C (sec): Die cooling time
D (°C): Mold temperature
E (m/s): Pushing speed—first stage
F (m/s): Pushing speed—second stage

The design matrix and three replicated test results are given as follows [Tariq et al. 2022]:

Run	A	B	C	D	E	F	Y_1	Y_2	Y_3
1	180	650	4.0	140	0.20	5.0	2.510	2.505	2.50
2	180	650	4.0	140	0.25	6.0	2.397	2.392	2.39
3	180	650	4.0	140	0.30	7.0	2.569	2.544	2.542
4	180	660	5.0	180	0.20	5.0	2.507	2.492	2.490
5	180	660	5.0	180	0.25	6.0	2.510	2.501	2.49
6	180	660	5.0	180	0.30	7.0	2.196	2.116	2.114
7	180	670	6.0	220	0.20	5.0	2.296	2.206	2.191
8	180	670	6.0	220	0.25	6.0	2.512	2.507	2.505
9	180	670	6.0	220	0.30	7.0	2.571	2.569	2.567
10	190	650	5.0	220	0.20	6.0	2.397	2.396	2.392
11	190	650	5.0	220	0.25	7.0	2.397	2.394	2.389
12	190	650	5.0	220	0.30	5.0	2.027	2.022	2.021
13	190	660	6.0	140	0.20	6.0	2.107	2.102	2.100
14	190	660	6.0	140	0.25	7.0	2.026	2.001	1.999
15	190	660	6.0	140	0.30	5.0	2.026	2.011	2.009
16	190	670	4.0	180	0.20	6.0	2.510	2.501	2.499
17	190	670	4.0	180	0.25	7.0	2.515	2.435	2.433
18	190	670	4.0	180	0.30	5.0	2.560	2.470	2.455
19	200	650	6.0	180	0.20	7.0	2.196	2.191	2.189
20	200	650	6.0	180	0.25	5.0	2.107	2.102	2.100
21	200	650	6.0	180	0.30	6.0	2.607	2.602	2.600
22	200	660	4.0	220	0.20	7.0	2.571	2.566	2.564
23	200	660	4.0	220	0.25	5.0	2.569	2.564	2.562
24	200	660	4.0	220	0.30	6.0	2.501	2.496	2.494
25	200	670	5.0	140	0.20	7.0	2.508	2.503	2.501
26	200	670	5.0	140	0.25	5.0	2.292	2.287	2.285
27	200	670	5.0	140	0.30	6.0	2.392	2.387	2.385

14

Electronic Fabrication

For power electronics of EVs, there has been an increasing shift toward applying DOE based on physics of failure for subsystem design before building up a fault tree analysis (FTA) for system reliability. This relies on an understanding of elastoplastic failure mechanisms in the thermomechanical environment with potential contamination of moisture, corrosion, temperature variation, and mechanical loads. A system analysis accounts for various loadings, fracture toughness, stress–strain energy, and material transformations (e.g., degradation) that lead to failure with a high level of detail. It is made possible with the use of finite element methods that can handle complex geometries and mechanisms such as creep, diffusion, oxidation, and fatigue, as well as probabilistic tools such as DOE, maximum-likelihood regression, and Monte Carlo simulations that can embrace the stochastic nature.

14.1. Electronic Packaging

Power electronic circuits with basic power semiconductor devices and advanced power electronic converters are fundamental to motor drive systems [Manias 2016]. These include analysis of modulation and output voltage, current control techniques, passive and active filtering, and the characteristics and gating circuits of different power semiconductor switches, such as bipolar junction transistors (BJTs), insulated gate bipolar transistors (IGBTs), metal-oxide-semiconductor field-effect transistors (MOSFETs), integrated gate commutated thyristors (IGCTs), and gate turn-off thyristors (GTOs). Electronic packaging is one discipline in the field of power electronic engineering that involves a wide variety of technologies, with the following functional requirements:

(a) Integrity of Circuit Design: Signal transmission and power need inherent with the circuit design are by default renowned for the robustness of unexpected environmental conditions. Signal integrity means the signal can be transmitted from the source to the receiver without distortion in a timely manner. Power integrity refers to the truthfulness of power delivery and ground in the circuit system.

(b) Inter-Component Connections: Active components include transistors, diodes, logic gates, processor IC, field programmable gate array (FPGA), and operational amplifiers. Passive devices refer to capacitors, inductors, and transformers. Testability may be instated in order to assure that all interconnections (e.g., pins) between connected components have been correctly soldered or fastened. Functional enclosures are used for lodging and situating the interconnection of ICs to form electronic systems.

(c) Packaging Materials: Potting, coating, staking material, and other encapsulants are applied in electronic devices to guard against environmental conditions such as moisture, contamination, chemicals, and radiation, which could otherwise degrade the product. Less expensive plastics are nowadays made suitable to package most electronic devices in a controlled environment with low moisture and moderate temperature cycling, while not requiring long-term storage. Electronic packaging polyimides (PIs) are widely used as an electronic insulation material. More expensive ceramic packages offer a greater performance in terms of thermal characteristics, moisture absorption, and endurance in a harsh working environment.

(d) Thermal Stability: The temperature control of EV propulsion system is crucial to its performance and operating efficiency. Heat dissipation and differential thermal expansion/contraction are well evaluated in order to warrant the thermal stability and component reliability. How material properties such as glass transition temperature, creep behavior, coefficients of thermal expansion, and degradation vary with respect to temperature and stress is of great concern.

(e) Electromagnetic Interference (EMI): EMI shielding refers to the shielding of radio- and/or microwave radiation so that the radiation cannot interfere with signal transmission. Metals and carbons are the main functional materials for EMI. Continuous carbon fiber composites and cement-based materials are preferred among structural shielding materials.

(f) System Reliability: Product reliability growth and analysis/test plans based on DOE and optimization techniques are needed for key areas, such as mount points and board constraints, which have a significant impact on the board strain magnitude. Mechanical joints including crimp connections, pinned joints, and bolted/screwed assemblies are expected not to degrade because of latent mechanical shock events and vibrations in their service lifetime.

The above functionalities are generally implemented in a PCB, which is an electronic assembly that uses conductors (e.g., copper) to make electrical connections between components. In general, PCBs are built from alternating layers of conductive copper with layers of electrically insulating material. The mechanical structure is strengthened with an insulating material laminated between layers of conductors. The inner copper layers are etched, leaving the intended traces of copper for connecting components in the circuit board. Then, both functional enclosures [e.g., vias in ball grid arrays (BGAs)] and protective features (e.g., thin-film encapsulants) are built into PCB assembly. It can be one-sided (components mounted on one surface) or double-sided (components mounted on both surfaces). A multi-layered PCB has conductors on internal layers that carry electrical signals between components, or the internal layers could be conductive plane layers. A multi-layered PCB may be single-sided or double-sided.

14.1.1. Hierarchy of Electronic Packaging

Electronic packaging provides housing and interconnection of ICs to form an electronic system. A hierarchy of electronics assembly can be described based on interconnection levels, including the chip, package, board, backplane, and pin-breakaway patterns. Ranging from micro- or even nano-scaled components to a commercial product, electronic assembly can be divided into the following six levels [Lopez-Buedo and Boemo 2019]:

1. Level 0: Fabricating inter-gate connections on a die
2. Level 1: Putting related dies together, forming single- or multiple-chip modules
3. Level 2: Packaging chip modules to PCB, forming PCB modules
4. Level 3: Interconnecting PCB modules, forming motherboard modules
5. Level 4: Assembling motherboard modules onto a rack, forming rack sub-assemblies
6. Level 5: Connecting rack sub-assemblies by a communication system, such as local area network (LAN)

Every level given above has a great impact on the reliability of automotive electronics [Liu et al. 2011, Ohring and Kasprzak 2014].

Surface mounting technology (SMT) stays as the mainstream of electronic packaging and interconnection, while chip scale package (CSP) is the trending evolutionary research to miniaturize the entire package size by putting more circuits in a smaller space. CSP, based on Institute for Printed Circuits (IPC)/Joint Electron Device Engineering Council (JEDEC) J-STD-012 definition, is a single-die, direct surface mountable package with an area of no more than 1.2 times the original die area.

14.1.2. Surface Mounting Assembly

SMT refers to a fully automated process for assembling in-line electronic components. SMT allows for reducing the size of a PCB assembly, which is in high demand. Components can be mounted on both sides of the board, maximizing the space and resources. As shown in Figure 14.1.2, a surface mounting process based on stencil printing can be divided into the following four steps:

(a) Alignment: It is to align the related components including PCB, pads, stencil, paste, and squeeze on a stencil printer. The first step in the stencil printing process is to decide where the components will be placed and how the PCB will function. Then, program machines to perform the tasks accordingly.

(b) Squeeze: Air pressure is applied to push the squeeze that would force the paste to move at the front of the squeeze blade. Automatic inspection by an optical device may be utilized to screen out the bad ones. Too much solder paste could result in short circuits, but not enough will lead to poor connections or a fragile board. It is suggested to perform a solder paste inspection before the component placement.

(c) Installation: Electronic parts, such as chips and plastic leaded chip carriers (PLCCs), are installed successively via a special-purpose chip shooter and a multi-purpose shooter. An automated optical inspection of the assembly is then conducted before reflow. Component types, values, and polarities are also tested here.

(d) Reflow: Solder paste is reflowed to perfect solder joints in a reflow furnace. Too little heat will result in poor connectivity, while high temperatures could damage the PCB.

The final step in a surface mounting assembly is testing the functionalities of the assembled PCB to ensure if the assembly is robust.

FIGURE 14.1.2 Surface mounting of electronic parts on a PCB [Tsai 2008].

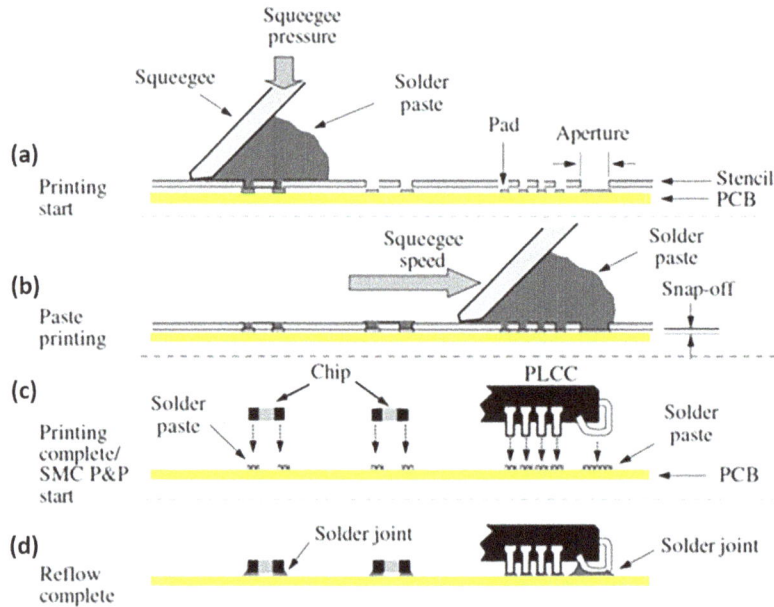

14.1.3. Interconnection in IC Packaging

As shown in Figure 14.1.3, interconnections in IC packaging can be decomposed into the following components:

1. Bonding Wires: A metallic wire connects (bonds) an IC to its packaging during a semiconductor fabrication. A bonding wire is usually approximated by three or four linear segments and parameters, which define the orientations of segments, are selected in compliance with standard Electronic Industries Alliance (EIA)/JEDEC, Standard No. 59. The design with four coplanar bonding wires in ground–signal–signal–ground (G–S–S–G) configuration is widely accepted [Van Quach et al. 2008] when all the metal is connected together.

2. Transmission Lines: A connecting line lies between two conductors to form a loop that has a certain length. It is also called a delay line. Several types of transmission lines are available, including microstrips, coplanar waveguides, and conductor-backed coplanar waveguide lines [Eisenstadt and Eo 1992].

3. Vias: A via is a conductor that provides a through-thickness bridging through the spacings between PCB layers. It is used for layer transition and thus is also called plated through via (PTN).

4. Solders: A solder connects the signal and power wires of an IC package to those of PCBs. Solder joints are the most widely used interconnection materials in electronic product packaging. Its shape is generally modeled as a polygonal or cylindrical column rather than a sphere for the simplicity of the analysis, but it still captures the capacitive effect of a solder.

FIGURE 14.1.3   Interconnecting wires in a typical IC package [Kim and Kong 2020].

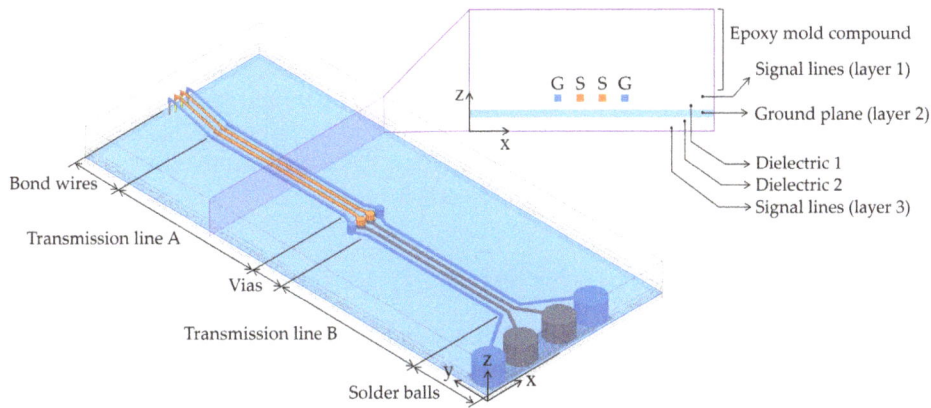

14.1.4. Design for PCB Assembly

With PCBs, guidelines include minimizing component variety, standardizing component packaging, using auto-insertable or placeable components, using a common component orientation and component placement to minimize soldering "shadows," selecting component and trace width that is within the process capability, using appropriate pad and trace configuration and spacing to assure good solder joints and avoid bridging, using standard board and panel sizes, using tooling holes, establishing minimum borders, and avoiding or minimizing adjustments.

14.2. Through-Thickness Interconnection

A via is an interconnect channel, through which an electrical conductor provides the vertical integration between adjacent PCB layers, as shown in Figure 14.1.2. As one of the surface mount technologies (SMTs), BGA packaging has been gaining industry-wide commitment as a potentially low-cost packaging technique for high-I/O devices such as automotive electronics and computers. The chip is "soldered" to the package substrate using the solder balls "bumps" that have been grown over the die pads. Flip chip, to get the chip soldered to the PCB, is now the preferred process for highly ICs with the following advantages [Lopez-Buedo and Boemo 2019]:

(a) Higher Circuit Density: not so bulky as wire bonding

(b) Mitigating Inductance (~0.1 nH): More power-ground pairs

(c) Improved Efficiency: Interconnection soldered at the same time

BGA can be designed to offer HD packaging with a comfortable distance between neighboring interconnects with some level of enhancement in electrical properties, e.g., lower electrical inductance. Widely applied lead-free solders include Sn95.5-Ag4.0-Cu (SAC405), Sn98.5-Ag1.0-0.5Cu (SAC105), and Sn96.5-Ag3.5. An electronic manufacturer uses an automated pick-and-place process to lay the BGA onto the PCB and then exposes the entire package including BGA and PCB pads to a reflow process at an elevated temperature. Each melt solder reflows and wets both pads and forms a metallurgical bond connecting the chip on the top side of the plastic grid ball array (PGBA) with the motherboard as an electric connector upon cooling.

Typically, ICs in mass production are made in batches of wafers or other semiconductors through processes like photolithography. Each wafer is diced into identical pieces, of which each is called a die that contains one copy of the distinct circuit by design. In a chip-on-board assembly process, a silicon die is mounted on the board via a soldering process (or with adhesive) and then electrically connected by bonding wires (gold or aluminum) of approximately 0.025 mm in diameter. Experimental data obtained from a durability test of electronic components on cycling to failure are generally right (positively) skewed and exhibit some long-life observations. It is likely because of thermomechanical stresses. DOE has been used for resolving problems like this when there are more than one acceleration factors in the durability test. Reliability based on a normal distribution is for sure not a proper approach anymore.

14.2.1. Scattering Parameters of PCB Vias

A via is an electric connector in a vertical transition, which builds a bridge between two adjacent layers in PCB holes and accomplishes the desired conduction between connected points.

Given that the signal routes are well-designed, a via transition can be the bottleneck in the end-to-end signal quality since it is a discontinuity in the electronic impedance of the transfer path [Fan et al. 2014]. In detecting the scattering parameters of PCB vias, a network analyzer is utilized to sweep the frequency of a sinusoidal waveform as it gets into a PCB under test, as shown in Figure 14.2.2.

FIGURE 14.2.2   Equivalent electric circuit for testing PCB vias in the frequency domain [LaMeres 1998].

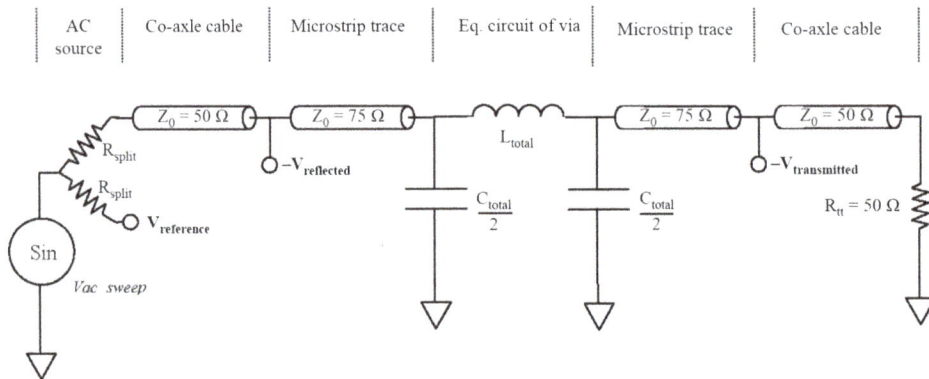

Courtesy of Brock LaMeres.

There are two detection points. The first detection device is located at the input via that measures the reflected signal (e.g., voltage or power). The second detection device is located at the output of the other end of the via, and it measures the transmitted signal (e.g., voltage or power). Given the predetermined incident voltage, the network analyzer can produce measurements known as S-parameters or Y-parameters [Gu et al. 2008]:

(a) S-Parameters (Differential Mode): Normalized incident and reflected traveling signals in each network port.

(b) Y-Parameters (Common Mode): Used to characterize a two-port network, based on input and output signals in operation.

As defined to clarify the quality of the electronic signal (voltage) crossing a single-ended transfer path between port 1 and port 2, four typical scattering parameters in the differential mode measured using the network analyzer [Hall and Heck 2009] are:

1. S_{21}: Parameter S_{21} is the input differential insertion loss, also called S_{dd21}. It is a measure of the signal coming out of port 2 relative to the signal stimulus entering port 1. It represents the transmission measured at port 2, referred to the incident signal at port 1. The smaller the $|S_{21}|$, the better the transmission truthfulness.

2. S_{11}: Parameter S_{11} is called input differential reflection loss, also called S_{dd11}. It is utilized to detect the reflection measured at the incident point (port 1), referred to the incident signal. If $|S_{11}| \gg 0$, there is no reflection loss. If $|S_{11}| = 0$, the load impedance is an open, short circuit, or purely reactive. The larger the absolute value, the better the design.

3. S_{12}: Parameter S_{12} is the output differential insertion loss, also called S_{dd12}.

4. S_{22}: Parameter S_{22} is the output differential reflection loss, also called S_{dd22}.

Generally speaking, a generic two-port network in a PCB comprising connectors and traces is symmetric as designed. Supposedly, S-parameters S_{11} and S_{22} have the same sign and value and S_{12} and S_{21} somehow correlate with each other. Thus, S-parameters S_{21} and S_{11} are commonly used to qualify the transmission and reflection of power signals when operating at a frequency where the skin effect is limited. Two charts of experimental test results done by LaMeres [1998] are demonstrated in Figure 14.2.3.

FIGURE 14.2.3 Typical $|S_{21}|$ and $|S_{11}|$ in the frequency range between 0 and 6 GHz [LaMeres 1998]. (a) $|S_{21}|$ (dB) and (b) $|S_{11}|$ (dB).

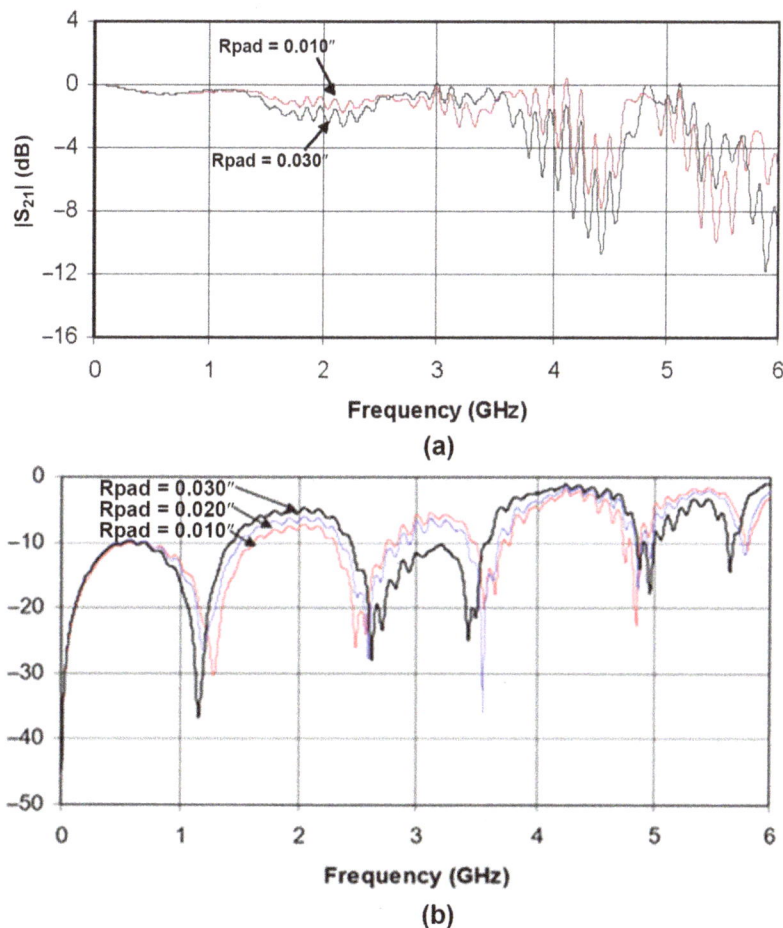

14.2.2. DOE for Examining Scattering Parameters of PCB Vias

A transmission line is a wire (or simply a pair of conductors) that must be analyzed according to the characteristics of high-frequency signal propagation. The signal transmission at high frequencies is more and more desirable. A study based on DOE, based on the three design factors listed in Table 14.2.1, was conducted to examine scattering parameters of PCB vias. The three design factors are schematically illustrated in Figure 14.2.4. Data of two signal responses measured at the frequency of 20 GHz [Asif 2019] are denoted as columns S_{21} and S_{11} in Table 14.2.2. For a low-loss system to operate at frequencies where the skin effect resistance is not significant, $|S_{21}|$ is small and $|S_{11}|$ is large.

TABLE 14.2.1 Factors and design levels for scattering parameters of PCB vias [Asif 2019].

Factor	Level (−1) (mil)	Level (+1) (mil)	Coded variable
A: Via diameter	2.9	8.7	a = (A − 5.8)/2.9
B: Via pitch from hole to hole	20	60	b = (B − 40)/20
C: Pad radius (anti-pad size)	25	50	b = (B − 37.5)/12.5

Note:
1 mil = 0.001″ = 0.0254 mm

FIGURE 14.2.4 Schematic drawing of vias and their surroundings [Asif 2019].

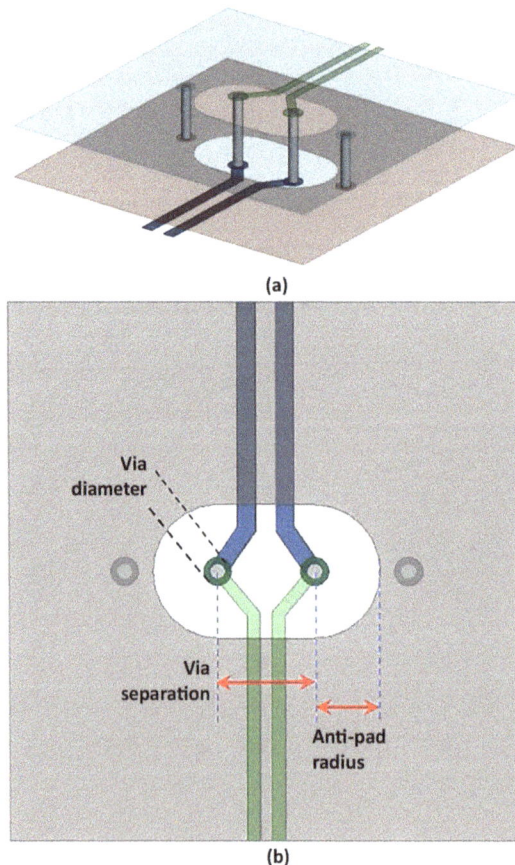

(a)

(b)

TABLE 14.2.2 Facial composite design based on 2^3 for examining scattering parameters of PCB vias [Asif 2019].

Run	Coded variable			Factors (mil = 0.001")			Scattering modes	
	a	b	c	A	B	C	S_{21} (dB)	S_{11} (dB)
1	−1	−1	−1	2.9	20	25	−0.892	−10.554
2	1	−1	−1	8.7	20	25	−0.532	−19.284
3	−1	1	−1	2.9	60	25	−1.702	−6.747
4	1	1	−1	8.7	60	25	−0.681	−15.781
5	−1	−1	1	2.9	20	50	−1.071	−9.000
6	1	−1	1	8.7	20	50	−0.457	−42.817
7	−1	1	1	2.9	60	50	−2.038	−5.647
8	1	1	1	8.7	60	50	−0.954	−10.737
9	−1	0	0	2.9	40	37.5	−1.712	−6.411
10	1	0	0	8.7	40	37.5	−0.689	−13.562
11	0	−1	0	5.8	20	37.5	−0.569	−15.939
12	0	1	0	5.8	60	37.5	−1.268	−8.379
13	0	0	−1	5.8	40	25	−0.841	−11.533
14	0	0	1	5.8	40	50	−1.134	−8.769
15	0	0	0	5.8	40	37.5	−1.039	−9.477

© SAE International.

Notes:
$|S_{21}|$: For detecting transmission loss; no transmission loss if $|S_{21}| = 0$
$|S_{11}|$: For detecting reflection loss, no reflection loss if $|S_{11}| \gg 0$

Multivariable linear regression analysis is employed to determine the correlations between the design factors and the scattering parameters. The analysis is conducted using the ANOVA based on the F-distribution as presented in Table 14.2.3. After removing the least insignificant factor or interaction one by one via the comparison of their resulting p-values, with a cutoff significance level (p-value) of 10.7%, an experimenter obtains the following two regression equations:

$$S_{21} \text{ (dB)} = -0.755 + 0.2562\, A - 0.0475\, B - 0.01829\, A^2 + 0.000320\, B^2$$

$$+ 0.002438\, A\, B - 0.000210\, B\, C$$

(14.2.1)

$$(R = 99.1\%, R_{adj} = 98.3\%, \text{ and } R_{pred} = 96.7\%)$$

and $S_{11} \text{ (dB)} = -15.97 + 0.702\, B + 0.243\, C - 0.01135\, B^2 - 0.1439\, A\, C$

$$+ 0.002100\, A\, B\, C$$

(14.2.2)

$$(R = 95.5\%, R_{adj} = 92.9\%, \text{ and } R_{pred} = 70.6\%)$$

According to the adjusted mean squares (column MS_{adj} in Table 14.2.3), factor B (via pitch) and its interactions with factors A and C (AB and AC) have decisive impacts on S_{21}, i.e., the transmission loss from point 1 to point 2. On the other hand, two-factor interaction AC and its further interaction with B (i.e., three-factor interaction ABC) have a dramatic impact on the S_{11}, i.e., the reflection.

TABLE 14.2.3 ANOVA for scattering parameters of PCB vias.

(i) S$_{21}$:	Source	SS$_{adj}$	DOF	MS$_{adj}$	F-value	p-Value
	Regression	3.03028	6	0.505046	69.56	0.0%
	A	0.08740	1	0.087400	12.04	0.8%
	B	0.13887	1	0.138869	19.13	0.2%
	A^2	0.06628	1	0.066282	9.13	1.7%
	B^2	0.04598	1	0.045978	6.33	3.6%
	AB	0.15990	1	0.159895	22.02	0.2%
	BC	0.13199	1	0.131985	18.18	0.3%
	Error	0.05809	8	0.007261		
	Subtotal	3.08836	14			
	Grand average	—	1			
	Total	—	15			
(ii) S$_{11}$:	Source	SS$_{adj}$	DOF	MS$_{adj}$	F-value	p-Value
	Regression	1059.72	5	211.94	18.52	0.0%
	B	37.79	1	37.79	3.30	10.3%
	C	65.28	1	65.28	5.70	4.1%
	B^2	68.65	1	68.65	6.00	3.7%
	AC	592.26	1	592.26	51.75	0.0%
	ABC	259.70	1	259.70	22.69	0.1%
	Error	103.01	9	11.45		
	Subtotal	1162.73	14			
	Grand average	—	1			
	Total	—	15			

© SAE International.

According to the adjusted mean squares, i.e., column MS$_{adj}$ in Table 14.2.3, factor C (anti-pad radius) is the least significant factor among the three design factors. Let factor C = 37.5 mil (=0.0375″ or 0.9525 mm, i.e., level 0). Equations (14.2.1) and (14.2.2) reduce to, respectively,

$$S_{21} \text{ (dB)} = -0.755 + 0.2562 \text{ A} - 0.055375 \text{ B} - 0.01829 \text{ A}^2 + 0.000320 \text{ B}^2$$

$$+ 0.002438 \text{ A B} \tag{14.2.3}$$

and $$S_{11} \text{ (dB)} = -6.8575 - 5.30625 \text{ A} + 0.702 \text{ B} - 0.01135 \text{ B}^2 + 0.07875 \text{ A B} \tag{14.2.4}$$

Plots of the response surface of scatter parameters S$_{21}$ and S$_{11}$ in terms of factor A (via diameter) and factor B (via pitch) are shown in Figure 14.2.5. It is shown that a large via diameter and a short via pitch are effective not only in the promotion of signal transmission (i.e., small |S$_{21}$|) but also in reduction in reflection loss (i.e., large |S$_{11}$|).

FIGURE 14.2.5   Response surface plots of S_{21} and S_{11} against via diameter (factor A) and via pitch (factor B). (a) S_{21} and (b) S_{11}.

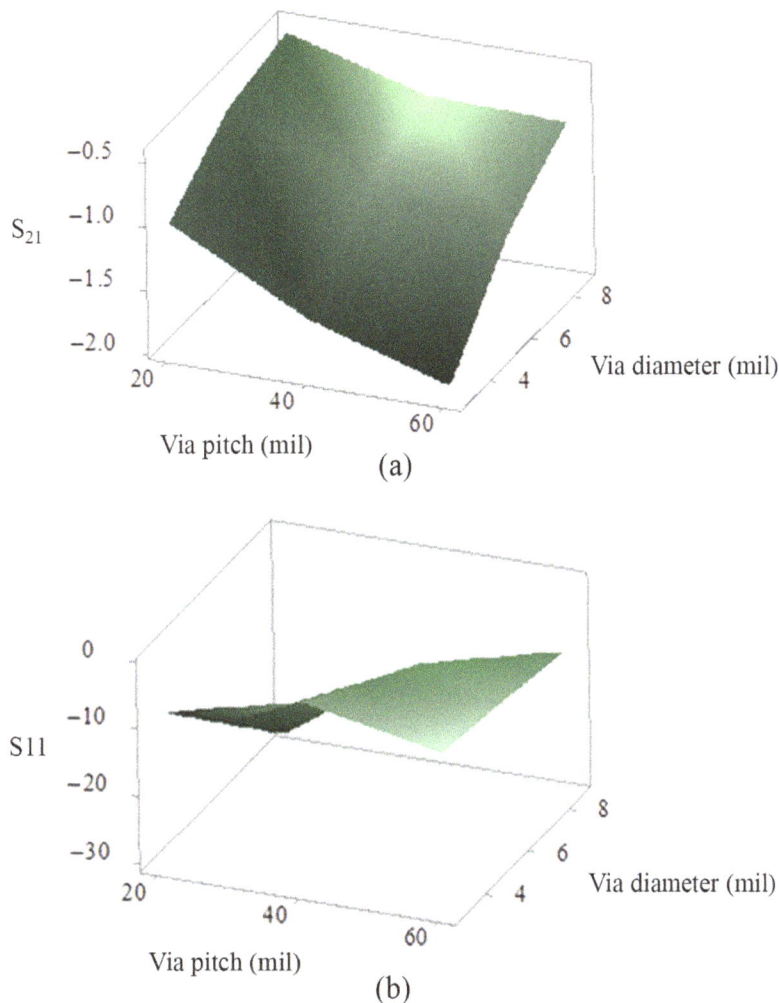

(a)

(b)

14.3. Durability of BGAs

Automotive electronics usually work under a thermomechanical loading environment that includes significant temperature variations and tremendous vibration levels over a wide range of frequencies. A thermomechanical cycling test, combining various temperature variations (ranging from −40 to 125°C) and vibration fluctuations, is routinely conducted concurrently for qualifying the durability of PCBs. For example, 15-min ramp and 15-min dwell at the two temperature extrema and 0.1-G^2/Hz stress level ranging from 100 to 1000 Hz are applied in the case study done by Qi et al. [2009]. Electrical resistances of daisy-chained packages were measured using a Fluke 75 multimeter in their study. The failed packages were observed from either the electrical open or extremely high resistance number (in the order of MΩ). A crack may initiate at the boundary edge in the joining area either between the component (Cu pad) and ball grid or at the boundary between the PCB and ball grid, as shown in Figure 14.3.1. It has the tendency to extend along with the intermetallic compound (IMC) layer. This is attributed to the stress (strain) singularity between two adjacent layers of dissimilar materials.

FIGURE 14.3.1 Cross-sectional view of a failed BGA joint under treatment 1 (i.e., A = −1, B = −1, and C = −1 in Table 14.3.2) [Qi et al. 2009].

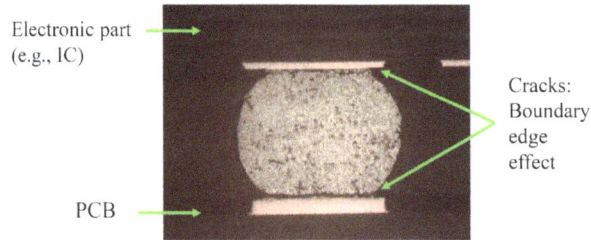

The process DOE is here illustrated using the example of the BGA packaging based on a full factorial design 2^3 as conducted by Qi et al. [2009]. Efforts are made to understand the product reliability problem under various thermomechanical loading conditions by varying the following three fabrication factors:

(A) Padding by Solder Mask: There are two types of land patterns used for surface mounting. They are non-solder mask-defined (NSMD) pads and SMD pads. Solder masks are generally photo-patterned thermoset polymers such as epoxies. The size of opening of a solder mask, formed by UV light, defines the contact area between the solder and the adjacent pad. Generally, the solder ball termination on a BGA package is formed on an SMD pad.

(B) Underfilling: Underfilling is to apply a polymer on top of the PCB and below the flip-chip layer after the solder reflow is completed. This encapsulates the bottom side of the silicon chip for protecting the fragile connections between the bottom of the flip chip and top of the PCB. Underfilling material is mostly epoxy or urethane.

(C) Conformal Coating: Transparent conformal coating material is applied to most electronic chips and wires to act as protection against moisture, dust, chemicals, and temperature extremes. Furthermore, conformal coating increases the voltage rating of a dense circuit assembly, as it can withstand a stronger electric field than otherwise ambient air. Conformal coating is more effective if surface contamination is removed beforehand, either by vapor degreasing or by semi-aqueous washing. The stencil printing method that applies wafer backside coating (WBC) adhesives is now in use at most mass production sites. Uniformity of coating thickness relies on the tooling precision.

Process levels for these three process factors are identified in Table 14.3.1 for experimental tests. The objective is to search for a robust solder joint configuration that would have a prolonged service life by varying these three process parameters.

TABLE 14.3.1 List of factors and design levels for experimental tests on BGA packaging [Qi et al. 2009].

Variables	Level (−)	Level (+)	Coded variable
A. Pad type	SMD	NSMD	SMD: a = A = −1 and NSMD: a = A = 1
B. Underfill	No	Yes	No: b = B = −1 and Yes: b = B = 1
C. Coating	Spray	Dip	Spray: c = C = −1 and Dip: c = C = 1

TABLE 14.3.2 Examining processing parameters for grid ball array joints using 2^3 design.

Treatment	a	b	c	ab	ac	bc	abc	t	ln(t)	ln(t_p)	t_p
1	−1	−1	−1	1	1	1	−1	5	1.609	1.5905	4.908
2	1	−1	−1	−1	−1	1	1	19	2.890	2.9085	18.34
3	−1	1	−1	−1	1	−1	1	1983	7.592	7.6105	2020.0
4	1	1	−1	1	−1	−1	−1	1432	7.267	7.2485	1406.5
5	−1	−1	1	1	−1	−1	1	2	0.693	0.7115	2.038
6	1	−1	1	−1	1	−1	−1	13	2.565	2.5465	12.77
7	−1	1	1	−1	−1	1	−1	1934	7.567	7.5485	1898.5
8	1	1	1	1	1	1	1	2176	7.685	7.7035	2216.8
Contrast	0.737	5.589	−0.212	−0.84	0.259	0.409	−0.037				
Effect	0.368	2.794	−0.106	−0.42	0.129	0.204	−0.019				
Error (σ_y)	0.019	0.019	0.019	0.019	0.019	0.019	0.019				
$|t_1|$	19.91	151	5.73	22.7	6.99	11.0	1				
$t_{1,10\%}$	3.078	3.078	3.078	3.078	3.078	3.078	3.078				
Valid	Yes	Yes	Yes	Yes	Yes	Yes	No				

where
t (h): Time to failure; data obtained from Qi et al. [2009]
ln(t): Natural log of time to failure
ln(t_p): Natural log of the predicted t_p value, i.e., time to failure
t_p (h): Predicted time to failure

© SAE International.

The effects of individual factors and their interactions are charted as a Pareto plot as shown in Figure 14.3.3. It is seen that factor B (underfill) has the largest impact on the crack initiation among all the factors in the given respective working ranges. Solders designed with underfill should be the choice in order to retard the crack initiation. The next influential factor is factor A (padding solder mask). The NSMD pad is the choice to mitigate the crack, because the solder ball is able to make a full connection and thus tends to have better solder adhesion than a SMD pad. Three two-factor interactions and one three-factor interaction (ABC) are also statistically effective, and their influences have to be taken into consideration in the quest of product reliability.

FIGURE 14.3.3   Pareto plot of natural logarithmic-transformed effects.

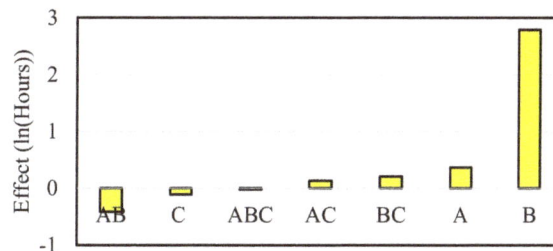

© SAE International.

14.3.1. Predictive Equation

One small effect, i.e., three-factor interaction ABC that is here defined as being less than 2% of the average (4.7335) of all the logarithmic-transformed data, is taken for granted as random error σ_y, which is used to measure the t-statistics associated with other potentially influential effects. Based on the t_1-statistic with the parting significance level of 10%, the calculated main and interactive effects and their statistical significances are demonstrated in Table 14.3.2. At a confidence level of 90% in light of t-distribution, the predictive equation for the lifespan of these ball grid joints can be expressed in terms of dimensionless coded variables as

$$\ln(t_p) = 4.7335 + 0.36825\ a + 2.79425\ b - 0.106\ c$$

$$- 0.42\ a\ b + 0.12925\ a\ c + 0.20425\ b\ c \qquad (14.3.1)$$

Based on the lognormal distribution, the estimated mean value for each treatment can be calculated as

$$t_{p.q} \approx \text{Mean} = E[t_p] = \exp(\mu_{tq} + \tfrac{1}{2}\ \sigma_y^2) \qquad (14.3.2)$$

of which subscript q is the number of each treatment, ranging from 1 to 8. The effect of the insignificant three-factor interaction (ABC) is taken as random error σ_T that is used to measure the t-statistic of the six remaining influential effects. It can be seen that all three main effects and two-factor interactions are statistically significant. Among all, factor B (underfill or not) has the dominating effect and it urges upon the engineer the necessity of designing a GBA joint with underfill. When a design with underfill, i.e., b = 1, Equation (14.3.1) reduces to

$$\mu_{tq} \approx \ln(t_p) = 7.527755 - 0.05175\ a + 0.09825\ c + 0.12925\ a\ c. \qquad (14.3.3)$$

$$\text{Thus,} \quad t_p = \exp(7.527755 - 0.05175\ a + 0.09825\ c + 0.12925\ a\ c + \tfrac{1}{2}\ \sigma_y^2). \qquad (14.3.4)$$

The response surface of t_p (hours) versus factors A and C in the coded scale is depicted in Figure 14.3.4. It shows that the ball grid joints made with underfill (b = 1) can enjoy a great lifespan between 1406 and 2216 h, when subjected to accelerated life testing.

On the other hand, the ball grid joints can have only a lifespan between 2.0 and 18.3 h if the ball grid joints are made with no underfill (b = −1).

FIGURE 14.3.4 Response surfaces of lifespan t_p (hours)—durability contrast due to factor B. (a) With underfill (B = 1); (b) without underfill (B = −1).

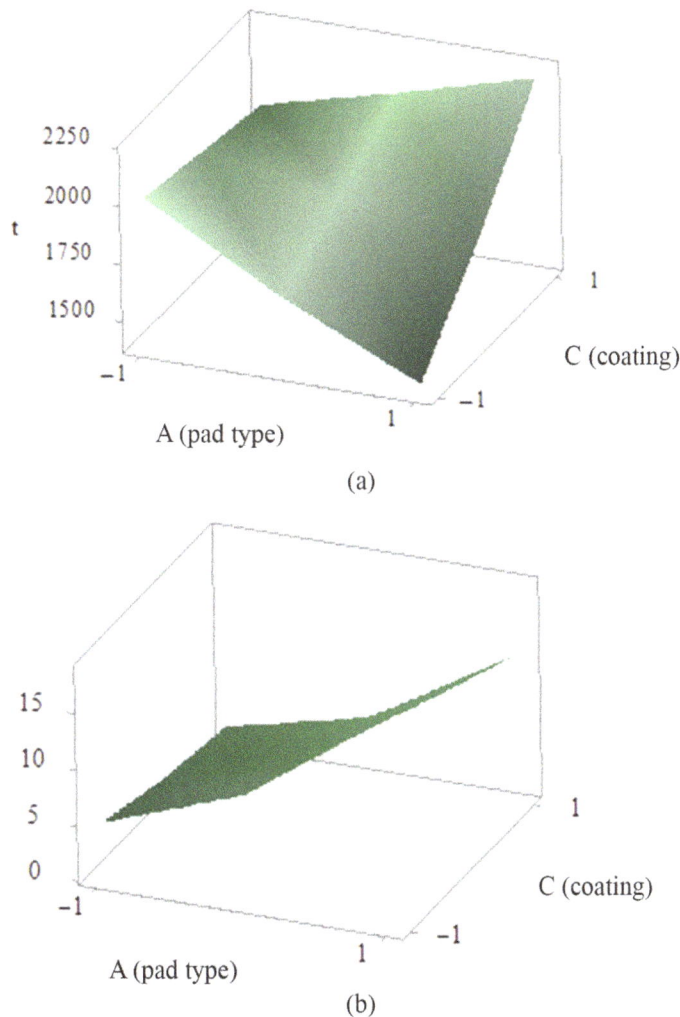

(a)

(b)

14.3.2. Diagnostic Checking

Diagnostic checking of model effectiveness and adequacy is carried out by the following two mechanisms: (a) correlation plot and (b) residual plot. The predicted values of $\ln(t_p)$ and t_p are listed in the last two columns in Table 14.3.2, from which residual $\varepsilon = t_p - t$. The correlation plot, i.e., physical test data (t) charted versus predicted values (t_p) is depicted in Figure 14.3.5(a), which yields a 99.97% correlation rendering the effectiveness of this study. The random pattern of the plot of residuals versus the predicted values as shown in Figure 14.3.5(b) evidences the adequacy of the predictive equation. The normal probability plot of all effects is presented in Figure 14.3.6, which also verifies the efficacy of the contributing main effects and interactions. The straight line mainly comprises the point of the nonsignificant factor, i.e., three-factor interaction ABC and the origin of the plot. Others that deviate from the dashed straight line mean they have "abnormal" effectiveness as the alternative hypothesis applies. The more the deviation the larger the effect, while the other two extremities, i.e., factor B and interaction AB, are marked down in the figure for the purpose of illustration.

FIGURE 14.3.5   Model checking for life prediction of grid ball array joints.

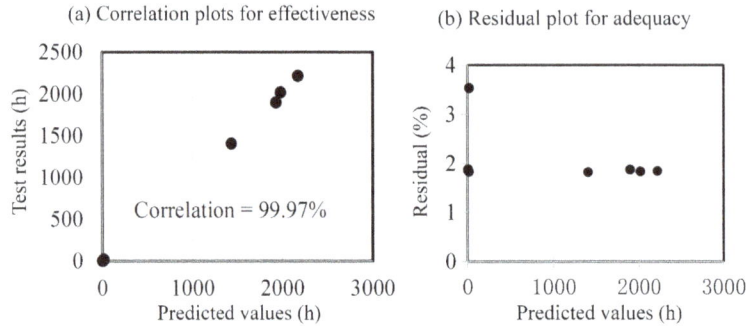

(a) Correlation plots for effectiveness

(b) Residual plot for adequacy

Correlation = 99.97%

FIGURE 14.3.6   Normal probability plot of factorial effects on BGA durability.

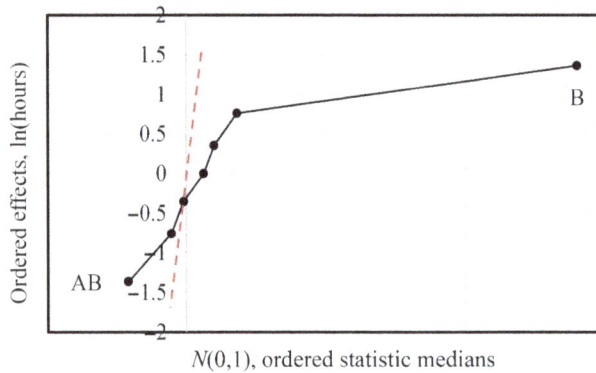

$N(0,1)$, ordered statistic medians

14.3.3. Study on Durability of BGA Packaging via Maximum-Likelihood Estimate

The maximum-likelihood method is here applied to estimating the lognormal parameters for a DOE of the three factors for estimating the life of BGA packaging. The regression model, without the three-factor interaction that is negligible as shown in Table 14.3.2, can be rewritten as

$$\mu_{yq} = \gamma_0 + \gamma_1 a_q + \gamma_2 b_q + \gamma_3 c_q + \gamma_{12} a_q b_q + \gamma_{13} a_q c_q + \gamma_{23} b_q c_q + \varepsilon_q. \tag{14.3.5}$$

Let $y = \ln(t)$. Substituting the above equation, without the residual, into Equation (8.4.14) leads to

$$L_{\ln}(\sigma_y, \gamma_0, \gamma_1, \gamma_2, \gamma_{12}, \gamma_{13}, \gamma_{23}) = \prod_{q=1}^{8} \frac{1}{t_q \, \sigma_y \, (2\pi)^{1/2}}$$

$$\exp\left\{-\frac{1}{2}\left[\frac{\ln(t_q) - (\gamma_0 + \gamma_1 a_q + \gamma_2 b_q + \gamma_3 c_q + \gamma_{12} a_q b_q + \gamma_{13} a_q c_q + \gamma_{23} b_q c_q)}{\sigma_y}\right]^2\right\}. \tag{14.3.6}$$

Thus, $\ln[L_{\ln}(\sigma_y, \gamma_0, \gamma_1, \gamma_2, \gamma_{12}, \gamma_{13}, \gamma_{23})] = \dfrac{1}{\sigma_y \, (2\pi)^{1/2}} \left(\displaystyle\sum_{q=1}^{8} \dfrac{1}{t_q} \right) -$

$$\dfrac{1}{2\sigma_y^2} \sum_{q=1}^{8} \left[\ln(t_q) - (\gamma_0 + \gamma_1 a_q + \gamma_2 b_q + \gamma_3 c_q + \gamma_{12} a_q b_q + \gamma_{13} a_q c_q + \gamma_{23} b_q c_q)\right]^2. \tag{14.3.7}$$

By taking partial differentiations of the logarithmic likelihood function given above with respect to the eight unknowns (i.e., γ_0, γ_1, γ_2, ..., and σ_y), one may maximize the logarithmic likelihood using the following eight equations:

$$\dfrac{\partial\{\ln[L_{\ln}()]\}}{\partial\gamma_0} = \dfrac{1}{\sigma_y^2} \sum_{q=1}^{8} [\ln(t_q) - (\gamma_0 + \gamma_1 a_q + \gamma_2 b_q + \gamma_3 c_q + \ldots)] = 0, \tag{14.3.8}$$

$$\dfrac{\partial\{\ln[L_{\ln}()]\}}{\partial\gamma_1} = \dfrac{1}{\sigma_y^2} \sum_{q=1}^{8} \{[\ln(t_q) - (\gamma_0 + \gamma_1 a_q + \gamma_2 b_q + \gamma_3 c_q + \ldots)] \, a_q\} = 0, \tag{14.3.9}$$

$$\dfrac{\partial\{\ln[L_{\ln}()]\}}{\partial\gamma_2} = \dfrac{1}{\sigma_y^2} \sum_{q=1}^{8} \{[\ln(t_q) - (\gamma_0 + \gamma_1 a_q + \gamma_2 b_q + \gamma_3 c_q + \ldots)] \, b_q\} = 0, \tag{14.3.10}$$

$$\dfrac{\partial\{\ln[L_{\ln}()]\}}{\partial\gamma_3} = \dfrac{1}{\sigma_y^2} \sum_{q=1}^{8} \{[\ln(t_q) - (\gamma_0 + \gamma_1 a_q + \gamma_2 b_q + \gamma_3 c_q + \ldots)] \, c_q\} = 0, \tag{14.3.11}$$

$$\dfrac{\partial\{\ln[L_{\ln}()]\}}{\partial\gamma_{12}} = \dfrac{1}{\sigma_y^2} \sum_{q=1}^{8} \{[\ln(t_q) - (\gamma_0 + \gamma_1 a_q + \gamma_2 b_q + \gamma_3 c_q + \ldots)] \, a_q b_q\} = 0, \tag{14.3.12}$$

$$\dfrac{\partial\{\ln[L_{\ln}()]\}}{\partial\gamma_{13}} = \dfrac{1}{\sigma_y^2} \sum_{q=1}^{8} \{[\ln(t_q) - (\gamma_0 + \gamma_1 a_q + \gamma_2 b_q + \gamma_3 c_q + \ldots)] \, a_q c_q\} = 0, \tag{14.3.13}$$

$$\dfrac{\partial\{\ln[L_{\ln}()]\}}{\partial\gamma_{23}} = \dfrac{1}{\sigma_y^2} \sum_{q=1}^{8} \{[\ln(t_q) - (\gamma_0 + \gamma_1 a_q + \gamma_2 b_q + \gamma_3 c_q + \ldots)] \, b_q c_q\} = 0, \tag{14.3.14}$$

and $\dfrac{\partial\{\ln[L_{\ln}()]\}}{\partial\sigma_y} = \dfrac{-1}{\sigma_y^2 \, (2\pi)^{1/2}} \left(\displaystyle\sum_{q=1}^{8} \dfrac{1}{t_q} \right) +$

$$\sum_{q=1}^{8} \left\{ \dfrac{1}{\sigma_y^3} [\ln(t_q) - (\gamma_0 + \gamma_1 a_q + \gamma_2 b_q + \gamma_3 c_q + \ldots)]^2 \right\} = 0, \tag{14.3.15}$$

where $\ln[L_{ln}()] = \ln[L_{ln}(\sigma_y, \gamma_0, \gamma_1, \gamma_2, \gamma_{12}, \gamma_{13}, \gamma_{23})]$. Substituting the values of a_q, b_q, and c_q that are coded variables, denoted by −1 and 1 in the design matrix (Table 14.3.2), into the above equations leads to eight simultaneous equations, which can be solved for eight unknowns as follows:

$$\gamma_0 = 4.7335, \gamma_1 = 0.36825, \gamma_2 = 2.79425, \gamma_3 = -0.106,$$

$$\gamma_{12} = -0.42, \gamma_{13} = -0.12925, \gamma_{23} = 0.20425, \text{ and } \sigma_y = 0.019.$$

These calculated values are the same as what was obtained in Table 14.3.2 using the traditional DOE based on the t-distribution. Note that the random error (σ_y) obtained from the maximum-likelihood method is different from what is calculated using the traditional DOE based on Student's t-distribution.

In conclusion, solder ball defects have been linked up with three key factors, namely solder paste material, tooling design such as stencil printing, and reflow-thermal profile.

14.4. **Reflow of Solders**

The solder paste reflows in a molten state, creating permanent solder joints. It is widely used for attaching surface mount components to PCBs. Solder balls may become defective in the reflow-soldering process because of explosive evaporation of solvents. Various models for characterizing rheological behaviors of solders are given in Chen et al. [2017]. The problem can be resolved with a proper cure–temperature curve that has good physical compatibility with the selected solder material, including proper preheat, soak, reflow ramp-up, and cooling durations. A typical curing curve is shown in Figure 14.4.1, of which the accumulated heat surpassing the melting point of solders is called the heating factor, formulated as [Alzameli et al. 2016] follows:

$$Q_\eta = \int_{t_2}^{t_5} (T - T_{melt})\, dt, \tag{14.4.1}$$

where

    Q_η (°C·s) is the heat factor
    T (°C) is the temperature
    T_{melt} (°C) is the melting point of solders
    t (sec) is the time

FIGURE 14.4.1 A typical curing curve for reflowing lead-free solders on FR-4 PCB [Tsai 2012].

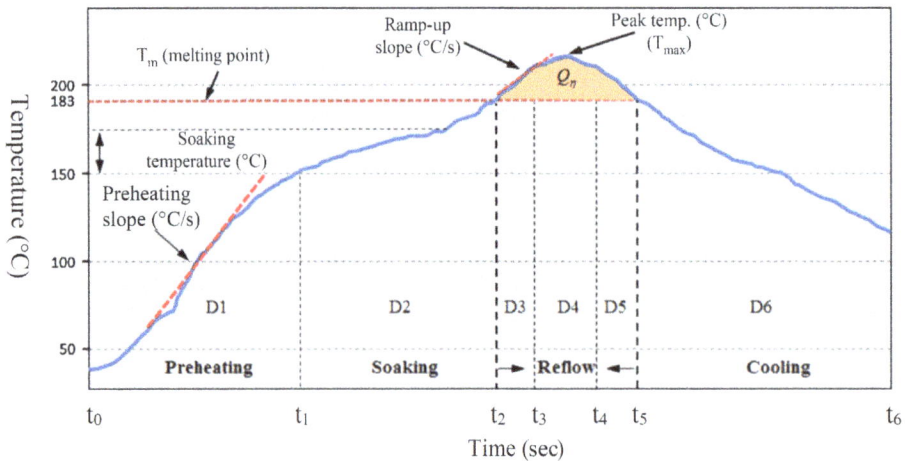

Two heating states are of great concern—preheating time duration (going from t_0 to t_1 in Figure 14.4.1) and ramp-up time duration (going from t_2 to t_3 in Figure 14.4.1). They are found to have a great impact on the number of defects with the solder balls. Design levels of these two factors are listed in Table 14.4.1, while the corresponding spherical composite design matrix based on factorial design 2^2 and test results are given in Table 14.4.2. The two extreme design levels are derived from $\lambda = k^{1/2} = 2^{1/2}$ (Chapter 5) for the spherical composite design.

TABLE 14.4.1 Factors and design levels for reflow analysis of solder balls [Chansa-Ngavej and Kasemsomporn 2008].

| Factor | Design levels | | | | | Coded variable |
	$-2^{1/2}$	-1	0	1	$2^{1/2}$	
A: Preheating time (sec)	0.7929	1	1.5	2	2.2071	$a = (A - 1.5)/0.5$
B: Ramp-up rate (°C/s)	17.574	30	60	90	102.43	$b = (B - 60)/30$

© SAE International.

TABLE 14.4.2 Spherical composite design for detecting defects in reflow of solder balls [Chansa-Ngavej and Kasemsomporn 2008].

Treatment	a	b	A	B	Y	ln(Y)	Y_p
1	-1	-1	1	30	79	4.3695	82
2	1	-1	2	30	244	5.4972	247
3	-1	1	1	90	76	4.3307	76
4	1	1	2	90	208	5.4765	230
5	$-2^{1/2}$	0	0.7929	60	47	3.8501	48
6	0	$-2^{1/2}$	1.5	17.574	191	5.2523	187
7	$2^{1/2}$	0	2.2071	60	239	5.4765	230
8	0	$2^{1/2}$	1.5	102.426	165	5.1060	169
9	0	0	1.5	60	7	1.9459	10.6
10	0	0	1.5	60	8	2.0794	10.6
11	0	0	1.5	60	10	2.3026	10.6
12	0	0	1.5	60	12	2.4849	10.6
13	0	0	1.5	60	14	2.6391	10.6
14	0	0	1.5	60	15	2.7081	10.6

© SAE International.

Notes:
Y: Number of defects from production tests
Y_p: Number of defects predicted using Equation (14.4.2)

When the regression is conducted using data Y, the lack of fit cannot meet the modeling requirement. Since the observation (Y) is the number of defects that is not a continuous variable, it may be transformed into its logarithmic domain for further data processing (Volume II, Chapter 8). The logarithmic-transformed data are listed as column ln(Y) in Table 14.4.2. The final model based on the regression using Minitab (Stat ➔ Regression ➔ Regression ➔ Fit Regression Model), after removing term AB (p-value > 5%), is

$$\ln(Y_p) = 16.680 - 12.60\ A - 0.1884\ B + 4.569\ A^2 + 0.001556\ B^2$$

$$(R = 99.1\%, R_{adj} = 98.6\%, \text{ and } R_{pred} = 98.6\%).\qquad(14.4.2)$$

The effectiveness and adequacy of the above predictive equation in the natural logarithmic domain are justified by the associated high correlations. The ANOVA that redeems the above equation is displayed in Table 14.4.3. Equation (14.4.2) can be rewritten as

$$Y_p = \exp(16.680 - 12.60 \, A - 0.1884 \, B + 4.569 \, A^2 + 0.001556 \, B^2). \tag{14.4.3}$$

It can be seen that both preheating time (factor A) and ramp-up rate (factor B) have strong influences on the number of solder defects and the impact is quite nonlinear as evidenced by Equation (14.4.3). The predictive values for all 14 test conditions based on the above equation are exhibited as column Y_p in Table 14.4.2.

TABLE 14.4.3 ANOVA for detecting defects in reflow of solder balls.

Source	SS_{adj}	DOF	MS_{adj}	F-value	p-Value
Regression	24.9075	4	6.2269	117.24	0.00%
A	7.9364	1	7.9364	149.43	0.00%
B	13.9368	1	13.9368	262.41	0.00%
A^2	9.6355	1	9.6355	181.42	0.00%
B^2	14.4778	1	14.4778	272.60	0.00%
Error	0.4780	9	0.0531		
Lack of fit	0.0099	4	0.0025	0.03	99.8%
Pure error	0.4681	5	0.0936		
Subtotal	25.3855	13			
Grand average	—	1			
Total	—	14			

© SAE International.

The optimal operating condition can be obtained using the following two partial derivatives:

$$\partial[\ln(Y_p)] / \partial A = -12.60 + 2 \, (4.569 \, A) = 0$$

and
$$\partial[\ln(Y_p)] / \partial B = -0.1184 + 2 \, (0.001556 \, B) = 0.$$

The above two equations lead to A = 1.379 and B = 60.54. Substituting these two values into Equation (14.4.3) yields $Y_p = 9.878$, which is the theoretical minimum, since $\partial^2[\ln(Y_p)]/\partial A^2 = 9.138 > 0$ and $\partial^2[\ln(Y_p)]/\partial B^2 = 0.003112 > 0$. A surface plot of the number of defects in terms of preheating time (sec) and ramp-up rate (°C/s) is given in Figure 14.4.2. The response surface is used to overview the number of defects subjected to two major parameters due to the thermal cure based on solder reflow.

FIGURE 14.4.2 Defects in reflow of solder balls vs. preheating time (factor A) and ramp-up rate (factor B).

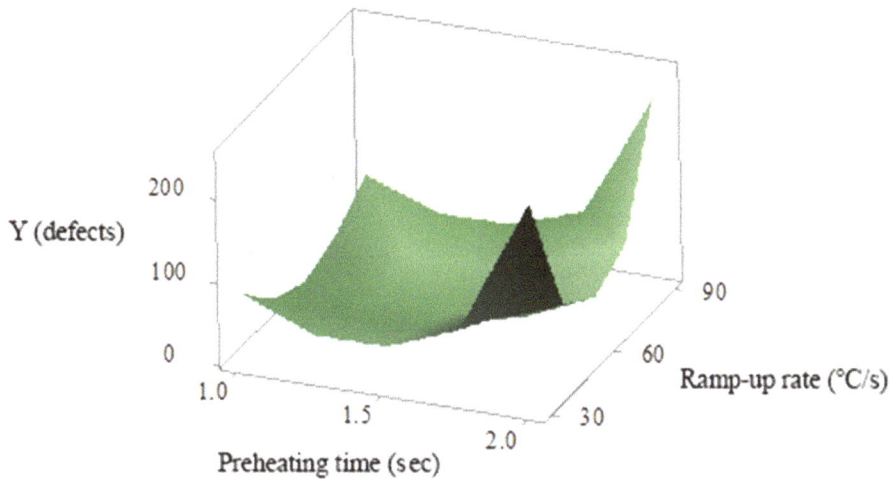

14.5. Deposition by Stencil Printing

As electronic products become smaller, the application of solder paste in an accurate and repeatable manner is a challenge. The stencil printing technique deposits a thin film of solder paste of a thickness between 75 and 200 µm, on which apertures are formed according to the solder pads on the PCB. Stencil printing process, as an SMT, has been in extensive use for mass production. The other deposition method is thin-film growth by molecular beam epitaxy (MBE), which is an epitaxial technology suited for the preparation of advanced structures with special composition and doping profiles controlled on a nanometer scale.

14.5.1. Stencil Printing Process

As an SMT process, a stencil printer process can be divided into the following stages:

1. Feeding the board on the conveyor into the chuck
2. Aligning and lifting the board against the stencil
3. Squeezing solder paste to the stencil by pressure
4. Rolling the paste
5. Aperture filling
6. Aperture emptying
7. Lowering the board to the conveyor and releasing it

A schematic drawing of stages (3)–(6) is exhibited in Figure 14.5.1. The force balance in quasi-static squeeze in a stencil printing process is shown in Figure 14.5.2.

Rheological characteristics of the paste are viable in determining print robustness, especially for printing smaller deposits. The paste ought to be tacky enough to hold the components in place and maintain its shape upon heating to reflow after it is deposited onto the pads. There are two potential nuisance factors, i.e., temperature and humidity, involved in the procedure.

FIGURE 14.5.1   Solder paste printing process: (a) Squeeze, (b) roll, (c) fill, and (d) empty [Kumar et al. 2013].

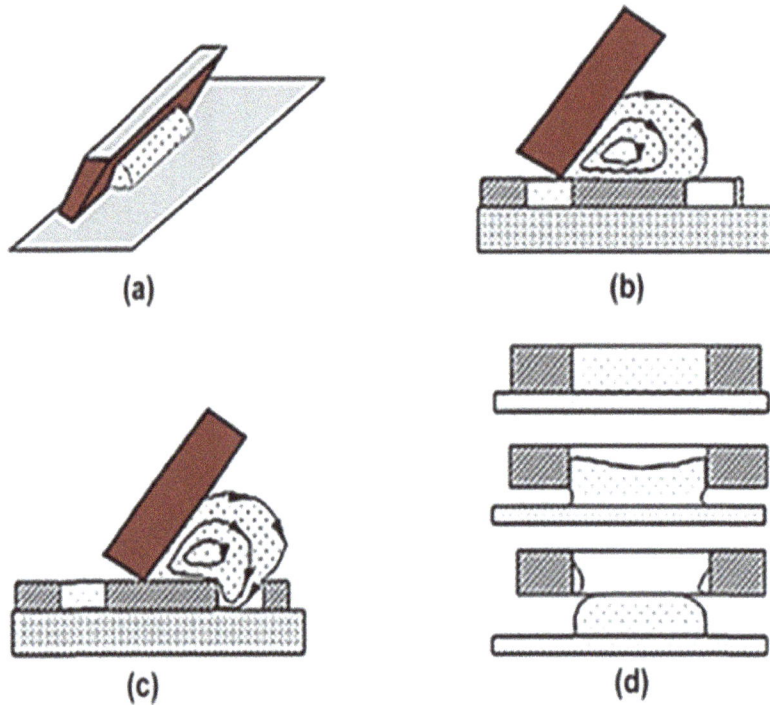

FIGURE 14.5.2   Force balance of paste in quasi-static squeeze during a stencil printing process [Kumar et al. 2013].

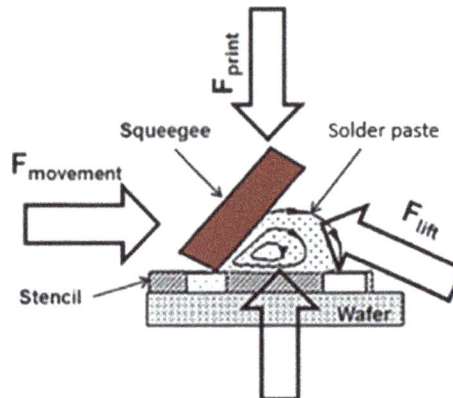

14.5.2. Diameter Variations of Sn-3Ag-0.5Cu Solder Paste

The stencil printing behavior of lead-free Sn-3Ag-0.5Cu solder paste for wafer-level bumping for sub-100-μm size solder bumps is investigated by Kumar et al. [2013]. Four operating parameters of interest and their design levels are listed in Table 14.5.1; other parameters are fixed at their respective levels as given in the table.

TABLE 14.5.1 Stencil printing of Sn-3Ag-0.5Cu solder—process parameters and their design levels [Kumar et al. 2013].

Factor	(1)	(2)	(3)	(4)	(5)
A: Squeegee load (N)	58.84	68.65	—	—	—
B: Squeegee speed (mm/s)	20	30	40	50	—
C: Location of solder (Figure 14.5.2)	1	2	3	4	5
D: Reflow (1: before, 2: after)	1	2	—	—	—
E: Snap-off levels (mm)	0 (contact printing)				
F: Paste material	Sn-3Ag-0.5Cu (fixed)				
G: Aperture size of screen pattern (μm)	30 μm (fixed)				
H: Humidity, relative	40% (fixed)				
I: Temperature (°C)	25°C (fixed)				
J: Wafer	Silicon (fixed)				
K: Stencil material	NiCo (fixed)				
L: Stencil thickness (mm)	Fixed				
M: Stencil aperture ratio (%)	Fixed				
N: Lead pitch (mm)	Fixed				

TABLE 14.5.2 ANOVA for modeling stencil printing of Sn-3Ag-0.5Cu solder.

Source	SS_{adj}	DOF	MS_{adj}	F-value	p-Value
Regression	1557.2	6	259.53	7.82	0.0%
A	315.3	1	315.29	9.50	0.3%
B	218.8	1	218.82	6.59	1.2%
BB	167.0	1	167.00	5.03	2.8%
AB	182.8	1	182.76	5.51	2.2%
ABB	136.5	1	136.50	4.11	4.6%
D	121.3	1	121.28	3.65	6.0%
Error	2423.2	73	33.19	—	—
Lack of fit	306.0	9	34.00	1.03	42.8%
Pure error	2117.2	64	33.08		
Subtotal	3980.4	79			
Grand average	—	1			
Total	—	80			

TABLE 14.5.3 Factorial design $2^2 \times 4 \times 5$ for modeling stencil printing process.

Run	A	B	C	D	Y (μm)
1	58.84	20	1	1	66.5
2	58.84	20	1	2	61.5
3	58.84	20	2	1	62.5
4	58.84	20	2	2	68
5	58.84	20	3	1	58
6	58.84	20	3	2	59.5
7	58.84	20	4	1	71
8	58.84	20	4	2	60.5
9	58.84	20	5	1	70
10	58.84	20	5	2	68.5
11	58.84	30	1	1	62.5
12	58.84	30	1	2	56
13	58.84	30	2	1	58.5
14	58.84	30	2	2	53
15	58.84	30	3	1	55
16	58.84	30	3	2	47
17	58.84	30	4	1	54.5
18	58.84	30	4	2	52
19	58.84	30	5	1	53
20	58.84	30	5	2	48
21	58.84	40	1	1	45
22	58.84	40	1	2	46
23	58.84	40	2	1	63.5
24	58.84	40	2	2	58
25	58.84	40	3	1	60
26	58.84	40	3	2	72
27	58.84	40	4	1	47
28	58.84	40	4	2	51
29	58.84	40	5	1	54
30	58.84	40	5	2	57
31	58.84	50	1	1	55
32	58.84	50	1	2	60
33	58.84	50	2	1	68
34	58.84	50	2	2	55
35	58.84	50	3	1	44
36	58.84	50	3	2	54
37	58.84	50	4	1	61
38	58.84	50	4	2	60.5
39	58.84	50	5	1	63
40	58.84	50	5	2	53
41	68.65	20	1	1	53.5
42	68.65	20	1	2	51.5

(*Continued*)

TABLE 14.5.3 (Continued) Factorial design $2^2 \times 4 \times 5$ for modeling stencil printing process.

Run	A	B	C	D	Y (μm)
43	68.65	20	2	1	55
44	68.65	20	2	2	45.5
45	68.65	20	3	1	55
46	68.65	20	3	2	45
47	68.65	20	4	1	55
48	68.65	20	4	2	48
49	68.65	20	5	1	53.5
50	68.65	20	5	2	45.5
51	68.65	30	1	1	49
52	68.65	30	1	2	51.5
53	68.65	30	2	1	48
54	68.65	30	2	2	42
55	68.65	30	3	1	50
56	68.65	30	3	2	55
57	68.65	30	4	1	45
58	68.65	30	4	2	50
59	68.65	30	5	1	62
60	68.65	30	5	2	50.5
61	68.65	40	1	1	51
62	68.65	40	1	2	49
63	68.65	40	2	1	58.5
64	68.65	40	2	2	51.5
65	68.65	40	3	1	54
66	68.65	40	3	2	39.5
67	68.65	40	4	1	55
68	68.65	40	4	2	47
69	68.65	40	5	1	57
70	68.65	40	5	2	56
71	68.65	50	1	1	48
72	68.65	50	1	2	51
73	68.65	50	2	1	56
74	68.65	50	2	2	59
75	68.65	50	3	1	46
76	68.65	50	3	2	47
77	68.65	50	4	1	54
78	68.65	50	4	2	63
79	68.65	50	5	1	60.5
80	68.65	50	5	2	51.5

Y (μm): Diameter of the printed paste; data obtained from Kumar et al. [2013]

Multivariable linear regression analysis is employed to determine the correlations between the design factors and the scattering parameters. The analysis is conducted using the ANOVA based on the F-distribution as presented in Table 14.5.2. After removing the least insignificant factor or interaction one by one via the comparison of their resulting p-values, with a cutoff significance level (p-value) of 10%, an experimenter obtains the following two regression equations:

$$Y_{before\_reflow} = 374.4 - 4.65 \ A - 15.21 \ B + 0.1883 \ B^2 + 0.2175 \ A \ B - 0.00266 \ A \ B^2 \ (D = 1)$$

and $$Y_{after\_reflow} = 371.9 - 4.65 \ A - 15.21 \ B + 0.1883 \ B^2 + 0.2175 \ A \ B - 0.00266 \ A \ B^2 \ \ (D = 2)$$

$$(R = 62.5\%, \ R_{adj} = 58.4\%, \ and \ R_{pred} = 52.2\%).$$

Although the ANOVA (Table 14.5.2) shows that the above two equations are effective, the adequacy is in doubt as evidenced by the three low correlations (R, R_{adj}, and R_{pred}) given above. It means that additional relevant factors have to be considered in the experiments. Additional parameters for stencil printing can be found in Table 14.5.3.

14.5.3. Volume of Dispensed Solder Paste

One more objective of stencil printing is to achieve a desired volume of solder paste with little dispersion. Half-fractional factorial design 2_V^{5-1} is applied in order to investigate the influences of five process parameters (Table 14.5.4) on the uniformity of the dispensed volume of solder paste, which is measured using a Synthetic Vision Systems View 8100 3D-inspection machine and then analyzed by a computer [Montgomery et al. 2000]. The final response data (Y) corresponding to the 16 treatments for a specific device of concern are given in Table 14.5.5, in which the order for conducting these experimental tests is randomized as listed.

TABLE 14.5.4 Stencil printing—process factors and design levels for fractional factorial design 2^{5-1} [Montgomery et al. 2000].

Factor	Level (−)	Level (+)	Dimensionless variate
A: Squeegee pressure (kPa)	55.16	109.84	a = (109.84 − 82.5)/27.34
B: Squeegee speed (mm/s)	12.7	76.2	b = (76.2 − 44.45)/31.75
C: Snap-off levels (mm)	−0.254	1.016	c = (1.016 − 0.381)/0.635
D: Snap-off speed (2 levels)	Low	High	d = −1 for low; d = 1 for high
E: Down-stop levels (mm)	0.762	2.032	e = (2.032 − 1.397)/0.635

TABLE 14.5.5 Fractional factorial design matrix 2_V^{5-1} for dispensed volume of solder pastes in stencil printing process.

(I) Main effects									
Run	Order	A	B	C	D	E	Y	S_q^2	Y_p
1	12	−1	−1	−1	−1	1	7393	252,521	7331
2	16	1	−1	−1	−1	−1	7157	297,391	7266
3	14	−1	1	−1	−1	−1	7350	198,574	7308
4	7	1	1	−1	−1	1	7247	167,252	7243
5	1	−1	−1	1	−1	−1	7900	296,660	8038
6	3	1	−1	1	−1	1	7672	290,064	7646
7	12	−1	1	1	−1	1	8452	236,594	8358
8	13	1	1	1	−1	−1	7984	282,082	7966
9	5	−1	−1	−1	1	−1	7172	277,828	7200
10	8	1	−1	−1	1	1	7210	269,614	7135
11	6	−1	1	−1	1	1	7101	141,906	7177
12	9	1	1	−1	1	−1	7141	197,578	7112
13	15	−1	−1	1	1	1	7208	297,610	7109
14	4	1	−1	1	1	−1	6729	545,387	6717
15	2	−1	1	1	1	−1	7374	155,282	7429
16	10	1	1	1	1	1	6981	181,592	7037
Contrast	−228.6	148.6	316.1	−529.9	57.1				
Effect	−114.3	74.3	158.1	−264.9	28.6	7379.4 (average)			
Error	505.5	505.5	505.5	505.5	505.5				

(II) 2-factor and 3-factor interactions									
Run	AB	AC	AD	BC	BD	CD	ABC	ABD	ACD	BCD
1	1	1	1	1	1	1	−1	−1	−1	−1
2	−1	−1	−1	1	1	1	1	1	1	−1
3	−1	1	1	−1	−1	1	1	1	−1	1
4	1	−1	−1	−1	−1	1	−1	−1	1	1
5	1	−1	1	−1	1	−1	1	−1	1	1
6	−1	1	−1	−1	1	−1	−1	1	−1	1
7	−1	−1	1	1	−1	−1	−1	1	1	−1
8	1	1	−1	1	−1	−1	1	−1	−1	−1
9	1	1	−1	1	−1	−1	−1	1	1	1
10	−1	−1	1	1	−1	−1	1	−1	−1	1
11	−1	1	−1	−1	1	−1	1	−1	1	−1
12	−1	−1	1	−1	1	−1	−1	1	−1	−1
13	1	−1	−1	−1	−1	1	1	1	−1	−1
14	−1	1	1	−1	−1	1	−1	−1	1	−1
15	−1	−1	−1	1	1	1	−1	−1	−1	1
16	1	1	1	1	1	1	1	1	1	1
Contrast	−2.4	−163.4	30.1	171.9	−79.1	−399.1	−36.1	24.4	−74.1	−32.4
Effect	−1.2	−81.7	15.1	85.9	−39.6	−199.6	−18.1	12.2	−37.1	−16.2
Error	505.5	505.5	505.5	505.5	505.5	505.5	505.5	505.5	505.5	505.5

Notes:
E = ABCD
Y: Dispensed volume of solder paste; data obtained from Montgomery et al. [2000]
S_q^2: Sample variance within treatment q, q = 1, 2, ..., and 16; three replications each

There are three replicated tests conducted for each treatment (run). Given 16 sample variances resulting from 16 treatments in a DOE, the unbiased estimate of overall sample standard deviation (i.e., random error) can be obtained from combining individual sample variances S_q^2 associated with treatment q ($1 \leq q \leq 16$) and their associated degrees of freedom ν_q ($1 \leq q \leq 16$) as

$$S = \left(\frac{\nu_1 S_1^2 + \nu_2 S_2^2 + \dots + \nu_{16} S_{16}^2}{\nu_1 + \nu_2 + \dots + \nu_{16}}\right)^{1/2} = \left(\frac{2 \times 252521 + \dots + 2 \times 181592}{2 + \dots + 2}\right)^{1/2} = 505.5.$$

Thus, no effect is valid if measured by the sample standard deviation given above on the condition that the t-ratio $\geq 90\%$ for each valid effect (main or interactive) is expected.

Assume that the sample standard deviation (i.e., random error) is ignored. Making an engineering sense of 1% of the average of the three measured data of each run and doing calculations of design contrasts and effects based on Student's t-distribution as presented in Table 14.5.5, one may obtain the following predictive equation of the dispensed volume of solder paste:

$$Y_p = 7379.44 - 114.3\,a + 74.3\,b + 158.1\,c - 264.9\,d - 81.7\,a\,c + 85.9\,b\,c - 199.6\,c\,d. \tag{14.5.1}$$

Factor D (snap-off speed) has the greatest influence on the dispensed volume of solder paste, and a high snap-off tends to reduce the dispensed volume of solder paste. Note that the predictive equation is only valid in the corresponding working ranges of all the variables given in Table 14.5.4.

14.5.4. Volume Dispersion of Dispensed Solder Paste

The next step is to look into the dispersion of the dispensed volume of solder paste given in Table 14.5.5. It is desired to have a low variation. Assume that the sample standard deviation (i.e., random error) is ignored. The analysis is conducted using the t-ratios as presented in Table 14.5.6. Making an engineering sense of 2% of the average of the 16 random errors (sample standard deviations) and doing calculations of design contrasts and effects based on Student's t-distribution as presented in Table 14.5.6, one may obtain the following predictive equation:

$$S_{qp} = 498.44 + 20.61\,a - 59.25\,b + 27.19\,c - 22.92\,e$$

$$+ 12.35\,a\,c + 15.01\,a\,d - 26.87\,b\,d - 22.89\,b\,c\,d. \tag{14.5.2}$$

Since alias equation E = ABCD holds, three-factor interaction BCD (last item in the above equation) and two-factor interaction AE are confounded with each other. More experimental tests are required to resolve it.

However, it can be also resolved from the optimal conditions. Assume that the two-factor interaction AE is statistically significant. Then, Equation (14.5.2) becomes

$$S_{qp} = 498.44 + 20.61\,a - 59.25\,b + 27.19\,c - 22.92\,e$$

$$+ 12.35\,a\,c + 15.01\,a\,d - 26.87\,b\,d - 22.89\,a\,e. \tag{14.5.3}$$

However, if Equation (14.5.3) is true, $\partial S_{qp}/\partial c = 27.19 + 12.35a = 0$ and $\partial S_{qp}/\partial e = -22.92 - 22.89a = 0$ would lead to two different values of a. Therefore, it is likely that Equation (14.5.2) is true, rather than that Equation (14.5.3) is true.

TABLE 14.5.6 Fractional factorial design matrix 2_V^{5-1} for random errors in stencil printing process.

(I) Main effects								
Run	Order	A	B	C	D	E	S_q	S_{qp}
1	12	−1	−1	−1	−1	1	502.5	510.4
2	16	1	−1	−1	−1	−1	545.3	542.7
3	14	−1	1	−1	−1	−1	445.6	445.7
4	7	1	1	−1	−1	1	409.0	386.3
5	1	−1	−1	1	−1	−1	544.7	540.1
6	3	1	−1	1	−1	1	538.6	530.2
7	12	−1	1	1	−1	1	486.4	475.3
8	13	1	1	1	−1	−1	531.1	557.0
9	5	−1	−1	−1	1	−1	527.1	534.1
10	8	1	−1	−1	1	1	519.2	534.8
11	6	−1	1	−1	1	1	376.7	361.8
12	9	1	1	−1	1	−1	444.5	454.2
13	15	−1	−1	1	1	1	545.5	563.8
14	4	1	−1	1	1	−1	738.5	705.5
15	2	−1	1	1	1	−1	394.1	391.6
16	10	1	1	1	1	1	426.1	441.7
Contrast	41.22	−118.5	54.38	−3.93	−45.85	—		
Effect	20.61	−59.25	27.19	−1.96	−22.92	498.44 (average)		

(II) Two-factor and three-factor interactions										
Run	AB	AC	AD	BC	BD	CD	ABC	ABD	ACD	BCD
1	1	1	1	1	1	1	−1	−1	−1	−1
2	−1	−1	−1	1	1	1	1	1	1	−1
3	−1	1	1	−1	−1	1	1	1	−1	1
4	1	−1	−1	−1	−1	1	−1	−1	1	1
5	1	−1	1	−1	1	−1	1	−1	1	1
6	−1	1	−1	−1	1	−1	−1	1	−1	1
7	−1	−1	1	1	−1	−1	−1	1	1	−1
8	1	1	−1	1	−1	−1	1	−1	−1	−1
9	1	1	−1	1	−1	−1	−1	1	1	1
10	−1	−1	1	1	−1	−1	1	−1	−1	1
11	−1	1	−1	−1	1	−1	1	−1	1	−1
12	−1	−1	1	−1	1	−1	−1	1	−1	−1
13	1	−1	−1	−1	−1	1	1	1	−1	−1
14	−1	1	1	−1	−1	1	−1	−1	1	−1
15	−1	−1	−1	1	1	1	−1	−1	−1	1
16	1	1	1	1	1	1	1	1	1	1
Contrast	−14.24	24.69	30.03	−13.89	−53.75	4.80	−13.28	−7.07	16.58	−45.78
Effect	−7.12	12.35	15.01	−6.95	−26.87	2.40	−6.64	−3.54	8.29	−2.89

Notes:
E = ABCD
Y: Dispensed volume of solder paste
S_q: Random error within treatment q, q = 1, 2, …, 16; three replications each

Four main effects of individual factors are statistically significant, as governed by Equation (14.5.2). Their impacts on the dispersion of dispensed volume of solder paste are described as follows:

1. Factor B: The squeeze speed is the leading influential factor. Increasing the speed will reduce the dispersion of dispensed volume.
2. Factor C: Snap-off level is the next influential factor. It reveals that contacting printing (low level) reduces the dispersion of dispensed volume.
3. Factor E: A higher down-stop level will reduce the dispersion of dispensed volume.
4. Factor A: A smaller squeeze pressure will reduce the dispersion of dispensed volume.

Four two-factor interactions are also statistically significant. The predicted values based on Equation (14.5.2) are listed in column S_{ip} in Table 14.5.2. The effectiveness of the model is validated by the high predictive correlation between the test data and the predicted values; $R_{pred} = 98.3\%$.

A Pareto frontier plot (Volume I, Chapter 6) of the predicted values of random errors (column S_{qp} in Table 14.5.6) versus the predicted values of dispensed volume of the solder paste (column Y_p in Table 14.5.5) is exhibited in Figure 14.5.3. It is used for balancing the desired dispended volume (the larger the better) and the associated random error (the smaller the better). The target dispensed volume of solder paste is 7177 with a sample standard deviation of 361.8, denoted by the red triangle in the figure.

FIGURE 14.5.3 Pareto Frontier plot of predicted random errors versus predicted values of dispensed volume of solder paste.

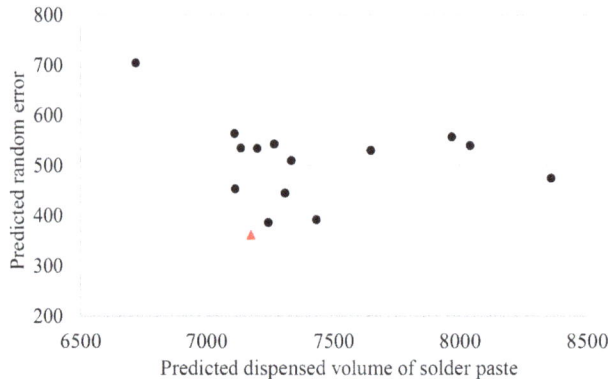

© SAE International.

14.6. System Identification of IC Interconnects

Signal interconnects, power interconnects, grounding, and thermal dissipation means (e.g., thermal through-silicon vias and microfluidic channels) compete for routing space in an IC package, as shown in Figure 14.6.1.

Full-wave simulation techniques using finite element methods in the dynamic domain provide a convenient assessment of IC interconnects in terms of scattering parameters in the wide-band frequency range [Polycarpou et al. 1997]. A system identification of IC interconnects in the design space using finite element simulations in combination with DOE was reported by Kim and Kong [2020]. It aims to ensure the system reliability of an IC against thermal warpage and delamination. Eleven design parameters (factors) are used in the study as depicted schematically in Figure 14.6.2, and their corresponding design levels are described in Table 14.6.1. Besides dielectric constants of the two dielectrics, other design factors are part dimensions and spacings.

FIGURE 14.6.1 Signal through-silicon vias, ground through-silicon vias, and microfluidic channels of a die in a 3D IC [Lee and Lim 2011].

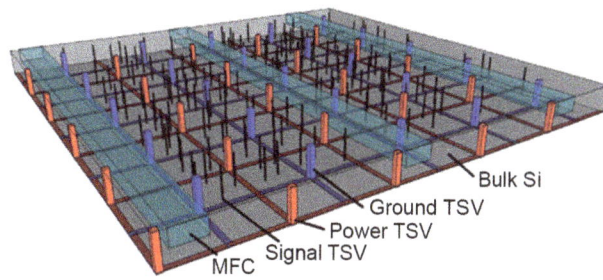

FIGURE 14.6.2 Topology of design factors for system identification of an IC interconnect [Kim and Kong 2020].

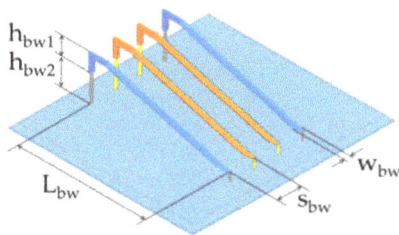

(a) Bonding wires

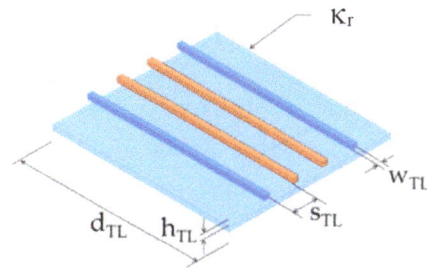

(b) Transmission wires

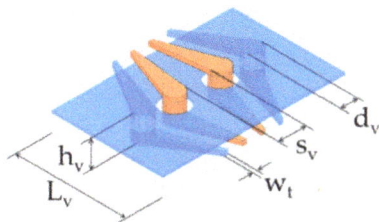

(c) Through-thickness vias

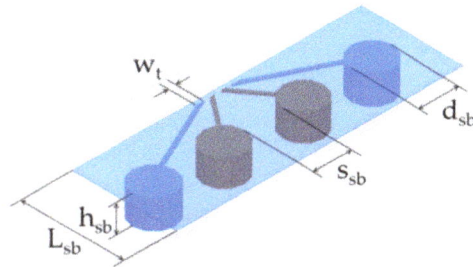

(d) Solder balls

TABLE 14.6.1 Factors and their design levels for system identification of an IC interconnect [Kim and Kong 2020] as shown in Figure 14.5.1.

Factor	(0)	(1)	(2)
A (μm): d_{sb} (diameter—solder)	240	300	360
B (μm): d_v (diameter—via)	35	65	95
C (μm): d_{TL} (length—transmission wire)	1000	2000	3000
D (μm): d_{Bw} (length—bonding wire)	350	700	1050
E (μm): h_{sb} (height—solder)	115	230	345
F (μm): s_{sb} (spacing—solder balls)	100	200	300
G (μm): s_{TL} (spacing—signal lines)	20	40	60
H (μm): s_v (spacing—vias)	37.5	75	112.5
J (μm): w_{TL} (width—signal lines)	10	20	30
K (μm): h_{TL} (dielectric thickness)	10	20	30
M: κ_r (relative dielectric constant)	3.44	4.30	5.16

κ_r: Dielectric constant; κ_r = permittivity of substance/permittivity of free space; air = 1.0006, PC = 2.3, FR4 = 4.4, alumina = 8.8

The equivalent electric circuit of an electric design, such as an IC interconnect, can be comprehended in terms of S-parameters based on its system RLGC model using the full-scale finite element simulations [Eisenstadt and Eo 1992]. A typical RLGC model in the telegrapher's equation is displayed in Figure 14.6.3, of which

- R is the resistance along the line
- L is the inductance along the line
- G is the conductance shunting the line
- and C is the capacitance shunting the line

FIGURE 14.6.3   Transmission parameters (RLGC model) of an interconnect [Eisenstadt and Eo 1992].

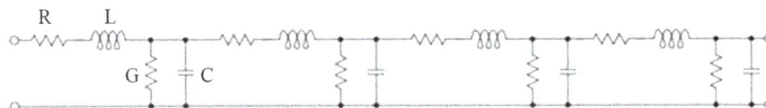

When applied to a two-port interconnect along a transmission path, the RLGC model can be related to its characteristic impedance (Z_0) and a propagation constant (γ). Based on the telegrapher's equations, they can be written as

$$Z_0 = [(R + j\,\omega\,L)\,/\,(G + j\,\omega\,C)]^{1/2}, \tag{14.6.1}$$

and $\quad \gamma = [(R + j\,\omega\,L)\,(G + j\,\omega\,C)]^{1/2}. \tag{14.6.2}$

Note that R (series resistance) and G (shunt conductance) are frequency-dependent, while L (inductance) and C (capacitance) are less frequency-dependent. In most transmission lines, the effects due to L and C tend to dominate because of the relatively low R and G. The characteristic impedance of high-frequency transmission lines is determined not only by the thickness of the substrate material but also by its dielectric constant that varies with respect to temperature.

Working in a dynamic mode, the LC model of a coupled line includes the following four elements:

1. Self-inductance L_{11} (nH)
2. Mutual inductance L_{12} (nH)
3. Self-capacitance C_{11} (pF)
4. Mutual capacitance C_{12} (pF)

The FEA used to identify the LC model is set up to have three conductor layers, while dielectrics 1 and 2 are placed above and below the ground plane (i.e., layer 2), respectively.

The nominal values of the dielectric constant and loss tangent of each dielectric are 4.3 and 0.009, respectively. The encapsulating epoxy mold compound, with a dielectric constant of 4.1 and a loss tangent of 0.022, is placed on top of dielectric 1 for package protection. Copper is used for all conductors. Finite element simulations, set up according to fractional factorial design 3_{III}^{11-8}, are given in Table 14.6.2, and so are the simulation results.

TABLE 14.6.2 Fractional factorial design 3_{III}^{11-8} for system identification of an IC interconnect [Kim and Kong 2020].

(1) Experimental treatments											
Run	**A/d_sb**	**B/d_v**	**C/d_TL**	**D/d_BW**	**E/h_sb**	**F/s_sb**	**G/s_TL**	**H/s_v**	**J/w_TL**	**K/h_TL**	**M/κ_T**
1	240	35	1000	350	115	100	20	37.5	10	10	3.44
2	240	65	2000	700	230	200	40	112.5	30	30	3.44
3	240	95	3000	1050	345	300	60	75	20	20	3.44
4	300	35	2000	1050	115	200	60	37.5	30	20	3.44
5	300	65	3000	350	230	300	20	112.5	20	10	3.44
6	300	95	1000	700	345	100	40	75	10	30	3.44
7	360	35	3000	700	115	300	40	37.5	20	30	3.44
8	360	65	1000	1050	230	100	60	112.5	10	20	3.44
9	360	95	2000	350	345	200	20	75	30	10	3.44
10	240	35	1000	350	230	200	40	75	20	20	4.30
11	240	65	2000	700	345	300	60	37.5	10	10	4.30
12	240	95	3000	1050	115	100	20	112.5	30	30	4.30
13	300	35	2000	1050	230	300	20	75	10	30	4.30
14	300	65	3000	350	345	100	40	37.5	30	20	4.30
15	300	95	1000	700	115	200	60	112.5	20	10	4.30
16	360	35	3000	700	230	100	60	75	30	10	4.30
17	360	65	1000	1050	345	200	20	37.5	20	30	4.30
18	360	95	2000	350	115	300	40	112.5	10	20	4.30
19	240	35	1000	350	345	300	60	112.5	30	30	5.16
20	240	65	2000	700	115	100	20	75	20	20	5.16
21	240	95	3000	1050	230	200	40	37.5	10	10	5.16
22	300	35	2000	1050	345	100	40	112.5	20	10	5.16
23	300	65	3000	350	115	200	60	75	10	30	5.16
24	300	95	1000	700	230	300	20	37.5	30	20	5.16
25	360	35	3000	700	345	200	20	112.5	10	20	5.16
26	360	65	1000	1050	115	300	40	75	30	10	5.16
27	360	95	2000	350	230	100	60	37.5	20	30	5.16

(Continued)

TABLE 14.6.2 (Continued) Fractional factorial design 3_{III}^{11-8} for system identification of an IC interconnect [Kim and Kong 2020].

(2) Test results				
Run	**L_{11}**	**L_{12}**	**C_{11}**	**C_{12}**
1	2.52	0.85	1.41	0.24
2	2.44	0.75	0.8	0.14
3	2.44	0.64	0.7	0.07
4	2.14	0.56	0.85	0.08
5	1.85	0.55	1.32	0.2
6	3.14	1.06	0.86	0.14
7	2.58	0.8	0.75	0.13
8	2.78	0.81	0.87	0.08
9	1.79	0.53	1.76	0.21
10	3.02	0.92	1.2	0.12
11	2.61	0.67	1.06	0.05
12	1.88	0.65	0.94	0.24
13	2.47	0.97	0.92	0.27
14	2.12	0.56	1.11	0.13
15	2.26	0.6	1.55	0.06
16	1.63	0.33	1.54	0.06
17	2.26	0.88	1.13	0.25
18	3.12	0.94	1.06	0.11
19	3.41	1.00	1.21	0.09
20	2.07	0.71	1.15	0.24
21	2.19	0.58	1.03	0.09
22	1.9	0.5	1.41	0.1
23	3.28	0.86	0.78	0.07
24	2.27	0.82	1.5	0.23
25	2.37	0.83	1.11	0.24
26	1.88	0.53	1.91	0.1
27	2.96	0.81	1.08	0.09

According to the ANOVA obtained from the regression using Minitab (Stat ➔ Regression ➔ Regression ➔ Fit Regression Model) after removing the terms with the least statistical significance (p-value > 10%) one by one, the final predictive equations are

$$L_{11p} = 2.846 - 0.001120\ A - 0.000178\ C - 0.000656\ D + 0.000906\ F + 0.01119\ G$$

$$- 0.02733\ J + 0.03217\ K \qquad (R = 97.1\%, R_{adj} = 96.1\%, R_{pred} = 94.2\%)$$

$$C_{11p} = 0.949 + 0.001583\ A - 0.000131\ C - 0.000186\ D - 0.00444\ G + 0.01400\ J$$

$$- 0.02511\ K + 0.1202\ M(R = 96.1\%, R_{adj} = 94.7\%, \text{and } R_{pred} = 92.1\%)$$

$$L_{12p} = 0.9983 - 0.000287\ A - 0.000093\ C - 0.000143\ D + 0.000356\ F - 0.001417\ G$$

$$- 0.010222\ J + 0.014667\ K \qquad (R = 98.1\%, R_{adj} = 97.3\%, \text{and } R_{pred} = 96.0\%)$$

and $\quad C_{12p} = 0.2707 - 0.004083\ G + 0.001722\ K \quad (R = 96.2\%, R_{adj} = 95.8\%, \text{and } R_{pred} = 95.2\%).$

The resulting correlation, adjusted correlation, and predictive correlation for each LC parameter are extremely high. Thus, these predictive equations are effective. Once the desired inductance and conductance are identified using these predictive equations, the related characteristic impedance (Z_0) and a propagation constant (γ) can be calculated using Equations (14.6.1) and (14.6.2). Nevertheless, since fractional factorial design 3_{III}^{11-8} comes with resolution III, the confounding between an individual main effect and its corresponding two-factor interaction has to be further resolved.

14.7. Thermomechanics of PCBs

Copper-clad laminates (CCLs) have been widely used as a bulk conductor for PCBs. The structural material that reinforces a PCB may be divided into the following two categories: fiber-reinforced composites and special material base (ceramic, aluminum alloys, etc.). Most fiber-reinforced composites, including FR-4 ($T_g \approx 140°C$) and FR-5 ($T_g \approx 160°C$), are generally epoxy (FR-4 matrix) reinforced by glass fibers. Standard FR-4 substrates have a relative dielectric constant of 4.5 approximately, although the weave pattern used in the substrate will affect the exact value of the dielectric constant.

The glass point, also called glass temperature (T_g), referring to the maximum temperature (°C) that a PCB can still be structurally stiff, is a key factor related to the durability of a PCB. Most FR-4 PCBs with T_g rated at 140°C are good enough for general applications like automotive interior systems, while FR-5 PCBs with T_g rated between 160 and 170°C are required for automotive systems in the working environment at an elevated temperature such as an engine compartment. PCBs that use two-layered composites of epoxy reinforced with woven glass as the substrate material are called CEM-1 PCBs, which hold great dimensional stability but are fragile.

Other matrices, namely base materials, have been in use are bismaleimide-modified triazine (BT) resin, PI resin, diphenyl ether (PPO) resin, maleic anhydride imide-styrene (MS) resin, polycyclic ester resin, and polyolefin resin. It can be shown in Figure 14.7.1 that vias on PI-based PCBs perform better than those on high-T_g epoxy-based PCB (FR-4, rated at 160°C) in the aspect of electric impedance, as evidenced by the thermal durability test [Salvado et al. 2017].

FIGURE 14.7.1   Electric resistance of vias on PCBs subjected to isothermal aging as stored at 190°C: epoxy versus PI [Salvado et al. 2017].

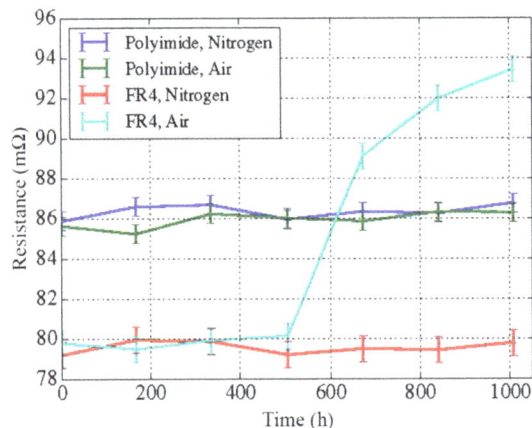

The demand of thin PCBs rises in pursuit of the allowable thermal conductivity of PCBs in the stencil printing of solder paste at room temperature and the reflow quality at a temperature above the melting point of solders, as depicted in Figure 14.7.2. The coefficient of thermal expansion along the z-axis (the thickness) of the dielectric material is usually designed for a low value to minimize dimensional changes with temperature and maintain the integrity of plated through holes (PTHs), which provide paths from the top to the bottom of the circuit board as needed for ground connections, as well as for interconnecting multilayer circuit boards.

FIGURE 14.7.2 Impact of PCB warpage on stencil printing of solder paste and reflow quality [Aravamudhan et al. 2016].

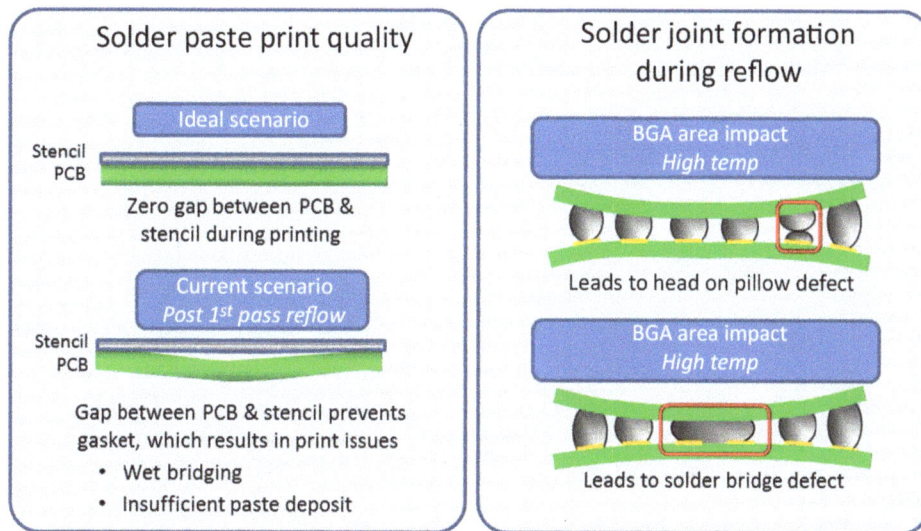

14.7.1. Thermal Conductivity of PCBs

Thermal conductivity (W/m/°C) is a measure of the ability of a substance to conduct heat. It is determined by the rate of heat flow through a unit thickness of the substance normally through a surface of unit area, subjected to a unit temperature gradient in a steady state. Thermal resistivity is the opposite of thermal conductivity. It may meet the need of PCB thermal conductivity with a combination of the following design configurations: horizontal thermal conduction through copper surfaces, vertical thermal conduction through an array of thermal vias, heat sinks, and/or microfluidic channels.

In pursuit of a path of good thermal conductivity right beneath high-power dissipating transistors, an optimization study on a thermal via in the related PCB was performed [Asghari 2006]. A DOE was set up to identify the potential effects of six design factors listed in Table 14.7.1. Thermal conductivities for the 16 treatments based on fractional factorial design 2_V^{5-1} obtained from simulation runs using a finite-volume computational fluid dynamics (CFD) software tool—Icepakreg—are listed in Table 14.7.2. The thermal resistance (°C/W) in the direction normal to the PCB in a steady state is employed as the heat dissipation index without considering the overall or individual through-package thickness. The measured data are listed in column Y in Table 14.7.2.

TABLE 14.7.1 Factors and design levels for thermal resistance of a PCB [Asghari 2006].

Factor	Level (−1)	Level (+1)	Coded variable
A: Diameter of PCB (circular)	14 mm	16 mm	a = (A − 15)/1
B: Pitch (no. of holes per inch)	25 holes	35 holes	b = (B − 30)/5
C: Copper size	2 mm	2.5 mm	c = (C − 2.25)/0.25
D: Barrel size	1 mm	1.4 mm	d = (D − 1.2)/0.2
E: Layers	4	6	e = (E − 5)/1

© SAE International.

TABLE 14.7.2 Fractional factorial design matrix 2_V^{5-1} (E = ABCD) for thermal resistance of PCBs.

(I) Main effects							
Treatment	**A**	**B**	**C**	**D**	**E = ABCD**	**Y (°C/W)**	**Y_p**
1	−1	−1	−1	−1	1	1.95	1.94
2	1	−1	−1	−1	−1	2.21	2.18
3	−1	1	−1	−1	−1	2.79	2.72
4	1	1	−1	−1	1	2.36	2.48
5	−1	−1	1	−1	−1	2.16	2.22
6	1	−1	1	−1	1	1.99	1.98
7	−1	1	1	−1	1	2.38	2.34
8	1	1	1	−1	−1	2.58	2.58
9	−1	−1	−1	1	−1	1.97	2.03
10	1	−1	−1	1	1	1.81	1.79
11	−1	1	−1	1	1	2.12	2.10
12	1	1	−1	1	−1	2.36	2.34
13	−1	−1	1	1	1	1.71	1.68
14	1	−1	1	1	−1	1.94	1.92
15	−1	1	1	1	−1	2.03	2.06
16	1	1	1	1	1	1.81	1.82
Contrast	−0.006	0.336	−0.121	−0.334	−0.239		
Effect	_−0.003_	0.168	−0.061	−0.167	−0.119	2.13563 (average)	
Error	0.0157	0.0157	0.0157	0.0157	0.0157		
$|t_8|$	0.198	10.67	3.85	10.60	7.580		
$t_{8,95\%}$	1.860	1.860	1.860	1.860	1.860		
Significant?	No	Yes	Yes	Yes	Yes		

© SAE International.

(*Continued*)

TABLE 14.7.2 **(Continued)** Fractional factorial design matrix 2_V^{5-1} (E = ABCD) for thermal resistance of PCBs.

(II) Two-factor and three-factor interactions										
Treatment	AB	AC	AD	BC	BD	CD	ABC	ABD	ACD	BCD
1	1	1	1	1	1	1	-1	-1	-1	-1
2	-1	-1	-1	1	1	1	1	1	1	-1
3	-1	1	1	-1	-1	1	1	1	-1	1
4	1	-1	-1	-1	-1	1	-1	-1	1	1
5	1	-1	1	-1	1	-1	1	-1	1	1
6	-1	1	-1	-1	1	-1	-1	1	-1	1
7	-1	-1	1	1	-1	-1	-1	1	1	-1
8	1	1	-1	1	-1	-1	1	-1	-1	-1
9	1	1	-1	1	-1	-1	-1	1	1	1
10	-1	-1	1	1	-1	-1	1	-1	-1	1
11	-1	1	-1	-1	1	-1	1	-1	1	-1
12	-1	-1	1	-1	1	-1	-1	1	-1	-1
13	1	-1	-1	-1	-1	1	1	1	-1	-1
14	-1	1	1	-1	-1	1	-1	-1	1	-1
15	-1	-1	-1	1	1	1	-1	-1	-1	1
16	1	1	1	1	1	1	1	1	1	1
Contrast	-0.046	0.163	0.029	-0.086	-0.114	-0.071	0.026	0.034	-0.034	-0.041
Effect	_-0.023_	_0.008_	_0.014_	-0.043	-0.057	-0.036	_0.013_	_0.017_	_-0.017_	_-0.021_
Error	0.0157	0.0157	0.0157	0.0157	0.0157	0.0157	0.0157	0.0157	0.0157	0.0157
$|t_8|$	1.468	0.516	0.913	2.738	3.611	2.262	0.833	1.071	1.071	1.310
$t_{8,95\%}$	1.860	1.860	1.860	1.860	1.860	1.860	1.860	1.860	1.860	1.860
Significant?	No	No	No	Yes	Yes	Yes	No	No	No	No

Y (°C/W): Thermal resistance in the direction normal to the PCB [Asghari 2006]

The analysis is conducted based on the Student t-distribution, instead of ANOVA based on the F-distribution carried out by Asghari [2006]. The pooled sample error is calculated using the eight small effects, underlined in Table 14.7.2, as

$$S = \left(\frac{\gamma_A^2 + \gamma_{AB}^2 + \gamma_{AC}^2 + \gamma_{AD}^2 + \gamma_{ABC}^2 + \gamma_{ABD}^2 + \gamma_{ACD}^2 + \gamma_{BCD}^2}{8} \right)^{1/2}$$

$$= 0.0157 \qquad\qquad\qquad (14.7.1)$$

Since all the three-factor interactions are invalid, the possibility for the four-factor interaction ABCD to be effective is remote. Thus, the predictive equation with p-value = 5% can be written as

$$Y_p = 2.13563 + 0.168\,b - 0.061\,c - 0.167\,d - 0.1194\,e$$

$$- 0.043\,b\,c - 0.057\,b\,d - 0.036\,c\,d \qquad\qquad (14.7.2)$$

The predicted values for the 16 treatments are listed in column Y_p (Table 14.7.2), which demonstrates that Equation (14.7.2) comes with a high predictive correlation, $R_{pred} = 98.8\%$. Four of the five main effects are statistically significant, and they are factors B, C, D, and E. The temperature rise can be mitigated by varying viable parameters:

1. Reduction in the number of holes (factor B)
2. Having bigger barrels (factor C)
3. Applying a larger top copper area (factor D)
4. Using the six-layered instead of the four-layered (factor E)

Some two-factor interactions are also significant, but their influences are generally mild as compared to factors B, D, and E.

14.7.2. PCB Warpage

It is imperative to have a uniform contact between the PCB and the stencil to achieve a consistent print process. A PCB with warpage can create a gap between the PCB and the stencil. This excessive gap turns into poor gasketing and yields unevenly dispensed (either excessive or insufficient) solder paste and wet bridging sometimes.

14.8. Spin Coating of Electronic Wafers

Spin coating is an easy and quick technique for applying thin films, ranging from a few nanometers to a few microns in thickness, onto substrates. When a solution of a material and a solvent is spun at high speeds, the centripetal force acts with the surface tension of the liquid coating to create an even covering. After any remaining solvent has evaporated, spin coating results in a thin solid film. For example, a spin coater is used to apply a photoresist to a silicon wafer. These process factors can be controlled by five machine settings [Anderson and Whitcomb 2006]:

(A) Spin Speed: For a given solution, the range of spin speeds available is a deciding factor that defines the range of thicknesses that can be achieved.

(B) Spin-Up Acceleration: The acceleration of the substrate toward the final spin speed can also affect the coated film properties, since the resin begins to dry while subjected to the radial force during the first part of the spin cycle.

(C) Volume of Resist: Most likely, 50% of the solvents in the resist (resin) are lost to evaporation in the first few seconds of the process.

(D) Spin Coating Duration: It is the time duration required to keep the substrate spinning until the film is fully dry.

(E) Cover of Exhaust: The exhaust cover (lid) is part of a spin coater system to minimize unwanted random turbulence during the spin operation.

Besides these operating factors, the vendor of resist is also taken into consideration as an additional factor in the study based on DOE by Anderson and Whitcomb [2006]. The six design factors and their corresponding design levels are listed in Table 14.8.1. Experimental tests, as listed in Table 14.8.2, are deployed according to a fractional factorial design 2_{IV}^{6-2} with the following two alias equations: E = BCD and F = ABC. The objective of the study is to explore the influences of these six factors on the coating thickness, termed Y (μm).

TABLE 14.8.1 Operating factors and their design levels for spin coating of electronic wafers [Anderson and Whitcomb 2006].

Factor	Level (−1)	Level (+1)	Coded variable
A. Speed, spin (rpm)	6650	7350	$a = (A − 7000)/350$
B. Acceleration, spin-up	5	20	$b = (B − 12.5)/7.5$
C. Volume of resist (CC)	3	5	$c = (C − 4)/1$
D. Time, spin coating (sec)	6	14	$d = (D − 10)/4$
E. Cover of exhaust	Off (E⁻)	On (E⁺)	Off ($e = −1$) and On ($e = 1$)
F. Vendor of resist	Supplier F⁻	Supplier F⁺	F⁻ ($f = −1$) and F⁺ ($f = 1$)

© SAE International.

TABLE 14.8.2 Fractional factorial design 2_{IV}^{6-2} (E = BCD and F = ABC) for spin coating of electronic wafers [Anderson and Whitcomb 2006].

Run	A	B	C	D	E	F	Y (μm)	Y_p (μm)
1	−1	−1	−1	−1	−1	−1	4475	4478.2
2	1	−1	−1	−1	−1	1	4630	4643.0
3	−1	1	−1	−1	1	1	4334	4336.2
4	1	1	−1	−1	1	−1	4534	4503.8
5	−1	−1	1	−1	1	1	4455	4490.0
6	1	−1	1	−1	1	−1	4478	4471.2
7	−1	1	1	−1	−1	−1	4222	4164.4
8	1	1	1	−1	−1	1	4523	4515.6
9	−1	−1	−1	1	1	−1	4440	4478.2
10	1	−1	−1	1	1	1	4664	4643.0
11	−1	1	−1	1	−1	1	4330	4336.2
12	1	1	−1	1	−1	−1	4515	4503.8
13	−1	−1	1	1	−1	1	4566	4490.0
14	1	−1	1	1	−1	−1	4456	4471.2
15	−1	1	1	1	1	−1	4115	4164.4
16	1	1	1	1	1	1	4467	4515.6

© SAE International.

The analysis is conducted using the ANOVA based on the F-distribution, as demonstrated in Table 14.8.3. After removing the least insignificant factor or interaction one by one via the comparison of their resulting p-values, an experimenter will obtain the following regression equation with a cutoff significance level (p-value) of 5%:

$$Y_p = 4450.3 + 83.1\,a - 70.3\,b - 40.0\,c + 45.9\,f + 46.6\,c\,f$$

i.e. $$Y_p = 4450.3 + 83.1\,[(A - 7000)/350] - 70.3\,[(B - 12.5)/7.5] - 40.0\,(C - 4)$$

$$+ 45.9\,F + 46.6\,(C - 4)\,F$$

$$= 3065.5 + 0.237\,A - 9.373\,B - 40.0\,C - 140.5\,F + 46.6\,C\,F. \tag{14.8.1}$$

This predictive equation is established with the following high correlations: $R = 96.9\%$, $R_{adj} = 95.3\%$, and $R_{adj} = 91.8\%$. Note that CF is confounded with AB (i.e., AB = CF), as the interaction is subjected to the alias equation F = ABC. It is imperative to resolve these ambiguities. With the aid of a follow-up study of eight runs,

Anderson and Whitcomb [2006] further verified that interaction CF is statistically significant. The predicted responses for the 16 treatments are listed in column Y_p, of which the maximum error is 1.38%.

The main effects of factors A, B, C, and F are statistically significant. A higher spin speed (factor A), less spin-up acceleration (factor B), and lowering the volume of resist (factor C) will increase the coating thickness. The resistance comes from supplier F+ is likely to provide higher coating thickness. The interaction between the volume of resist and the vendor also has a significant influence on the coating thickness.

TABLE 14.8.3 ANOVA for spin coating of electronic wafers.

Source	SS_{Adj}	DOF	MS_{Adj}	F-value	p-Value
Regression	283,572	5	56,714	30.47	0.0%
A	110,556	1	110,556	59.40	0.0%
B	78,961	1	78,961	42.42	0.0%
C	25,600	1	25,600	13.75	0.4%
F	33,672	1	33,672	18.09	0.2%
AB = CF	34,782	1	34,782	18.69	0.2%
Error	18,613	10	1861		
Subtotal	302,185	15			
Grand average		1			
Total	—	16			

© SAE International.

14.9. CVD and Epitaxial Deposition

CVD has been widely used for creating thin films, called topological insulators such as polymer coatings, used in part of semiconductor packaging. It is one of the essential unit operations in semiconductor manufacturing because of the ability to deposit thin, smooth films conformally onto submicron-scaled features. CVD oxide is a linear growth process where a precursor gas deposits a thin film onto a wafer in a reactor. As the technology proceeds, one may have the spatial actuation and sensing capabilities necessary to control deposition patterns, e.g., film uniformity in single-wafer combinatorial CVD experiments [Kleijn and Werner 1993]. The pressure and flow rates of these chemicals during the process play an important role in the film uniformity and growth rate.

Epitaxial deposition, or epitaxy, is referred to as the arrangement of atoms in a crystal form upon a crystal substrate. With the MBE technique, layer-by-layer atomically thin films may work as topological insulators for electronics at a temperature lower than CVD but under ultra-high vacuum conditions (i.e., 10^{-8}–10^{-12} Torr). It is possible to dissect sharp interfaces between the layers and to accomplish comparatively easily an exact alloy composition or doping. Recently, topological insulators have become a class of materials that innovate the phase of quantum matter, in which time-reversal-symmetry-protected electrical conduction is confined to the surfaces and edges. The key performance of the process is the deposition rate.

Both CVD and MBE processes are subjected to material compatibility. A significant mismatch between the lattice sizes of the growing film and the supporting material (crystal) may gather elastic energy in the growing film. As the residual strain energy is lowered accordingly, the film breaks into isolated islands at some critical film thickness and the in-plane tensile strength of the film is reduced.

14.9.1. CVD for Amorphous Silicon (a-Si) Thin Film

CVD is a "coating" process whereby solid material is deposited from a vapor by means of a chemical reaction that occurs on or near a heated substrate surface. When molecules are in their radical forms, they react with other radicals and re-combines to form new materials. As a vapor-phase epitaxy, the coating material is vaporized inside a vacuum chamber and begins to uniformly settle on the substrate and excessive byproducts and energies are released from the new materials back into the freestream in the CVD process. The operation involves mainly growing solid thin films on the substrate surface, and it is also used to produce powders and single-crystal materials. Three different measures can be taken to comprehend the CVD performance. They are film thickness (nm = 10^{-9} m), non-uniformity (%), and in-plane stresses (MPa).

Current CVD reactor designs typically consist of a cooled-wall reaction (vacuum) chamber, in which one reactor has several wafers. The growth process proceeds at a relatively low temperature and has a much higher growth rate when compared to thermal oxide. Reactor design components including chamber, wafer position and rotation, pumping, heating, and gas inlet are configured to achieve a uniform deposition with cylindrical symmetry onto the wafer, as well as fulfilling uniform film growth. Innovative design like a programmable CVD reactor has the potential for real-time control of gas-phase composition across the wafer surface for realizing possible novel operating modes.

TABLE 14.9.1 Design factors and their design levels for plasma-enhanced CVD [Patel and Metzler 2017].

Factor	Level (−1)	Level (0)	Level (+1)
A: Silane (SiH_4) flow rate (sccm)	200	500	800
B: Argon flow rate (sccm)	200	500	800
C: Chamber pressure (10^{-3} Torr)	1200	1500	1800
D: High-frequency power (W)	75	150	225
E: Low frequency power (W)	0	0	0
F: Temperature, electrode (°C)	350°C	350°C	350°C
G: Wafer size	4″	4″	4″
H: Capacitor	Auto	Auto	Auto

Notes:
sccm: Standard cubic centimeters per minute
10^{-3} Torr (miniTorr): 0.1333223684211 Pa (1 Torr = 133.3223684211 Pa)

A DOE was conducted by Patel and Metzler [2017] to study the influences of the four factors listed in Table 14.9.1 on the film deposition rate (nm/min) of a CVD process. Experimental tests were developed to create amorphous silicon (a-Si) thin films deposited by plasma-enhanced CVD (PECVD) using the Oxford Plasma Lab 100 system based on fractional factorial design 3_{III}^{4-2} with two alias equations, i.e., C = AB and D = A^2B. The responses are recorded in column Y of Table 14.9.2.

TABLE 14.9.2 Fractional factorial design 3_{III}^{4-2} (C = AB and D = A^2B) for film deposition rate of electronic wafers by plasma-enhanced CVD [Patel and Metzler 2017].

Run	A	B	C	D	Y	$\sigma_{11,uts}$	$\sigma_{22,uts}$	Y_p
1	200	200	1200	75	46.635	399	367.9	46.65
2	200	500	1500	150	59.90	297.2	277.3	60.17
3	200	800	1800	225	52.85	310.8	313.2	52.55
4	500	200	1500	225	93.15	717.2	707.2	93.40
5	500	500	1800	75	69.45	296.9	291.3	69.16
6	500	800	1200	150	87.4	378.2	417.1	87.03
7	800	200	1800	150	89.35	567.6	563.9	88.98
8	800	500	1200	225	111.85	657.8	650.4	111.44
9	800	800	1500	75	74.25	337.1	329.5	74.24

where
Y (nm/min): Deposition rate, i.e., film thickness growth per unit time
σ_{11} (MPa): Ultimate tensile strength along the major axis (in-plane) of the film
σ_{22} (MPa): Ultimate tensile strength perpendicular to the major axis

Multivariable linear regression analysis is employed to determine the correlations between the design factors and the scattering parameters. The analysis is conducted using the ANOVA based on the F-distribution as presented in Table 14.9.3. After removing the least insignificant factor or interaction one by one via the comparison of their resulting p-values, with a cutoff significance level (p-value) of 10%, an experimenter can obtain the following regression model:

$$Y_p = 3.42 + 0.18516\ A + 0.06366\ B - 0.019019\ C + 0.3733\ D$$

$$- 0.000121\ A^2 - 0.000072\ B^2 - 0.000744\ D^2$$

$$(R = 99.99\%,\ R_{adj} = 99.94\%,\ \text{and}\ R_{pred} = 99.19\%), \tag{14.9.1}$$

of which high correlations confirm the effectiveness and adequacy of the predictive equation for the film growth of the PECVD.

TABLE 14.9.3 ANOVA for film deposition rate of electronic wafers by plasma-enhanced CVD.

Source	SS_{Adj}	DOF	MS_{Adj}	F-value	p-Value
Regression	3590.37	7	512.910	1071.76	2.4%
A	539.21	1	539.207	1126.71	1.9%
B	63.73	1	63.733	133.17	5.5%
C	195.34	1	195.339	408.17	3.1%
D	95.97	1	95.971	200.54	4.5%
A^2	235.92	1	235.915	492.96	2.9%
B^2	83.48	1	83.485	174.45	4.8%
D^2	35.04	1	35.042	73.22	7.4%
Error	0.479	1	0.479		
Subtotal	3590.85	8			
Grand average	—	1			
Total	—	9			

© SAE International.

Note that the value of film deposition rate (Y) redeems "the higher the better." Every factor manifests its own individual main effect on the film growth. The film deposition rates of the nine experimental treatments predicted using Equation (14.9.1) are listed in the last column, i.e., column Y_p, in Table 14.9.2. Taking partial differentiations with respect to factors A, B, C, and D, one has

$$\partial Y_p / \partial A = 0.18516 - 0.000242\ A = 0 \quad \text{i.e.}\ A = 765.1$$
$$\partial Y_p / \partial B = 0.06366 - 0.000144\ B = 0 \quad \text{i.e.}\ B = 442.1$$
$$\partial Y_p / \partial C = -0.038038 \quad\quad\quad\quad \text{Less pressure} \rightarrow \text{Higher deposition rate}$$
$$\text{and} \quad \partial Y_p / \partial D = 0.3733 - 0.001488\ D = 0 \quad \text{i.e.}\ D = 250.9$$

An experimental treatment with A = 765.1 sccm, B = 442.1 sccm, and D = 250.9 W will maximize the film deposition rate.

The film deposition rate will increase with a decreasing chamber pressure that is limited by the law of physics. Given that A = 765.1 sccm, B = 442.1 sccm, and D = 250.9 W, the variation of film deposition rate versus factor C (chamber pressure) in the experimental range given in Table 14.9.1 is plotted in Figure 14.9.1. Note that the maximization of film deposition rate is valid only within the design levels in Table 14.9.1. Any extrapolation is to be taken cautiously.

FIGURE 14.9.1   Film deposition rate versus chamber pressure under optimal conditions.

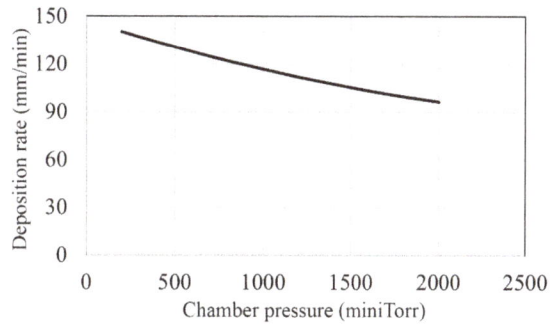

14.9.2. CVD for Diamond-Like Carbon (DLC) Coating

DLC, consisting of an intensive network of HD carbon crystallites, has been increasingly applied as thin-film polycrystalline coatings due to its excellent mechanical properties, such as high wear resistance, low friction, good optical transmittance, with enhanced material modulus and hardness [Kalita et al. 2022]. These DLC coatings are synthesized over silicon (Si 100) substrates via a thermal CVD process. Process parameters of interest and their design levels are listed in Table 14.9.4. The design matrix and resulting data obtained from experimental tests are listed in Table 14.9.5. The goal is to reduce the coefficient of friction (column F) and wear rate (column W) simultaneously.

TABLE 14.9.4   Process parameters of interest and their design levels for a thermal CVD process of DLC coating [Kalita et al. 2022].

Factor	Level−1	Level 1	Coded variable
A (sccm): H_2 flow rate	40	80	$a = (80 - 60)/20$
B (sccm): C_2H_2 flow rate	4	8	$b = (8 - 6)/2$
C (°C): CVD chamber temperature	750	850	$c = (850 - 750)/50$
D (sccm): Ar	Fixed		
E (sccm): N_2	Fixed		

where
sccm: Standard cubic centimeters per minute
C (°C): Rates of temperature increase and decrease are 3 and 5°C/min, respectively

TABLE 14.9.5 To examine CVD process of DLC coating using a composite design as extended from factorial design 2^3 [Kalita et al. 2022].

Run	a	b	c	A	B	C	F	W (μm²)
1	−1	−1	−1	40	4	750	0.140	0.000330
2	1	−1	−1	80	4	750	0.074	0.000132
3	−1	1	−1	40	8	750	0.210	0.000720
4	1	1	−1	80	8	750	0.160	0.000380
5	−1	−1	1	40	4	850	0.074	0.000350
6	1	−1	1	80	4	850	0.094	0.000128
7	−1	1	1	40	8	850	0.146	0.000650
8	1	1	1	80	8	850	0.142	0.000320
9	−1.682	0	0	25	6	800	0.240	0.000840
10	1.682	0	0	95	6	800	0.125	0.000250
11	0	−1.682	0	60	2.5	800	0.060	0.000120
12	0	1.682	0	60	9.5	800	0.185	0.000560
13	0	0	−1.682	60	6	700	0.159	0.000450
14	0	0	1.682	60	6	900	0.086	0.000310
15	0	0	0	60	6	800	0.127	0.000280
16	0	0	0	60	6	800	0.078	0.000230
17	0	0	0	60	6	800	0.119	0.000290
18	0	0	0	60	6	800	0.113	0.000249
19	0	0	0	60	6	800	0.089	0.000298
20	0	0	0	60	6	800	0.105	0.000259

where
Coefficient of friction = tangential force/normal force
W (μm²): Wear rate = true volume/scratch length; obtained using a nano-scratch tester (Model-NHTX 55-0019, CSM instruments) with a Berkovich diamond indenting tip (B-I93, 20 lm radius of curvature) loaded at 0.040 N

The data analysis is carried out using Minitab (Stat → Regression → Regression → Fit Regression Model). The "most insignificant" factor or interaction that comes with the largest p-value is weeded out one by one until all p-values are less than 10%, which is here defined as the parting significance level based on the F-distribution. In light of the ANOVA presented in Table 14.9.6, the final predictive equations for the coefficient of friction and wear rate based on the factorial design are, respectively,

$$F = 1.391 - 0.02115\ A - 0.001333\ C + 0.000057\ A^2 + 0.001437\ B^2 + 0.000017\ A\ C$$

$$(R = 94.8\%,\ R_{adj} = 92.9\%,\ \text{and}\ R_{adj} = 85.2\%), \tag{14.9.2}$$

and $\quad W = 7645 - 28.16\ A + 61.6\ B - 16.25\ C + 0.2111\ A^2 + 4.38\ B^2 + 0.00985\ C^2 - 0.781\ A\ B$

$$(R = 98.8\%,\ R_{adj} = 98.2\%,\ \text{and}\ R_{adj} = 95.1\%), \tag{14.9.3}$$

High correlations confirm the effectiveness and adequacy of the above predictive equations for the coefficient of friction and wear rate of the DLC films.

A Pareto frontier plot of the wear rate versus coefficient of friction is shown in Figure 14.9.2. Potential solutions to the selection of process parameters can be one of the three points identified (diamonds in red) in the figure.

TABLE 14.9.6 ANOVA for characterizing DLC film prepared by CVD.

(i) Friction:	Source	SS$_{Adj}$	DOF	MS$_{Adj}$	F-value	p-Value
	Regression	710,909	7	101,558	72.89	0.000
	A	92,178	1	92,178	66.16	0.000
	B	4415	1	4415	3.17	0.100
	C	16,557	1	16,557	11.88	0.005
	AA	114,347	1	114,347	82.07	0.000
	BB	4912	1	4912	3.53	0.085
	CC	15,593	1	15,593	11.19	0.006
	AB	7812	1	7812	5.61	0.036
	Error	16,719	12	1393	—	—
	Lack of fit	13,306	7	1901	2.78	0.139
	Pure error	3413	5	683		
	Subtotal	727,628	19			
	Grand average	—	1			
	Total	—	20			
(ii) Wear:	Source	SS$_{Adj}$	DOF	MS$_{Adj}$	F-value	p-Value
	Regression	710,909	7	101,558	72.89	0.000
	A	92,178	1	92,178	66.16	0.000
	B	4415	1	4415	3.17	0.100
	C	16,557	1	16,557	11.88	0.005
	AA	114,347	1	114,347	82.07	0.000
	BB	4912	1	4912	3.53	0.085
	CC	15,593	1	15,593	11.19	0.006
	AB	7812	1	7812	5.61	0.036
	Error	16,719	12	1393	—	—
	Lack of fit	13,306	7	1901	2.78	0.139
	Pure error	3413	5	683		
	Subtotal	727,628	19			
	Grand average	—	1			
	Total	—	20			

© SAE International.

FIGURE 14.9.2 Pareto frontier plot of wear rate versus coefficient of friction of DLC films prepared by CVD.

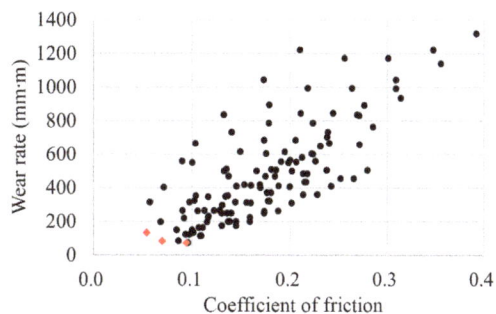

© SAE International.

14.9.3. CVD for Graphene Film

Investigation of how the process parameters of a CVD for growing graphene films on the film quality in light of crystallinity and number of layers was conducted by Shanmugam et al. [2016] using a DOE. It will not be rephrased here.

14.9.4. MBE

MBE consists essentially of atoms or clusters of atoms, which are produced by heating up a solid source. Gaseous atoms migrate in an ultra-high-vacuum environment and impinge on a hot substrate surface, where they can diffuse and get integrated into the growing film. Detailed information on theoretical background and experimental facility can be found in Herman and Sitter [2013], Henini [2018], Asahi and Horikoshi [2019]. It is an epitaxial technology developed for the preparation of advanced structures with composition and doping profiles on a nanometer scale for the manufacture of semiconductor devices, including transistors for cell phones and Wi-Fi. Miniature devices, e.g., thin gallium arsenide (GaAs)-based Hall effect sensors [Mohades-Kassai and Soufi 2000] and topological insulators [Shoemaker et al. 1991] can be also fabricated using the MBE technique. Topological insulators are a class of materials that host new phases of quantum matter in which time-reversal-symmetry-protected electrical conduction is confined to the surfaces and edges. Various techniques of DOE have been applied for identifying the key factors in each individual MBE process [Lee et al. 2000, Uddin et al. 2010].

Here is an example of thin films grown on GaAs and InP substrates by use of MBE for the study of topological insulators. The cross-facet variation in film thickness is of great concern. Data of film thickness at six different facets obtained from the adapted epitaxial layer growth experiment conducted at AT&T [Shoemaker et al. 1991] are given in Table 14.9.7. The goal is to find out how the film deposition thickness across these facets varies with the following four controllable factors: deposition time duration, deposition temperature (°C), nozzle position, and susceptor-rotation method.

The low (−1) and high (+1) levels of the three factors identified for this study by the expert team at AT&T are given in Table 14.9.7. The next step is to assign the low (−1) and high (+1) levels directly to the extremities of each factor (attribute or variable). For a continuous variable factor such as the temperature given here, the nominal value (1215°C) of the deposition temperature can be described with a "0," while "−1" represents 1210°C and a "+1" represents 1220°C. The discrete variable (i.e., nozzle location) can be directly assigned "low (−1)" and "high (+1)" to either of them arbitrarily. The deposition time is a continuous variable, but it is considered as a piece of confidential information and no details have been released. Herein, it can be just identified as "short" and "long" for the purpose of analysis.

TABLE 14.9.7 Factors and levels used for adapted epitaxial layer growth tests.

Factor	Description	Level (−1)	Level (+1)	Coded variable
A:	Deposition time duration	High	Low	$a = A$
B:	Deposition temperature (°C)	1210	1220	$b = (B - 1215)/5$
C:	Nozzle position	2	6	$c = C$
D:	Susceptor-rotation method	Continuous	Oscillating	$d = D$
E:	Wafer code	Fixed		
F:	Arsenic flow rate	Fixed		
G:	HCl etch temperature	Fixed		
H:	HCl flow rate	Fixed		
I:	Location of wafer on susceptor	Either top or bottom; taken as a noise factor		
J:	Facet	Facets 1, 2, 3, 4, 5, and 6 are measured		

Experimental data resulting from physical tests based on design matrix 2^4 are denoted by Y and listed in Table 14.9.8. The corresponding variances of the 16 treatments are recorded in column S_q^2 in the table. Since there is no replication for variance (i.e., N = 1), the "not-so-significant" main effects (i.e., a and d) and two-factor interactions (i.e., "ac," "ad," "bd," and "cd"), and all three-factor interactions (i.e., abc, abd, acd, and bcd) as underlined in Table 14.9.8, are pooled together to assess the sample random error for the variance as follows:

$$\text{Error} = [(\gamma_a^2 + \gamma_d^2 + \gamma_{ac}^2 + \gamma_{ad}^2 + \gamma_{bd}^2 + \gamma_{cd}^2 + \gamma_{abc}^2 + \gamma_{abd}^2 + \gamma_{acd}^2 + \gamma_{bcd}^2) / 10]^{1/2}$$

$$= \{[(-0.043)^2 + (0.029)^2 + \ldots + (-0.059)^2 + (-0.048)^2] / 10\}^{1/2}$$

$$= 0.0039$$

Note that each effect, either main or interactive, possesses one degree of freedom with the variance of reference. The predictive equation of the variance of film thickness (S_q^2) at a confidence level of 10% ($t_{10,10\%}$) is formulated as follows:

$$S_{qp}^2 = 0.24 - 0.0119\,b - 0.0158\,c - 0.0097\,a\,b + 0.0058\,a\,d + 0.0096\,b\,c$$

$$- 0.0059\,a\,c\,d - 0.0124\,a\,b\,c\,d \qquad \qquad (14.9.4)$$

Coded variables a, b, c, and d are the four dimensionless coded variables derived from physical variables A, B, C, and D, respectively. The predicted values of the film thickness variance using the above equation are listed in column S_{qp}^2. The correlation between S_q^2 and S_{qp}^2 is 95.1%.

After model checking, one can conclude that two main effects, as well as three two-factor, one three-factor, and one four-factor interactions, are statistically significant. On main effects, the variance reduces with an increasing deposition temperature (factor B) and having the nozzle positioned at 6 (factor C). Let b = 1 (i.e., B = 1220°C) and c = 1 (i.e., Position 6), and the predictive equation reduces to

$$S_{qp}^2 = 0.2219 - 0.0097\,a - 0.0125\,a\,d.$$

The response surface plot of the predicted variance versus deposition time duration (factor A) and susceptor-rotation method (factor D) is depicted in Figure 14.9.3. It can be seen that the minimum value of S_{qp}^2 is 0.1996 given that a = 1 and d = 1. This predicted value is correlated well with the data obtained from physical tests $S_q^2 = 0.192$ (run 16), with an error of 3.96%. In summary, the minimum variance can be achieved using A = low deposition time duration, B = high deposition temperature (1220°C), C = nozzle positioned at 6, and D = oscillating susceptor rotation.

TABLE 14.9.8 Factorial design 2^4 for variability of adapted epitaxial layer growth tests.

(I) Main effects (q = 1, 2, 3, ..., 16)

Run	A	B	C	D	Thickness (µm)	$y_{ave,q}$	S_q^2	$S_{q\_p}^2$
1	−1	−1	−1	−1	14.506, 14.153, 14.134, 14.339, 14.953, 15.455	14.59	0.270	0.267
2	1	−1	−1	−1	12.886, 12.963, 13.669, 13.869, 14.145, 14.007	13.59	0.291	0.288
3	−1	1	−1	−1	13.926, 14.052, 14.392, 14.428, 13.568, 15.074	14.24	0.268	0.268
4	1	1	−1	−1	13.758, 13.992, 14.808, 13.554, 14.283, 13.904	14.05	0.197	0.201
5	−1	−1	1	−1	14.629, 13.940, 14.466, 14.538, 15.281, 15.046	14.65	0.221	0.229
6	1	−1	1	−1	14.059, 13.989, 13.666, 14.706, 13.863, 13.357	13.94	0.205	0.224
7	−1	1	1	−1	13.800, 13.896, 14.887, 14.808, 14.469, 13.973	14.40	0.222	0.219
8	1	1	1	−1	13.707, 13.623, 14.210, 14.042, 14.881, 14.378	14.14	0.215	0.225
9	−1	−1	−1	1	15.050, 14.361, 13.916, 14.431, 14.968, 15.293	14.67	0.269	0.268
10	1	−1	−1	1	14.249, 13.990, 13.065, 13.143, 13.708, 14.255	13.72	0.272	0.286
11	−1	1	−1	1	13.327, 13.457, 14.368, 14.405, 13.932, 13.552	13.84	0.220	0.220
12	1	1	−1	1	13.605, 13.190, 13.695, 13.259, 14.428, 14.223	13.90	0.229	0.249
13	−1	−1	1	1	14.274, 13.904, 14.317, 14.754, 15.188, 14.922	14.56	0.227	0.205
14	1	−1	1	1	13.775, 14.586, 14.379, 13.775, 13.382, 13.382	13.88	0.253	0.249
15	−1	1	1	1	13.723, 13.914, 14.913, 14.808, 14.469, 13.973	14.30	0.250	0.244
16	1	1	1	1	14.031, 14.467, 14.675, 14.252, 13.658, 13.578	14.11	0.192	0.200
Contrast	−0.0087	−0.0238	−0.0316	0.0057	—		—	—
Effect	−0.0043	−0.0119	−0.0158	0.0029	—		—	—
Error	0.0039	0.0039	0.0039	0.0039	—		—	—
t_{10}	1.115	3.056	4.044	0.734	—		—	—
$t_{10,90\%}$	1.372	1.372	1.372	1.372	—		—	—
Significant	No	Yes	Yes	No	—		—	—

(II) Interactions

Run	AB	AC	AD	BC	BD	CD	ABC	ABD	ACD	BCD	ABCD
1	1	1	1	1	1	1	−1	−1	−1	−1	1
2	−1	−1	−1	1	1	1	1	1	1	−1	−1
3	−1	1	1	−1	−1	1	1	1	−1	1	−1
4	1	−1	−1	−1	−1	1	−1	−1	1	1	1
5	1	−1	1	−1	1	−1	1	−1	1	1	−1
6	−1	1	−1	−1	1	−1	−1	1	−1	1	1
7	−1	−1	1	1	−1	−1	−1	1	1	−1	1
8	1	1	−1	1	−1	−1	1	−1	−1	−1	−1
9	1	1	−1	1	−1	−1	−1	1	1	1	−1
10	−1	−1	1	1	−1	−1	1	−1	−1	1	1
11	−1	1	−1	−1	1	−1	1	−1	1	−1	1
12	1	−1	1	−1	1	−1	−1	1	−1	−1	−1
13	1	−1	−1	−1	−1	1	1	1	−1	−1	1
14	−1	1	1	−1	−1	1	−1	−1	1	−1	−1
15	−1	−1	−1	1	1	1	−1	−1	−1	1	−1
16	1	1	1	1	1	1	1	1	1	1	1
Contrast	−0.0193	−0.0072	0.0116	0.0192	−0.0047	0.0070	−0.0015	0.0034	−0.0117	−0.0095	−0.0248
Effect	−0.0097	−0.0036	0.0058	0.0096	−0.0023	0.0035	−0.0008	0.0017	−0.0059	−0.0048	−0.0124
Error	0.0039	0.0039	0.0039	0.0039	0.0039	0.0039	0.0039	0.0039	0.0039	0.0039	0.0039
t_{10}	2.477	0.919	1.492	2.466	0.600	0.901	0.192	0.438	1.503	1.211	3.178
$t_{10,10\%}$	1.372	1.372	1.372	1.372	1.372	1.372	1.372	1.372	1.372	1.372	1.372
Significant	Yes	No	Yes	Yes	No	No	No	No	Yes	No	Yes

where
$y_{ave,q}$: Sample mean (average)
S_q^2: Sample variance, where S_q is the sample standard deviation of each treatment
q: Subscript for treatment order, q = 1, 2, 3, ..., and 16

FIGURE 14.9.3 Response surface plot of predicted variance versus deposition time duration (factor A) and susceptor-rotation method (factor D).

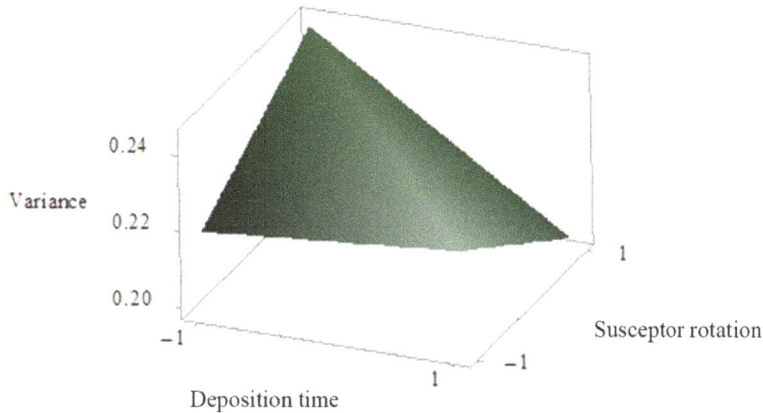

14.10. Insulation of Windings in Electric Motors

Windings in electric motors are insulated with resin or varnish that protects windings from shorting electrically and contamination and makes the windings more mechanically rigid as well. PI, PA-imide (PAI), and polyetherimide (PEI) are frequently used as base materials for insulating the electric windings of electric motors and other high-power machines.

PI and PI/nanofillers nanocomposites are often used as insulating material for turn-to-turn insulation of inverter-fed traction motors because of their excellent electrical, thermal, and mechanical characteristics. The influences of nanofillers, such as alumina [aluminum oxide (Al_2O_3)], boron nitride (BN), silica [silicon dioxide (SiO_2)], titania (TiO_2), graphenes, and carbon nanotubes (CNTs), on the thermomechanical and electrical properties of PI nanocomposites for insulation applications are summarized in Ogbonna et al. [2020]. The performance of PI/Al_2O_3 nanocomposites with different percentages of nanoparticles of Al_2O_3) is presented in Figure 14.10.1.

FIGURE 14.10.1 Lifetime PI/Al_2O_3 nanocomposites as insulation material under repetitive impulsive voltages [Luo et al. 2014].

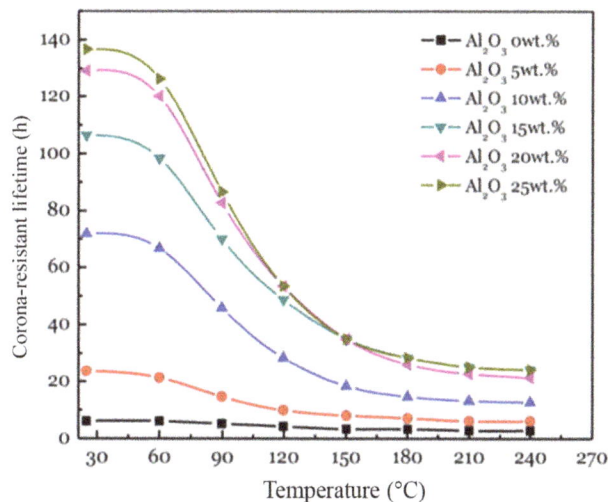

It is well known that the insulation degradation in rotating machines, such as permanent magnet synchronized motors (PMSMs), can be greatly affected by many operating factors, such as heat, tension, vibration, oxidation, hydrolysis, voltage spikes, voltage harmonics [pulse width modulation (PWM)], partial discharge (PD)-induced dielectric breakdown, tracking, gas, water, acids, radiation, and material flaws (e.g., cracking). These stress factors may complicate the operation of electric motors and the related inverters [Du and Robertson 2017], and it is hard to have a comprehensive model for insulation lifespan prediction. Studies based on DOE aiming at modeling the lifespan of insulation materials may throw new light on the causes of this failure mode.

14.10.1. Stress Factors Induced of Induction Motors as Powered PWM

When an induction motor is driven with PWM supply, the voltage and the current applied to the motor are non-sinusoidal. Spikes and harmonics of PWM voltages generate higher electrical and thermal stresses in the stator insulation of induction motors. These two extra loads function as two extra stress factors, in addition to the original stress factor originates only from theoretical sinusoidal loads. Spikes and harmonics may result in the premature failure of PWM-controlled induction motors than those loaded with purely sinusoidal loads [Arora and Aware 2019]. The contribution to the life reduction in induction motors can be assorted to the following three stress factors:

1. Theoretical Sinusoidal Voltage: Applied voltage V_p in the induction motor is expected to comply with its 50/60-Hz sinusoidal supply under normal operation conditions without PWM. Given that the switching voltage varies with respect to PWM, the stress factor subject to sinusoidal voltage (harmonics) is measured as

$$\Lambda_{sine} = V_{pwm} / V_p. \qquad (14.10.1)$$

2. Voltage Spikes at Induction Motor Induced by PWM: For example, the insulation systems of inverter-fed motors are exposed to transient surge voltages with steep fronts generated by the switching of IGBT [Khanna 2003]. The spike magnitude factor (Λ_{sine}) is measured as

$$\Lambda_{spike} = V_{spike} / V_p, \qquad (14.10.2)$$

of which V_{spike} is the amplitude of the voltage spike, as shown in Figure 14.10.2. For example, consider an induction motor that is supplied with a 50-Hz power source. The number of spikes per second is defined for sampled spikes that have a magnitude higher than 1000 V and a rise time less than 2 sec. In practice, the average of 20 replicated random data points is recorded [ABB 1997].

3. Harmonics Induced by PWM Operations at the Induction Motor End: Even-order harmonics cancel out each other. This is also true for third-order harmonics (third, sixth, ninth, etc.), because it is a three-phase power supply system. More worrisome are the fifth-, seventh-, eleventh-, thirteenth-, and higher-order harmonics, which are worrisome. The amplitude of the harmonics produced by a variable-frequency drive (VFD) is greatest for the fifth-, seventh-, and eleventh-order harmonics, and it drops quickly with the thirteenth-order and higher-order harmonics. For a non-sinusoidal

voltage waveform exerted on an induction motor as a generalized case, the Fourier decomposition of the applied voltage V(t) is given as

$$V(t) = \sum_{n=1}^{N} V_{pn} \sin(n\, \omega_1 t + \psi_n),$$

$$(14.10.3)$$

where

V_{pn} is the voltage amplitude in the nth harmonic mode
ω_1 is the fundamental frequency
ψ_n is the phase angle in the nth harmonic mode
N is the total number of harmonic modes considered; $n = 1, 2, \ldots, N$

Assume that N harmonic modes are taken into consideration and the damaging is related to the applied energy as usual. The stress factor due to harmonics is then evaluated using the root-mean-square value as

$$\Lambda_{harmonics} = \left[\sum_{n=1}^{N} (n\, V_n / V_1)^2 \right]^{1/2}.$$

$$(14.10.4)$$

The operation of an induction motor at increasing switching frequencies fed by PWM in variable-speed duty cycles calls for a good understanding of the whole power system as well as the interactions among the related parts, including power line, PWM frequency of inverter, induction motor, and the applied load.

FIGURE 14.10.2   Voltage spikes induced by PWM [Arora and Aware 2019].

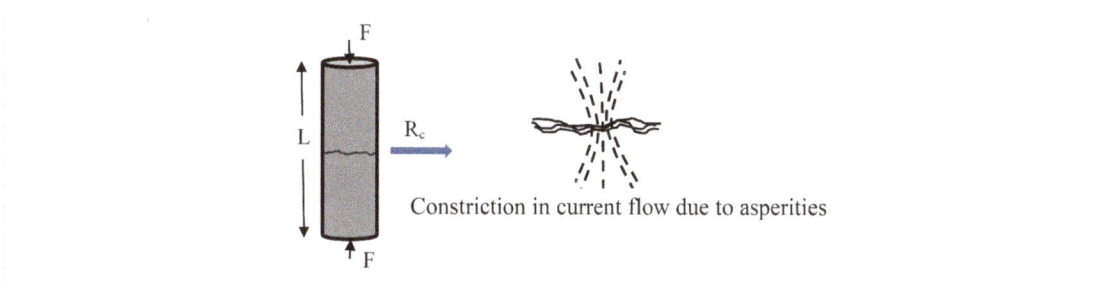

Constriction in current flow due to asperities

Reprinted with permission. © Taylor & Francis Group.

Experimental tests are conducted to explore these three stress factors with a three-phase induction motor rated at 3 hp at the speed of 1440 rpm, supplied with a sinusoidal power source of 440 V at 50 Hz [Arora and Aware 2019]. Relative stress levels under various switching frequencies under PWM are exploited, as listed in Table 14.10.1.

TABLE 14.10.1 Relative stress factors at various switching frequencies [Arora and Aware 2019].

f (kHz)	A = Λ_{sine} (per sec)	B = Λ_{spike} (peaks)	C = $\Lambda_{harmonics}$ (peaks)
Sine (normal)	0	1	1
1	18	1.74	34
5	25	2.12	159
8	36	2.19	244
11	75	2.26	309
15	100	2.43	375

A: Stress factor subject to theoretical sinusoidal voltage (harmonics)
B: Frequency of spikes, i.e., number of peaks per second in spiking; for 1 sec, the number of spikes (KF) with magnitude higher than 1000 V is counted, while the average of 20 randomly recorded samples is taken
C: Frequency of harmonic peaks, i.e., number of harmonic peaks per second

The material damage, or called degradation, done at various switching frequencies may vary with respect to the voltage amplitude and the related phase angle as nonproportional loadings applied in thermomechanics [Chiang 2019]. Overheating is generally the root cause that results in most failures of electric motors. It is due mainly to the deteriorating effect of heat on the motor insulation system. The tactic is herein to introduce quadratic interpolation functions of material damage as used in the shape functions for formulating quadratic elements in the finite element methods. Therefore, the life expectancy (second), denoted by L, is assumed to be represented by a response surface model as

$$L = \gamma_0 + \gamma_1 A + \gamma_2 B + \gamma_3 C + \gamma_{11} A^2 + \gamma_{22} B^2 + \gamma_{33} C^2$$

$$+ \gamma_{12} A B + \gamma_{13} A C + \gamma_{23} B C + \varepsilon, \tag{14.10.5}$$

As listed in Table 14.10.2, the average of lives of ten insulation sheets for each experimental condition (individual switching frequency f_s) according to the applied PWM voltage is given in column L.

TABLE 14.10.2 Life tests of insulators at various switching frequencies [Arora and Aware 2019].

f_s	L_1	L_2	L_3	L_4	L_5	L_6	L_7	L_8	L_9	L_{10}	L
0	533	533	532	518	536	517	536	517	513	517	525.2
1	386	377	415	377	408	416	403	393	392	410	397.3
5	330	293	311	310	280	300	315	303	315	288	304.5
8	272	248	275	251	240	285	268	258	265	267	262.9
11	238	222	227	216	260	222	235	202	233	255	231.0
15	183	182	217	222	190	202	205	197	211	196	200.5

where
f_s (kHz): Switching frequency of sine waves
L_i (sec): Life of the ith (i = 1, 2, ..., 10) specimen tested at the switching frequency
L (sec): Average of ten test specimens

TABLE 14.10.3 ANOVA for life expectancy of insulators of electric motor windings at various switching frequencies.

Source	SS_{Adj}	DOF	MS_{Adj}	F-value	p-Value
Regression	738,192	3	246,064	1244.50	0.0%
B	3716	1	3716	18.79	0.0%
C	5903	1	5903	29.86	0.0%
B^2	1110	1	1110	5.61	2.1%
Error	11,072	56	198	—	—
Lack of fit	265	2	132	0.66	52.0%
Pure error	10,808	54	200	—	—
Subtotal	749,265	59			
Grand average	—	1			
Total	—	60			

TABLE 14.10.4 Design matrix for insulators at various switching frequencies.

Run	A	B	C	L	L_p
Sine (normal)	0	1	1	525.2	525.6
1 kHz	18	1.74	34	397.3	394.5
5 kHz	25	2.12	159	304.5	307.0
8 kHz	36	2.19	244	262.9	264.2
11 kHz	75	2.26	309	231.0	231.5
15 kHz	100	2.43	375	200.5	197.7

where
L (sec): Life at each switching frequency; data obtained from Arora and Aware [2019]
L_p (sec): Predicted life at each switching frequency using Equation (14.10.6)

The ANOVA based on the F-distribution resulting from the life data (second) is demonstrated in Table 14.10.3. After removing the least significant effect (factor or interaction) one by one via the comparison of their resulting p-values, an experimenter will obtain the following regression model with a cutoff significance level (p-value) of 10%:

$$L_p = 805.8 - 350.5 \ B - 0.4653 \ C + 70.8 \ B^2$$

$$(R = 99.26\%, R_{adj} = 99.22\%, \text{ and } R_{adj} = 99.15\%). \tag{14.10.6}$$

The three high correlations given above assure that the above equation is effective as a predictive means in characterizing the life expectancy. The low correlation of lack of fit (52.0% in Table 14.10.3) tell that the experimental tests are well conducted.

The reduction in life expectancy is proportional to the increasing number of spikes per second (factor B), but it varies quadratically with the increasing number of harmonics per second (factor C). The damage done by the sinusoidal vibration is negligible. The predicted life expectancy corresponding to the six nonuniformly stepwise-increasing treatments is listed in column L_p (Table 14.10.4). A response surface plot of life expectancy versus factors B and C is exhibited in Figure 14.10.3.

FIGURE 14.10.3 Response surface of life expectancy in terms of number of spikes and number of harmonic peaks.

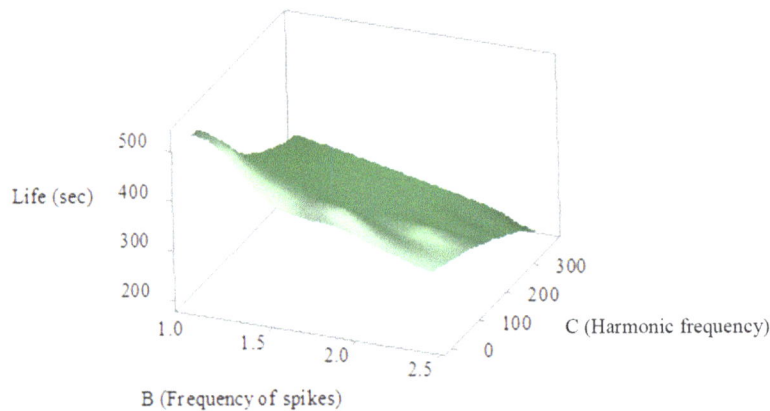

14.10.2. PDs

PDs are minor electrical sparks that may occur across the insulation of medium- and high-voltage electrical devices such as power transformers and bushings, switchgear, motors, and generators.

Practical discharges are likely to induce dielectric breakdown. There are several design problems that cause PDs, including overvoltage caused by impedance mismatch (e.g., between the supply cable and motor) and uneven potential distribution in the stator of an electric motor. PDs may also originate from sharp corners of conductors, discontinuity in coating, and voids formed during manufacture, which occur often in windings, splices, and terminations, especially for conductors with rectangular cross sections [Wang et al. 2013]. PDs are also closely related to the degree of insulation aging.

14.11. Electric Connectors

A connector is made up of two separate sub-assemblies, known as plug and receptacle, which intermate to conduct the electricity, e.g., connect wires with pin-to-socket contacts. For example, a charger for EVs provides a multi-contact electrical connection as a sub-assembly within the interconnect wiring system [SAE J1772]. With crimp or solder contacts two basic connector families are:

1. Circular or Cylindrical Connectors: They are configured with a circular interface to smoothly connect and disconnect circuits using the following mechanical-coupling mechanisms: bayonets, threads, ball detent (push/pull), breech lock, and spring rack/panel. Circular connectors offer rugged solutions, as engineered for reliable performance in a wide variety of harsh environmental applications.

2. Rectangular Connectors and Rack/Panel Connectors: They are configured to maximize the robust contact in a restricted space, or under limited environmental constraints, using the following mechanical-coupling mechanisms: flange mounting-jackscrew, center drive jackscrew, guide pin-flange mounting, flange mating, cam lock, center engagement device, and spring-loaded-flange mounting. Rack and panel connectors feature socket insert style and socket contact type.

In order to warrant proper connections of electric connectors, manufacturers also offer connector accessories to facilitate contacts, such as lubricants, crimp tools, insert/extraction tools, sealing plugs, plastic cap plugs, and basic strain-relief clamps. By placing a film of lubricant between mating surfaces, proper lubrication lowers the operating efforts and reduces mechanical wear. A lubricated electric connector also shows a propensity for mitigating corrosion with appropriate additives in the lubricant [Mraz 2004].

High power density and low contact resistance command a fine prospect for EV chargers. Where there is contention around universal plug technology, there is critical mass from global automakers:

(a) North America and Europe: Combined Charging System (CCS), revised from SAE J1772.

(b) Japan: CHAdeMO, used by Nissan and Mitsubishi EVs.

(c) China: GB/T.

Home EV charging is better off with a 240 V (Level 2) home charger.

14.11.1. Electric Contact Resistance and Impedance

The electrical contact resistance and impedance are used for describing the slowdown of electric conductivity in electric connectors such as switching devices, relays, cable joints, grounded electrodes, and other connections. Top six contributing design factors are

1. Electric voltage (including amplitude and frequency)

2. Electric current

3. Surface topology (e.g., surface roughness)

4. Mechanics of contact (e.g., contact pressure and contact patch)

5. Working temperature [Talukder et al. 2022]

6. Electric transport (e.g., copper and aluminum are different inherently)

For a simple case of electric contact shown in Figure 14.11.1, the electric transport of two practical surfaces in contact with each other consists of at least three components:

$$R_T = R_{B1} + \left[\sum \left(\frac{1}{R_C} + \frac{1}{R_{FT}} + \frac{1}{R_{air}} + \ldots \right) \right]^{-1} + R_{B2}$$

$$\text{and} \quad R_C \approx \tfrac{1}{2} (\rho_1 + \rho_2) / D_C, \tag{14.11.1}$$

where
 R_T (Ω) is the total resistance
 R_{B1} (Ω) is the bulk resistance of the first (top) cylinder; e.g., $R_{B1} \approx \rho_1 L_1/A_1$
 R_{B2} (Ω) is the bulk resistance of the second (bottom) cylinder; e.g., $R_{B2} \approx \rho_2 L_2/A_2$
 R_C (Ω) is the constriction resistance, subjected to asperities of surfaces in contact
 R_{FT} (Ω) is the film tunneling resistance, e.g., oxidation
 R_{air} (Ω) is the air trapped in the tunnel
 ρ_1 and ρ_2 (Ω·m) are the electric resistivities of the upper and lower rods, respectively
 L_1 and L_2 (m) are the lengths of upper and lower rods, respectively (Figure 14.11.1)
 A_1 and A_2 (m²) are the cross-sectional areas of upper and lower rods, respectively (Figure 14.11.1)
 D_C (m) is the equivalent diameter of a circular asperity that accounts for all contact asperities collectively

FIGURE 14.11.1 Constriction in current flow subjected to asperities between two cylindrical rod surfaces.

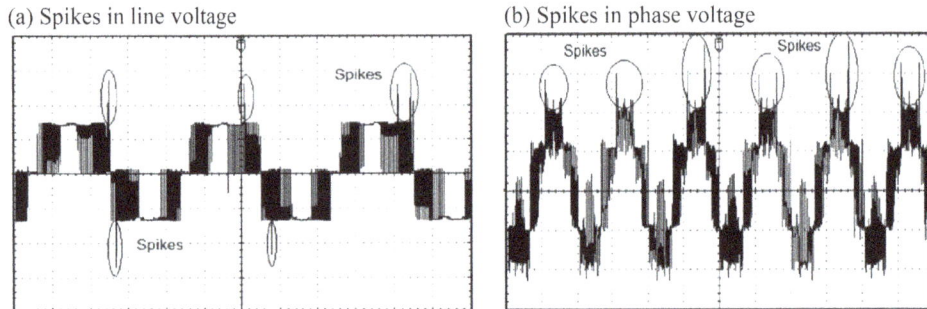

(a) Spikes in line voltage

(b) Spikes in phase voltage

© SAE International.

Material properties of the two frequently used materials for characterizing electrical contact behaviors are given as follows:

Material	E	ν	k	ρ	C_p	α	σ
Cu disk	130	0.3	400	8960	384.4	0.004	1.7×10^{-8}
Al pin	70	0.3	235	2375	904	0.0043	2.65×10^{-8}

where
E (GPa): Young's modulus
ν: Poisson's ratio
k (W/m/°K): Thermal conductivity
ρ (kg/m³): Density
C_p (J/kg/°K): Thermal capacity
α (°K^{-1}): Coefficient of thermal expansion
σ (Ω-m): Electric resistivity

Electric contact resistance and impedance change as the interfacial fretting degradation occurs, and this is generally recognized as one of the major failure modes of electric connectors. The major cause of fretting damage is the relative motion in the interfacial contact between electric plugs and receptacles, whereas material transfer and deformation develop. The relative motion that triggers fretting can be induced by the differential of thermal expansion (contraction), vibration, and/or corrosion.

There is still a need for having a generalized DOE to simultaneously explore the functionalities and the interactions of related design, material, and operating parameters, including insertion forces, withdrawal forces, contact forces/Hertz stresses (normal and shear) in operation, working temperature, plating thickness, surface roughness, aging, and number of mating cycles.

14.11.2. Power Loss due to Electric Contact Resistance in EV Battery Pack

Power loss due to electric contact resistance in an EV battery pack's electric transport at the interface between electrodes and current collectors (bars), as shown in Figure 14.11.2, can be a significant waste of energy. The reduction of surface roughness may be achieved by a precise machining process. The values for mean surface roughness (R_a) for copper collectors and brass electrode brackets used in EV battery packs are generally expected to be less than 0.35 μm.

Pouch cells provide detached tabs, but hard-casing cells provide a certain casing area for electrical connection in the production of high-voltage battery assemblies for EVs. Resistance spot/projection welding,

USW, pulsed arc [micro-tungsten inert gas (TIG)] welding, and laser beam welding can be used for connecting battery cells [Lee et al. 2013, Das et al. 2018]. Each welding technique has its own characteristics depending on the material properties, contact geometry, and contact pressure applied [Brand et al. 2015].

| **FIGURE 14.11.2** | Interface between the current collector and electrode [Taheri et al. 2011]. |

(a) Physical topology

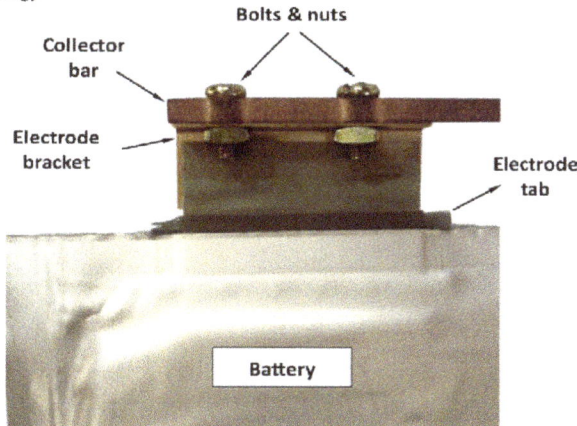

Bolts & nuts

Collector bar

Electrode bracket

Electrode tab

Battery

(b) Schematic drawing of electric contact resistance test on EV battery cell

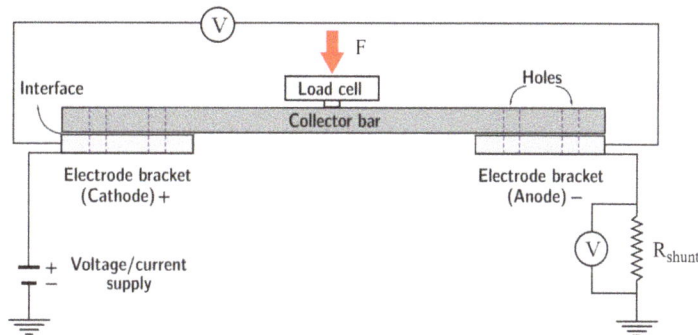

V

F

Interface

Load cell

Holes

Collector bar

Electrode bracket (Cathode) +

Electrode bracket (Anode) −

V

R_{shunt}

+
−   Voltage/current supply

14.11.3. Variation of Electric Contact Resistance with Respect to Contact Pressure

Bolt-nut joints are applied for increasing the contact pressure around the holes where the mean surface roughness is significantly high, e.g., 10 μm or higher. The waviness and the worse surface roughness may appear near the holes simultaneously. The electric contact resistance decreases with increasing contact pressure as demonstrated in Figure 14.11.3. Designers may resort to the DOE based on finite element models, as presented in Volume II, Chapter 11, Section 11.4 (Sealing), to optimize the contact pressure for minimizing electric contact resistance.

FIGURE 14.11.3 Variation of electric contact resistance versus contact pressure [Taheri et al. 2011].

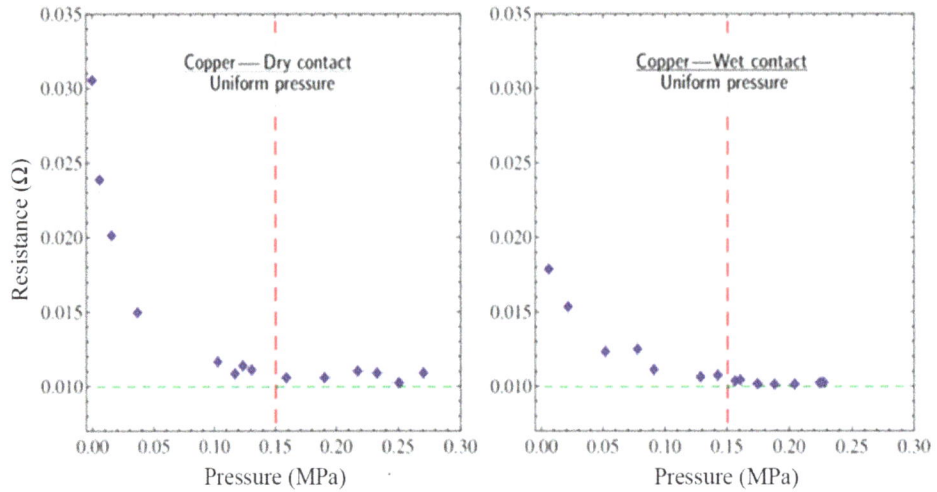

14.11.4. Fretting Wear Resistance and Lifetime of Electrical Contacts

Copper alloys are widely used base material for electrical conductors, but they are prone to corrosion, including oxidation. Electrical contacts are thus plated with a layer of coating material that can be tin, silver, gold, palladium, nickel, or a diverse alloy of these metals. In general, gold plating is slightly better than silver plating and silver plating is much better than tin plating. Quantitative correlations between fretting wear resistance as a function of lifetime of electrical contacts for different types of plating based on physical tests were presented by Song et al. [2012]. A typical chart of electric contact resistance as a function of number of fretting cycles is given in Figure 14.11.4.

FIGURE 14.11.4 Contact resistance as a function of fretting wear cycles [Song et al. 2012].

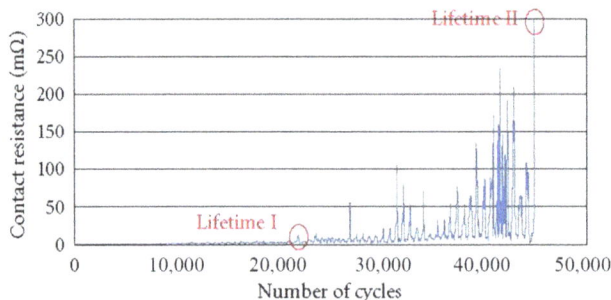

14.11.5. Mating Efforts of Connectors

As an electrical connector serves to couple two circuit devices, the mating effort consists of four elements: contact interface, contact finish, contact spring, and connector housing. Operating efforts for assembling a pair of mating electric connectors, including insertion forces, pull-out forces, and normal contact forces, can vary much from batch to batch. Tolerance management, control of surface roughness, match of thermal expansions, and design for vibration resistance are essential to the avoidance of fretting. A case study on sliding contact based on DOE using FEAs was published by Kallmeyer [2019].

The mating effort can be illustrated by automotive door harnesses [Krugh et al. 2016]. The inner door harness typically connects devices such as door lights, switches, and safety features to the main harness and is readily used by the vehicle occupants. During the automotive door assembly, the inner door wiring harness is clipped in place onto the inner door panel by a line associate that is then connected to the main door wiring harness and attached to the outer door panel. The impacts of operating variables on the defect propensity in a manual assembly process are ranked as follows: connector width, work height, connector height, engagement height, male pigtail, and female pigtail [Krugh et al. 2016].

14.11.6. Tolerance Analysis for Pin Insertion in Electric Connectors

A statistical tolerance analysis is conducted here to find out the gap required for mating "DC module pin window" to "BP pin," when they are in the engaged position. Based on the data listed in Table 14.11.1, the nominal dimension is

$$Y_{nominal} = A - C + E - G = 11.64 - 11.78 + 0.86 - 0.30 = 0.420.$$

Assume that all the dimensions are normally distributed. The overall variance σ_Y^2 can be calculated using the methodology of statistical tolerance analysis by the root-sum squares method (Volume II, Chapter 7), as follows.

Thus, $Y = Y_{nominal} \pm T_Y = 0.4200 \pm 0.157242,$

$$T_Y^2 = (T_A^2 + T_C^2 + T_E^2 + T_G^2) + T_B^2 + T_D^2 + T_F^2 + T_H^2 + T_J^2$$

$$= (0.05^2 + 0.075^2 + 0.025^2 + 0.025^2) + 0.025^2 + 0.05^2 + 0.04^2 + 0.025^2 + 0.01^2$$

$$= (0.15724)^2$$

Thus, $Y = Y_{nominal} \pm T_Y = 0.4200 \pm 0.157242,$

which can be rewritten as

$$Y = \mu \pm 3S = 0.4200 \pm 3\,(0.052414).$$

where
    T_Y is the tolerance for Y
    S is the natural standard deviation; S = 0.052414
    μ is the mean value

If no interference is allowed, Y > 0 is required to have a successful assembly. The process capability can be calculated using Equation (4.6.7) with lower control limit (LCL) = 0 [while upper control limit (UCL) is not specified] as

$$C_{pk} \cong \frac{\min (UCL - \mu, \mu - LCL)}{3S} = \frac{0.42 - 0}{0.157242} = 2.671$$

Note that a process with $C_{pk} > 1.67$ is accepted in the automotive industry.

TABLE 14.11.1 Dimensioning and tolerancing for BP pin insertion [Molex 2006].

No.	Factor: dimension and feature	Dimension (mm)	Level (−1)	Level (+1)
1.	A: DC module width	11.64 ± 0.05	11.59	11.69
2.	B: DC module width true position	0 ± 0.025	−0.025	0.025
3.	C: BP shroud width	11.780 ± 0.075	11.705	11.855
4.	D: BP shroud width true position	0 ± 0.05	−0.050	0.050
5.	E: DC window lead	0.860 ± 0.025	0.835	0.885
6.	F: DC window lead true position	0 ± 0.04	−0.040	0.040
7.	G: BP pin lead width	0.300 ± 0.025	0.325	0.275
8.	H: BP pin lead true position	0 ± 0.025	−0.025	0.025
9.	J: BP pin true position	0 ± 0.100	−0.100	0.100

14.11.7. Mechanics of Electrical Contact Resistance of Cable Wires

Cable wires shall be mounted in a way that the electric contact resistance is reduced and mechanical strains and vibrations are mitigated. For example, a latching mechanism may be utilized to prevent each connector from losing contact with the mating piece. These include ribs, stiffeners, hold-down clamps/brackets, spring retaining clips, clinching, adhesives, rubber pads, and encapsulation materials. Some compliance in the contact zone shall be added to compensate for thermal expansion/contraction.

Electric contact resistance refers to the contribution to the total electric resistance that is attributed to the contacting interfaces of a joint. Here is an experimental design employed to characterize the electrical contact resistance as a function of contact force and contact area. Two circular conductor wires (1.80 mm in diameter each) are pressed against each other on a test bench, as exhibited in Figure 14.11.7. Three crossing angles (90, 60, and 45°) of these two wires are exercised. The electrical contact resistance (mΩ) as a function of applied force (N) and contact area (mm²) is measured at an increasing compressive force, as shown in Table 14.11.2. The test procedure is addressed in detail in Zeroukhi et al. [2014].

FIGURE 14.11.7 Imprint measurement of contact area with two copper wires [Zeroukhi et al. 2014].

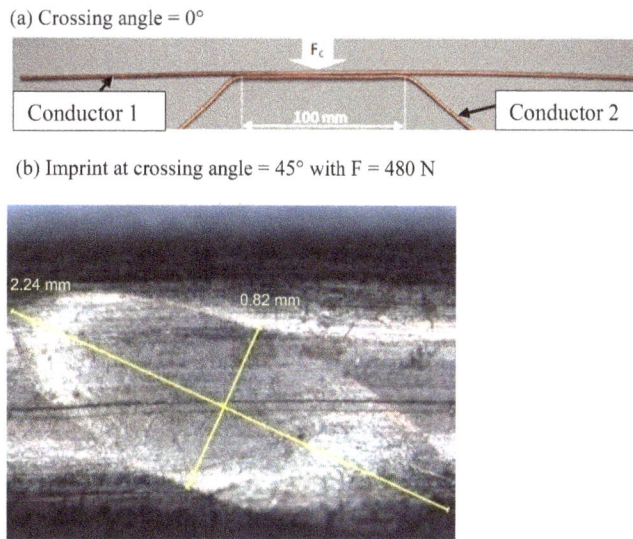

(a) Crossing angle = 0°

(b) Imprint at crossing angle = 45° with F = 480 N

The data analysis is carried out using Minitab (Stat → Regression → Regression → Fit Regression Model). The "most insignificant" factor or interaction that comes with the largest p-value is weeded out one by one until all p-values are less than 10%, which is here defined as the parting significance level based on the F-distribution. In light of the ANOVA presented in Table 14.11.3, the final predictive equation for the electric contact resistance based on the factorial design is

$$R = 2.599 - 0.004086 \, B - 0.6573 \, C + 0.001275 \, B \, C$$

$$(R = 97.7\%, R_{adj} = 97.1\%, \text{ and } R_{pred} = 95.0\%).$$

According to the model correlation, adjusted model correlation, and predictive correlation (i.e., R, R_{adj}, and R_{pred}) shown above, it is concluded that the regression model is effective and adequate.

The response surface of contact electrical resistance as a function of contact force (factor B) and contact area (factor C) is plotted in Figure 14.11.7. Two major findings are as follows:

1. Main effects of both contact force and contact area are statistically significant.

2. In the meanwhile, the interaction between them (i.e., BC) is also viable. With a small contact area, a large force is required to reduce the electrical contact resistance (Figure 14.11.8).

This can be applied to compacting processes for different wire strands. As for a case study, the compacting process for a simple 1 × 6 wire cable is exhibited in Figure 14.11.9.

On the other hand, the contact-stress level would lead to a potential failure as disclosed in Chiang [1996]. The targeted stress level can be higher than the yield strength, but it has to be well below the ultimate von Mises strength.

TABLE 14.11.2 Measured electrical contact resistance (mΩ) as a function of the applied force and contact area [Zeroukhi et al. 2014].

Run	A (°)	B (N)	C (mm²)	P (MPa)	R (mΩ)
1	90	101.5	0.221	459.28	2.328
2	90	202.14	0.478	422.89	1.5
3	90	300.75	0.724	415.40	0.984
4	90	401.37	1.021	393.11	0.785
5	90	481.12	1.287	373.83	0.684
6	60	101.5	1.027	98.83	1.703
7	60	202.14	1.979	102.14	0.945
8	60	300.75	2.859	105.19	0.688
9	60	401.37	4.134	97.09	0.479
10	60	481.12	4.971	96.79	0.413
11	45	101.5	0.865	117.34	1.587
12	45	202.14	2.009	100.62	0.814
13	45	300.75	3.318	90.64	0.561
14	45	401.37	4.366	91.93	0.309
15	45	481.12	5.77	83.38	0.276

where
A (deg): Aligned angle between test wires
B (N): Contact force in compression
C (mm²): Contact area between the two test wires
P (MPa): Contact pressure
R (mΩ): Electrical resistance in 10^{-3} Ohm

TABLE 14.11.3 ANOVA for investigating electrical contact resistance of cable wires.

Source	SS_{Adj}	DOF	MS_{Adj}	F-value	p-Value
Regression	4.6674	3	1.55581	78.79	0.0%
B	1.5665	1	1.56645	79.33	0.0%
C	1.0010	1	1.00102	50.69	0.0%
BC	0.7019	1	0.70190	35.55	0.0%
Error	0.2172	11	0.01975		
Subtotal	4.8846	14			
Grand average	—	1			
Total	—	15			

FIGURE 14.11.8 Response surface of contact electric resistance (10^{-3} Ω) as a function of contact force (N) and contact area (mm²).

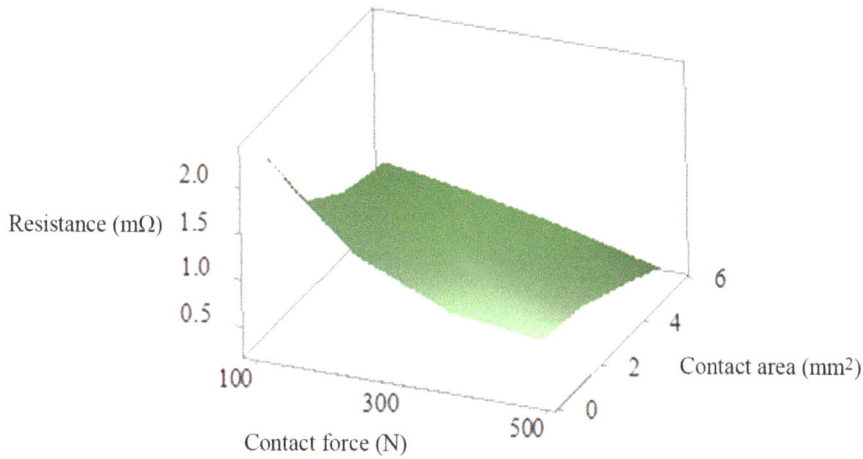

FIGURE 14.11.9 Deformation of 1 × 6 wire cable in a compacting process by FEA [Zeroukhi et al. 2014, Figure 11].

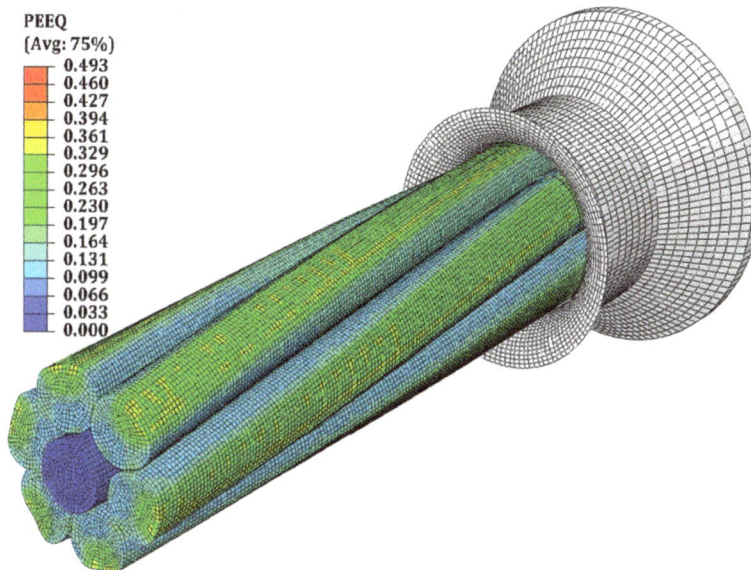

14.11.8. Adhesive Bonding between Electrodes and Current Collectors

There are three different types of failure modes in the adhesive bonding between a battery electrode and its adjacent current collector: (a) fracture of the adhesive layer, (b) fracture of metallic foils of the electrode, and (c) interfacial delamination. Delamination refers to the debonding between two layers of dissimilar materials (e.g., adhesive and current collector or adhesive and electrode), of which a weak stress singularity exists.

The concern about the adhesive joint between an electrode and its current collector is twofolds:

1. Thermomechanical Load: The adhesive of the electrode (cathode and anode) may experience a significant amount of thermomechanical stresses as subjected to repeated charge–discharge cycles. This could result in pronounced internal stresses in the adhesive layer, as well as the interface between the adhesive and its adjacent current collector or electrode [Luo et al. 2018]. This thermomechanical load in combination with external vibration loads may cause severe damage to the battery.

2. Electrical Contact Resistance: An imperfection or material variation (e.g., dendrite formation) of the electrode, current collector adhesive layer between them often yields an increasing electrical contact resistance [Yamada et al. 2020]. Should a delamination occur, a profound increase in electrical contact resistance is expected.

14.11.9. Vehicle Grounding System and the Related Grounded Electrodes

Vehicle grounding means the connection of earth ground to non-current-carrying conductive materials such as conduit, cable trays, junction boxes, enclosures, and electric motor frames, negative terminals of batteries, and even sheet metals in the vehicles. It means the connection of earth ground to the objects mentioned above either solidly or with a current-limiting device. The study involves grounding impedance assessment techniques for various vehicle grounding systems [Wen et al. 2021].

The neutral point of a circuit has to be grounded, too. The causes of a ground fault in a circuit may include loose connections, worn wire insulation, or miswiring that causes the hot wire to directly connect with another pathway to ground. A ground fault is an unwanted connection that may cause damage to parts and even systems.

14.11.10. Electrochemical Stability at Electrical Contact Surface

One main barrier of electric transport is oxidation. A thin film of metal oxide can be formed for most metallic alloys based on aluminum, copper, gold, nickel, platinum, silver, and tungsten, when exposed to the air for less than 10 min at room temperature.

14.11.11. Nano-Electric Connectors

An experimental investigation was reported to measure electrical and thermal conductivities of individual silver nanowires [Wang et al. 2018]. Theoretically, it has been known that miniaturized wires could lead to a reduction in both thermal and electrical conductivities when the cross-sectional dimension of a metallic nanowire is comparable to or even smaller than the electron mean free path (MFP).

TABLE 14.11.4 ANOVA for assessing electric conductivity of nanosilver wires [Wang et al. 2018].

Run	A	B	C	D	E	Y
1	2-P	6.49	4.88	93.2	298	2.10
2	2-P	6.49	4.88	93.2	196	3.0
3	2-P	6.49	4.88	93.2	123	4.2
4	2-P	6.49	4.88	93.2	298	2.36
5	2-P	6.49	4.88	93.2	196	3.8
6	2-P	6.49	4.88	93.2	123	4.5
7	2-P	27.23	27.23	227	298	1.26
8	4-P	15.49	14.67	97.0	298	3.51
9	4-P	15.49	14.67	97.0	196	6.0
10	4-P	15.49	14.67	97.0	123	8.0
11	4-P	15.49	14.67	97.0	298	3.04
12	4-P	15.49	14.67	97.0	196	5.5
13	4-P	15.49	14.67	97.0	123	7.0
14	4-P	11.0	9.0	107	298	3.5

where
A: Method of measurement
B (μm): Total wire length of measurement, i.e., length L in Figures 1 and 2 of Wang et al. [2013]
C (μm): Suspended length, i.e., length l_1 in Figures 1 and 2 of Wang et al. [2013]
D (nm): Wire diameter
E (°K): Temperature
Y (10^7 S/m): Electrical conductivity

The data analysis is carried out using Minitab (Stat → Regression → Regression → Fit Regression Model). The "most insignificant" factor or interaction that comes with the largest p-value is weeded out one by one until all p-values are less than 10%, which is here defined as the parting significance level based on the F-distribution. In light of the ANOVA presented in Table 14.11.5, the final predictive equation for the electrical conductivity based on the data available is

$$Y = 1.32 + 2.005\ B - 1.306\ C - 0.02668\ D - 0.001604\ B\ E$$

$$(R = 98.7\% \text{ and } R_{adj} = 98.1\%).$$

Thus, the influences of individual main effects of total wire length of measurement (factor B), suspended length (factor C), and wire diameter (factor D) on the electrical conductivity are statistically significant. The difference between P-2 and P-4 measurement methods (factor A) cannot be disclosed using the limited data available as given in Table 14.11.4, although these two measurement methods are reported to be different. A balanced rotatable DOE is needed to resolve the confusion.

The two-factor interaction between the temperature and total wire length of measurement has a dominant impact on the electrical conductivity according to Table 14.11.5, but the temperature variation has no individual main effect on the electrical conductivity by itself. This finding cannot be unveiled with a one-factor-at-a-time approach to the problem as presented by Cheng et al. [2015].

TABLE 14.11.5 ANOVA for assessing electric conductivity of nanosilver wires.

Source	SS$_{Adj}$	DOF	MS$_{Adj}$	F-value	p-Value
Regression	46.3345	4	11.5836	84.03	0.0%
B	1.6650	1	1.6650	12.08	0.7%
C	0.8658	1	0.8658	6.28	3.4%
D	3.8047	1	3.8047	27.60	0.1%
B E	22.4300	1	22.4300	162.72	0.0%
Error	1.2406	9	0.1378		
Lack of fit	0.1063	3	0.0354	0.19	90.1%
Pure error	1.1342	6	0.1890		
Subtotal	47.5751	13			
Grand average	—	1			
Total	—	14			

© SAE International.

14.12. Advanced Semiconductors

Wide-bandgap power semiconductor devices such as SiC and gallium nitride (GaN) have recently gained more and more applications in EVs because they are able to operate at a high temperature in a wide frequency range for the required missions [Davis 2009]. Nominal material properties of typical semiconductors are listed in Table 14.12.1. It would depend on the ability to exploit the unique advantages of SiC and GaN devices in light of their high-power capabilities running at a high reliability level.

TABLE 14.12.1 Nominal material properties of EV semiconductors at room temperature.

Property (crystals)	Si	GaAs	SiC-4H	GaN
Electronic properties				
Bandgap at 300°K (eV)	1.11	1.43	3.26	3.5
Breakdown electric field (10^6 V/m)	30	40	300	330
Electron mobility (m^2/V/s)	0.14	0.85	0.09	0.15
Hole mobility (m^2/V/s)	0.06	0.04	0.004	0.004
Dielectric constant	11.8	12.8	10.0‖/9.7⊥	10.4‖/9.5⊥
Saturation drift velocity (10^6 m/s)	0.1	0.2	0.27	0.27
Thermomechanical properties				
Density (g/cm^3)	2.33	5.32	4.6	6.15
Melting point (°C)	1410	1238	2830	1700
Hardness (Mohrs)	7	4–5	9.2–9.3	10.2
Young's modulus (GPa)	150	—	410	323
Tensile strength* (MPa)	170	—	34.5–138	—
Compressive strength (GPa)	3.33	—	3.9	10–15
Poisson's ratio	—	—	0.14	0.18
Specific heat (J/kg/°C)	710	327	580	—
Thermal conductivity* (W/m/°K)	92	55	120	130
Coeff. of thermal expansion (/°C)	7.5×10^{-6}	5.7×10^{-6}	4.7×10^{-6} (a) 4.3×10^{-6} (c)	5.6×10^{-6} (a) 3.2×10^{-6} (c)
Fracture toughness (MPa·m$^{1/2}$)	0.8	—	4.6	2
Maximum operating temperature (°C)	150	460	1200	400

© SAE International.

Notes:
Si, GaAs, and SiC(4H)/GaN are typical first, second, and third generations of semiconductors, respectively. Data are given for crystals in their standard state, except otherwise noted.

Both SiC and GaN devices are in favor for EVs and 5G devices, in which the voltage rating of different electrical systems may demand a wide range of voltage [Li et al. 2018]:

(a) SiC semiconductors are able to work at a voltage up to 1200 V as applied in IGBT, junction-gate field-effect transistors (JFETs), and MOSFETs. The capability of SiC-MOSFETs to operate at voltages between 650 and 1700 V makes them excellent for traction inverters, DC–DC converters, and onboard chargers (OBCs) [Frenzel 2019]. SiC exhibits the highest corrosion resistance among all the advanced ceramics. SiC inverters are crucial to increasing the charge speed of EVs, as electric motors migrate from a 400-V system to the more versatile 800-V system. Tesla is the first manufacturer to equip its EVs (Model 3) with SiC-MOSFETs.

(b) GaN semiconductors are able to work at a voltage up to 650 V as applied in high-electron-mobility transistors (HEMTs). GaN semiconductors are able to operate at higher temperatures while still maintaining their characteristics (up to 400°C). For the EV industry, power GaN providers have produced 650 V GaN-based OBCs and DC/DC converters. GaN devices can be also used for low-voltage applications ranging from 12 to 42 V mainly for electrical systems in on-ground vehicles. GaN is gaining increasing use in optoelectronics such as lidar systems for autonomous vehicles.

Both SiC and GaN can conduct current from a few amperes to tens of amperes at the peak voltage and thus fulfill higher energy conversion efficiency in vehicle-based battery chargers. SiC operates at higher voltages than GaN, but SiC requires a high gate drive voltage. Super junction MOSFETs are gradually being replaced by both GaN and SiC. It is important to identify applications in which the energy savings or other technical advantages achievable with SiC and GaN and sufficient to justify the cost increase. GaN semiconductors take off in automotive areas like economical OBCs and direct current (DC)–DC converters for EVs. SiC appears to be the favorite for more high-power applications.

14.12.1. Sintering of SiC Nanocrystalline Powders

Plasma pressure compaction (P^2C) involves resistive heating of the powder compact by passing an electric current through it [Srivatsan et al. 2002]. Rapid resistive heating in combination with high pressure results in full densification, as SiC grains grow. The four important process factors and their design levels are given in Table 14.12.2, as identified by Bothara et al. [2009]. Test results based on fractional factorial design 3^{4-2} are listed in Table 14.12.3, which shows that nanocrystalline powders may yield a theoretical density higher than 98% (i.e., 98% of their theoretical density of 3.26 g/cm^3) with proper control of these four process factors.

Because most ceramics including SiC are weak in fracture toughness, a further study for improving its fracture toughness is conducted here with the DOE data given in Bothara et al. [2009]. Note that the correlation between the hardness and fracture toughness is strong (95.9%), and so is the correlation between the density and fracture toughness (96.1%).

TABLE 14.12.2 Design parameters and levels for sintering study of nanocrystalline SiC fabricated using plasma pressure compaction [Bothara et al. 2009].

Factor	Level 1	Level 2	Level 3
A. Sintering temperature (°C)	1600	1700	1800
B. Holding time (min)	0	30	60
C. Applied pressure (MPa)	10	30	50
D. Heating rate (°C/min)	20	60	100

TABLE 14.12.3 Design matrix 3^{4-2} for sintering study of nanocrystalline SiC fabricated using plasma pressure compaction [Bothara et al. 2009].

Run	A	B	C	D	K_{IC}	H_V	ρ (%)	$K_{IC,pred}$
1	1600	0	10	20	1.9	2.1	62.5	1.78
2	1600	30	50	100	4.6	18.4	98.1	4.60
3	1600	60	30	60	4.6	21.1	99.8	4.43
4	1700	0	30	60	5.3	21.5	100	5.35
5	1700	30	50	20	5.4	23.1	99.8	5.32
6	1700	60	10	100	4.8	19.6	98.4	4.87
7	1800	0	30	100	4.8	18.2	99.1	4.97
8	1800	30	10	60	4.8	18.1	99.7	4.85
9	1800	60	50	20	4.4	19.3	99.0	4.57

Notes:
K_{IC} (MPa·m$^{1/2}$): Fracture toughness in opening mode by indentation (load = 1 kg)
H_V (GPa): Vickers hardness by indentation (ASTM E384)
ρ (%): Percentage of its theoretical density (3.26 g/cm^3)
K_{IC} (MPa·m$^{1/2}$): Fracture toughness in opening mode, predicted by DOE model

Multivariable linear regression analysis is employed to express the fracture toughness in terms of these four process factors. The analysis is conducted using ANOVA based on F-distribution as presented in Table 14.12.4. After removing the least insignificant factor or interaction one by one via the comparison of their resulting p-values, with a cutoff significance level (p-value) of 10%, an experimenter can obtain the following regression model for fracture toughness in opening mode:

$$K_{IC} = -311.1 + 0.3520\ A + 0.8509\ C - 0.000098\ A^2 + 0.000182\ B^2 - 0.001663\ C^2$$

$$- 0.000428\ A\ C \qquad (R = 99.8\% \text{ and } R_{adj} = 99.1\%). \tag{14.12.1}$$

The high model and adjusted model correlations, R and R_{adj}, prove that the regression model given above is effective. Since it is a 3^{4-2} design matrix with only nine treatments, the predictive correlation is not theoretically applicable. Nevertheless, after being adjusted for the predictive accuracy using the fracture toughness resulting from treatment 2, the constant (intercept) of Equation (14.12.2) is modified as

$$K_{IC} = -312.03 + 0.3520\ A + 0.8509\ C - 0.000098\ A^2 + 0.000182\ B^2 - 0.001663\ C^2$$

$$- 0.000428\ A\ C. \tag{14.12.2}$$

Given that $\partial K_{IC}/\partial B = 0.000364$, it can be seen that the fracture toughness increases steadily as the holding time persists. When the holding time is kept at 30 min (B = 30), Equation (14.12.2) reduces

$$K_{IC} = -311.8662 + 0.3520\ A + 0.8509\ C - 0.000098\ A^2 - 0.001663\ C^2 - 0.000428\ A\ C.$$

The response surface plot of fracture toughness in opening mode as a function of sintering temperature (factor A) and holding pressure (factor C) is shown in Figure 14.12.1. Setting $\partial K_{IC}/\partial A = 0$ and $\partial K_{IC}/\partial C = 0$ leads to the peak fracture toughness $K_{IC} = 5.63$, at A = 1720.8°C and C = 34.4 MPa, given that B = 30 min.

TABLE 14.12.4 ANOVA for sintering study of nanocrystalline SiC fabricated using plasma pressure compaction.

Source	SS_{Adj}	DOF	MS_{Adj}	F-value	p-Value
Regression	8.47234	6	1.41206	77.28	1.3%
A	2.14112	1	2.14112	117.18	0.8%
C	3.58032	1	3.58032	195.94	0.5%
A^2	1.93389	1	1.93389	105.84	0.9%
B^2	0.65207	1	0.65207	35.69	2.7%
C^2	0.86712	1	0.86712	47.45	2.0%
AC	2.82084	1	2.82084	154.38	0.6%
Error	0.03654	2	0.01827		
Subtotal	8.50889	8			
Grand average	—	1			
Total	—	9			

© SAE International.

FIGURE 14.12.1 Response surface plot of fracture toughness in opening mode as a function of sintering temperature and holding pressure.

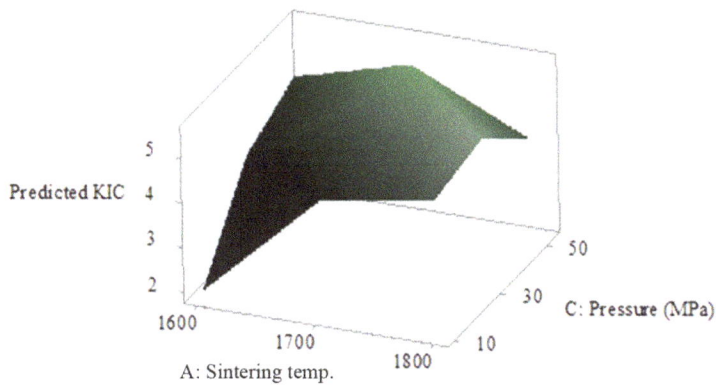

© SAE International.

References

ABB (1997), "Effects of AC Drives on Motor Insulation," ABB Technical Guide No. 102, ABB Industrial Systems, Inc.

Alzameli, K. et al. (2016), "Optimization of Thermal Profile Process in Assembly Line of Printed Circuit Boards (PCB) Using Design of Experiments," *Proceedings of the 2016 Int'l Conference on Industrial Engineering and Operations Management*, Detroit, MI, September 23–25, 2016.

Anderson, M. J. and Whitcomb, P. J. (2006), "Design of Experiments for Coatings," Revised August 8, 2006, https://www.researchgate.net/publication/253843410; uploaded by Mark Anderson on December 23, 2014; Retrieved December 26, 2018.

Aravamudhan, S. et al. (2016), "Multi-Faceted Approach to Minimizing Printed Circuit Board Warpage in Board Assembly Process," *Proceedings of SMTA Int'l*, Rosemont, IL, September 25–29, 2016.

Arora, T. G. and Aware, M. V. (2019), "Life Model for PWM Controlled Induction Motor Insulation Using Design of Experiments Method," *Electric Power Components and Systems*, 47(1-2), pp. 153-163.

Asahi, H. and Horikoshi, Y. (2019), *Molecular Beam Epitaxy*, John Wiley & Sons, New York.

Asghari, T. A. (2006), "PCB Thermal via Optimization Using Design of Experiment," *ITHERM 2006*, San Diego, CA, 2006, pp. 224-228.

Asif, A. A. (2019), "Design of Experiment Methods in High Speed Signal via Transition in Printed Circuit Board," *International Journal of Innovations in Engineering and Technology*, 13(4), pp. 1-7.

Bothara, M. G. et al. (2009), "Design of Experiment Approach for Sintering Study of Nanocrystalline SiC Fabricated Using Plasma Pressure Compaction," *Science of Sintering*, 41, pp. 125-133.

Brand, M. J., Schmidt, P. A., Zaeh, M. F., and Jossen, A. (2015), "Welding Techniques for Battery Cells and Resulting Electrical Contact Resistances," *Journal of Energy Storage*, 1, pp. 7-14.

Cao, B. et al. (2018), "How to Optimize Materials and Devices via Design of Experiments and Machine Learning: Demonstration Using Organic Photovoltaics," *ACS Nano*, 12, pp. 7434-7444.

Chansa-Ngavej, C. and Kasemsomporn, J. (2008), "Optimization of Reflow-Thermal Profile by Design of Experiments with Response Surface Methodology for Minimizing Solder-Ball Defects," *Proceedings of the 2008 IEEE ICMIT*, Bangkok, 2008, pp. 412-417.

Chen, G. et al. (2017), "A Critical Review of Constitutive Models for Solders in Electronic Packaging," *Advances in Mechanical Engineering*, 9(8), pp. 1-21.

Cheng, Z. et al. (2015), "Temperature Dependence of Electrical and Thermal Conduction in Single Silver Nanowire," *Scientific Reports*, 5, p. 10718.

Chiang, Y. J. (1996), "Characterizing Simple-Stranded Wire Cables under Axial Loading," *Finite Elements in Analysis and Design*, 24, pp. 49-66.

Chiang, Y. J. (2019), *Mechanics and Design for Product Life Prediction*, Chongqing University Press; ISBN: 978-7-5689-1917-6.

Czitrom, V. and Spagon, P. D. (1997), *Statistical Case Studies for Industrial Process Improvement*, ASA-SIAM Series on Statistics and Applied Probability, ASA-SIAM, Philadelphia, PA.

Das, A., Li, D., Williams, D., and Greenwood, D. (2018), "Joining Technologies for Automotive Battery Systems Manufacturing," *World Electric Vehicle Journal*, 9, p. 22.

Davis, S. (2009), "SiC and GaN Vie for Slice of the Electric Vehicle Pie," Power Electronics, November 1, 2009.

Du, R. and Robertson, P. (2017), "Cost Effective Grid-Connected Inverter for a Micro Combined Heat and Power System," *IEEE Transactions on Industrial Electronics*, 64 (7), pp. 5360-5367.

Eisenstadt, W. R. and Eo, Y. (1992), "S-Parameter-Based IC Interconnect Transmission Line Characterization," *IEEE Transactions on Components, Hybrids, and Manufacturing Technology*, 15, pp. 483-490.

Fan, J., Hardock, A., Rimolo-Donadio, R., Müller, S., Kwark, Y. H., and Schuster, C. (2014), "Signal Integrity: Efficient, Physics-Based via Modeling: Return Path, Impedance, and Stub Effect Control," *IEEE Electromagnetic Compatibility Magazine*, 3, pp. 76-84.

Frenzel, L. (2019), "Superior Gate Drivers Make SiC MOSFETs the Top High-Power Switching Devices," Electronics Design, August 22, 2019.

Gu, X., Ruehli, A. E., and Ritter, M. B. (2008), "Impedance Design for Multi-Layered Vias," *Proceedings of the 2008 IEEE-EPEP Electrical Performance of Electronic Packaging*, San Jose, CA, October 27–29, 2008, pp. 317-320.

Hall, S. H. and Heck, H. L. (2009), *Advanced Signal Integrity for High-Speed Digital Designs*, John Wiley & Sons, Inc., Hoboken, NJ.

Henini, M. (2018), *Molecular Beam Epitaxy*, 2nd Edition, Elsevier; ISBN: 978-0128121368.

Herman, M. A. and Sitter, H. (2013), *Molecular Beam Epitaxy*, Springer; ISBN: 3540190759.

Kalita, K. et al. (2022), "Parametric Optimization of CVD Process for DLC Thin Film Coatings: A Comparative Analysis," *Sådhanå*, 47, p. 57.

Kallmeyer, R. (2019), "Optimization of a Connector," *RDO-Journal*, 1, pp. 16-20.

Khanna, V. K. (2003), *Insulated Gate Bipolar Transistor (IGBT): Theory and Design*, Wiley-IEEE Press, Piscataway, NJ; ISBN: 978-0-471-23845-4.

Kim, M. and Kong, S. (2020), "Efficient Approach for Electrical Design and Analysis of High-Speed Interconnect in Integrated Circuit Packages," *Electronics*, 9(2), p. 303.

Kleijn, C. R. and Werner, C. (1993), *Modeling of Chemical Vapor Deposition of Tungsten Films*, Springer Basel AG, 139 pages; ISBN: 978-3034877435.

Krugh, M. et al. (2016), "Prediction of Defect Propensity for the Manual Assembly of Automotive Electrical Connectors," *Procedia Manufacturing*, 5, pp. 144-157.

Kumar, S. et al. (2013), "Stencil Printing Behavior of Lead-Free Sn-3Ag-0.5Cu Solder Paste for Wafer Level Bumping for Sub-100 μm Size Solder Bumps," *Metals and Materials International*, 19(5), pp. 1083-1090.

LaMeres, B. J. (1998), "Characterization of a Printed Circuit Board Via," MS Thesis, University of Colorado at Colorado Springs.

Lee, Y. and Lim, S. (2011), "Co-Optimization and Analysis of Signal, Power, and Thermal Interconnects in 3-D ICs," *IEEE Transactions on Computer-Aided Design of Integrated Circuits and Systems*, 30(11), pp. 1635-1648.

Lee, K. K. et al. (2000), "Using Statistical Experimental Design to Investigate the Role of the Initial Growth Conditions on GaN Epitaxial Films Grown by Molecular Beam Epitaxy," *Journal of Vacuum Science and Technology, B*, 18(3), pp. 1448–1452.

Lee, S. et al. (2013), "Characterization of Joint Quality in Ultrasonic Welding of Battery Tabs," *Journal of Manufacturing Science and Engineering*, 135(2), p. 021004.

Li, K., Evans, P., and Johnson, M. (2018), "SiC/GaN Power Semiconductor Devices: A Theoretical Comparison and Experimental Evaluation under Different Switching Conditions," *IET Electrical Systems in Transportation*, 8(1), pp. 3-11.

Liu, J. et al. (2011), *Reliability of Microtechnology Interconnects, Devices and Materials*, Springer, New York.

Lopez-Buedo, S. and Boemo, E. (2019), "Electronic Packaging Technologies," Universidad Autonoma de Madrid, Retrieved November 23, 2019, sergio.lopez-buedo@uam.es.

Luo, Y., Wu, G., and Liu, J. (2014), "Investigation of Temperature Effects on Voltage Endurance for Polyimide/Al_2O_3 Nanodielectrics," *IEEE Transactions on Dielectrics and Electrical Insulation*, 21(4), pp. 1824-1834.

Luo, H. et al. (2018), "Adhesion Strength of the Cathode in Lithium-Ion Batteries under Combined Tension/Shear Loadings," *RSC Advances*, 8, pp. 3996-4005.

Manias, S. (2016), *Power Electronics and Motor Drive Systems*, Academic Press, London, UK; ISBN: 978-0128117989.

Mohades-Kassai, A. and Soufi, H. R. (2000), "Design and Fabrication of High Accuracy GaAs Hall Effect Sensor Grown by Molecular Beam Epitaxy," *IEEE Proceedings of the 12th International Conference on Microelectronics*, Tehran, Iran, 2000.

Molex (2006), "Connector System Tolerance Analysis, Revision A," Molex Confidential, June 28, 2006.

Montgomery, D. C. et al. (2000), "Using Statistically Designed Experiments for Process Development and Improvement: An Application in Electronics Manufacturing," *Robotics and Computer Integrated Manufacturing*, 16, pp. 55-63.

Mraz, S. (2004), "Giving Electric Connectors the Slip: New Grease for Automotive Connectors Prevents Resistance Buildup," Machine Design, February 5, 2004.

Ogbonna, V. E. et al. (2020), "A Review on Polyimide Reinforced Nanocomposites for Mechanical, Thermal, and Electrical Insulation Application: Challenges and Recommendations for Future Improvement," *Polymer Bulletin*, 79, pp. 663–695.

Ohring, M. and Kasprzak, L. (2014), *Reliability and Failure of Electronic Materials and Devices*, 2nd Edition, Elsevier, 734 pages; ISBN: 978-0120885749.

Patel, R. and Metzler, M. (2017), "Optimization of Plasma Enhanced Chemical Vapor Deposition (PECVD) of Amorphous Silicon (aSi) Using Oxford Instruments System 100 with Taguchi L9 Based Design of Experiments (DOE)," Tool Data, Paper 43.

Polycarpou, A. C., Tirkas, P. A., and Balanis, C. A. (1997), "The Finite-Element Method for Modeling Circuits and Interconnects for Electronic Packaging," *IEEE Transactions of Microwave Theory and Techniques*, 45, pp. 1868-1874.

Qi, H., Osterman, M., and Pecht, M. (2009), "Design of Experiments for Board Level Solder Joint Reliability of PBGA Package under Various Manufacturing and Multiple Environmental Loading Conditions," *IEEE Transactions on Electronics Packaging Manufacturing*, 32(1), pp. 32-40.

Salvado, O. A. et al. (2017), "Evaluation of Printed-Circuit Boards Materials for High Temperature Operation," *HiTEN, IMAPS*, Cambridge, UK, July 2017.

Scholer, R. (2013), DC Charging and Standards for Plug-in Electric Vehicles, SAE 2013-01-1475.

Shanmugam, R. et al. (2016), "A Design of Experiments Investigation of the Effects of Synthesis Conditions on the Quality of CVD Graphene," *Materials Research Express*, 3, p. 125601.

Shoemaker, A. C., Tsui, K., and Wu, C. F. (1991), "Economical Experimentation Methods for Robust Design," *Technometrics*, 33(4), pp. 415-427.

Song, J., Koch, C, and Wang, L. (2012), "Correlation between Wear Resistance and Lifetime of Electrical Contacts," *Advances in Tribology*, 2012, Article ID 893145.

Srivatsan, T. S., Ravi, B. G., Petraroli, M., and Sudarshan, T. S. (2002), "The Microhardness and Microstructural Characteristics of Bulk Molybdenum Samples Obtained by Consolidating Nanopowders by Plasma Pressure Compaction," *International Journal of Refractory Metals and Hard Materials*, 20(3), pp. 181-186.

Szczepanski, M., Malec, D., Maussion, P., Petitgas, B., and Manfé, P. (2017), "Prediction of the Lifespan of Enameled Wires Used in Low Voltage Inverter-Fed Motors by Using the Design of Experiments (DoE)," *IEEE Transactions on Industry Applications*, 56(3).

Taheri, P. et al. (2011), "Investigating Electric Contact Resistance Losses in Lithium-Ion Battery Assembly for Hybrid and Electric Vehicles," *Journal of Power Sources*, 196, pp. 6525-6533.

Talukder, S. et al. (2022), "Analytical Modeling and Simulation of Electrical Contact Resistance for Elastic Rough Electrode Surface Contact Including Frictional Temperature Rise," *AIP Advances*, 12(2), p. 025204.

Tsai, T. N. (2008), "Modeling and Optimization of Stencil Printing Operations: A Comparison Study," *Computers & Industrial Engineering*, 54, pp. 374-389.

Tsai, T. N. (2012), "Thermal Parameters Optimization of a Reflow Soldering Profile in Printed Circuit Board Assembly: A Comparative Study," *Applied Soft Computing*, 12, pp. 2601-2613.

Uddin, G. M. et al. (2010), "Analysis of Molecular Beam Epitaxy Process for Growing Nanoscale Magnesium Oxide Films," *Journal of Manufacturing Science and Engineering*, 132(3), p. 030913.

Van Quach, M., Devnani, N. S., and Manley, R. B. (2008), "Device and Method for Reducing Cross-Talk in Differential Signal Conductor Pairs," U.S. Patent 7609125B2, Avago Technologies Enterprise IP Singapore Pte Ltd. System, April 17, 2008.

Wang, P., Wu G., Gao, B., Luo, Y., and Cao, K. (2013), "Study of Partial Discharge Characteristics at Repetitive Square Voltages Based on UHF Method," *Science China Technological Sciences*, 56(1), pp. 262-270.

Wang, J. et al. (2018), "Effect of Electrical Contact Resistance on Measurement of Thermal Conductivity and Wiedemann-Franz Law for Individual Metallic Nanowires," *Scientific Reports*, 8, p. 4862.

Wen, S.-W. et al. (2021), "Study of Ground Impedance Measurement Techniques for Vehicle Ground Systems," Retrieved November 20, 2021, https://www.artc.org.tw/.

Yamada, M. et al. (2020), "Review of the Design of Current Collectors for Improving the Battery Performance in Lithium-Ion and Post-Lithium-Ion Batteries," *Electrochem*, 2, pp. 124-159.

Zeroukhi, Y. et al. (2014), "Dependence of the Contact Resistance on the Design of Stranded Conductors," *Sensors*, 14, pp. 13925-13942.

Problems

P14.1: An experimental design was set up to explore the impact of three processing parameters, i.e., donor percentage by weight (factor A), total solution concentration (factor B), and bulk heterojunction (BHJ) spin-cast speed (factor C) on the power conversion efficiency (Y) of organic photovoltaics. The resulting data are given as follows [Cao et al. 2018]:

Run	a	b	c	A	B	C	Y ± S$_D$ (%)	N	H (nm)
1	−1	−1	0	20	20	1500	6.32 ± 0.38	5	73
2	1	−1	0	27	20	1500	7.21 ± 1.23	11	77
3	−1	1	0	20	25	1500	6.83 ± 0.48	6	126
4	1	1	0	27	25	1500	6.96 ± 0.42	6	131
5	−1	0.2	−1	20	23	1000	7.77 ± 2.25	4	109
6	1	0.2	−1	27	23	1000	6.87 ± 0.96	4	136
7	−1	0.2	1	20	23	2000	6.43 ± 1.22	8	76
8	1	0.2	1	27	23	2000	7.65 ± 1.84	7	88
9	0.4286	−1	−1	25	20	1000	7.43 ± 0.82	4	115
10	0.4286	1	−1	25	25	1000	6.88 ± 1.24	8	135
11	0.4286	−1	1	25	20	2000	7.32 ± 2.20	7	104
12	0.4286	1	1	25	25	2000	7.21 ± 2.24	8	126
13	0.4286	1	0	25	23	1500	7.40 ± 0.37	7	129

where
 A (%): Donor percentage by weight; a = (A − 23.5)/3.5
 B (%): Total solution concentration; b = (B − 22.5)/2.5
 C (rpm): BHJ spin-cast speed; c = (C − 1500)/500
 Y ± S$_D$ (%): Power conversion efficiency and its associated standard deviation
 N: Number of specimens, i.e., devices used in the corresponding treatment
 H (nm): Thickness of BHJ, which is a mixture of two or more phase-segregated materials that is the major part of a
   photovoltaic device

1. Find out the predictive equations for the power conversion efficiency, including linear terms, quadratic terms, and two-factor interactions.

2. Find out the predictive equations for the thickness of BHJ, including linear terms, quadratic terms, and two-factor interactions.

P14.2: An experimental design based on two-parameter Weibull statistics was set up to observe the lifespan of enameled wires (enamel for magnet wires: PEI base + PET overcoat) used in low-voltage inverter-fed motors [Szczepanski et al. 2017]. The shape factor (β) was assumed to be a constant, and the scale factor (η) obtained from accelerated life tests is listed as follows:

Run	A	B	C	A	B	C	η (h)
1	−1	−1	−1	2.0	5	30	7.25
1	−1	−1	1	2.0	5	100	3.29
1	−1	1	−1	2.0	5	30	5.00
1	−1	1	1	2.0	5	100	2.38
1	−1	−1	−1	2.25	10	30	3.00
1	−1	−1	1	2.25	10	100	1.31
1	−1	1	−1	2.25	10	30	2.04
1	−1	1	1	2.25	10	100	0.87

where
   A (kV): Voltage amplitude; coded variable a = (A − 2.125)/0.125
   B (kHz): Frequency; coded variable b = (A − 7.5)/2.5
   C (°C): Temperature; coded variable c = (C − 65)/35
   η (h): Scale factor of the anticipated 2-parameter Weibull statistics

How do these free factors affect the lifespan of the enamel of magnet wires?

P14.3: In a CVD process, an experiment based on an incomplete inscribed central composite design was set up to examine the impact of chamber pressure in kPa (factor A) and the ratio of gaseous reactants, H_2/WF_6 (factor B), on the uniformity in % (i.e., response U) and stress (i.e., response S) in MPa of deposition layers. Note that 1 standard temperature and pressure (STP) air pressure = 101.325 kPa at 20°C. The uniformity is estimated from sheet resistance measurements made at 49 different locations across a wafer. Data resulting from the experiment are given as follows [Czitrom and Spagon 1997]:

Run	A	b	A (kPa)	B (H_2/WF_6)	U (%)	S (MPa)
1	−0.707	−0.707	2.02	3.17	8.6	6.66
2	0.707	−0.707	9.18	3.17	3.4	7.58
3	−0.707	0.707	2.02	8.83	6.9	7.27
4	0.707	0.707	9.18	8.83	5.1	8.33
5	−1	0	0.5333	6	7.3	6.49
6	1	0	10.67	6	4.6	8.04
7	0	−1	5.60	2	6.3	7.16
8	0	1	5.60	10	5.4	8.19
9	0	0	5.60	6	6.2	7.78
10	0	0	5.60	6	6.4	7.69
11	0	0	5.60	6	5.0	7.90

Note that a = (A − 5.6)/5.066 and b = (B − 6)/4. They are coded variables of factors A and B, respectively.

(a) Which two treatments (runs) are missing in the above design matrix to make it a complete inscribed composite design?

(b) Derive the predictive equations for both the uniformity (U) and stress level (S). Use these two equations to make a U–S plot (dot plot), i.e., a Pareto frontier plot of two independent factors. The rule of "the smaller the better" fits for either uniformity U or stress S.

15

3D Printing and Additive Manufacturing

Three-dimensional printing, generally called additive manufacturing, has been transforming how products are designed and produced. 3D printing builds a 3D object one layer at a time from a computer-aided design (CAD) model, i.e., adding material to make a product instead of removing material. Besides the procedural parameters in a 3D printing operation, another major challenge is how to tune up physicochemical properties of materials applied and how to quantify the design parameters. It aims at reaching or exceeding the product quality of the counterparts that are produced with traditional manufacturing processes. Because of the inherent complexity, DOE turns out to be an excellent tool for exploiting the reliability of 3D-printed products.

15.1. Additive Manufacturing via 3D Printing

Additive manufacturing means to convert a 3D CAD model into a stereolithography (SLA or STL) model [Hull 1986], i.e., build job file, which is executable step by step usually on a 3D printing machine. Basically, it creates the solid parts in a "layer-upon-layer" manner with molten molecules or powders, followed by certain transformations such as curing, joining, and solidification. Commonly applied additive manufacturing techniques include:

1. Fused Deposition Modeling: The process is also called fused filament fabrication (FFF). Fused deposition modeling printers use a molten filament material (e.g., thermoplastics), which is heated to its melting point and then extruded, layer by layer, to create a 3D object. This is the most applied 3D printing method. It converts a 3D CAD model into a STL or SLA lattice surface model and then creates a solid part. Before an object can be printed, its CAD file is converted to a format that a 3D printer can follow. Z-axis is usually positioned to be perpendicular to the deposit surface, x-axis is in line with the traversing beam, and y-axis is in the plane and perpendicular to x-axis.

2. Powder Bed Fusion: This is to fuse small particulates/particles of plastic, metal, ceramic, or glass powders into a 3D object by a high-energy source. This includes selective laser sintering (SLS), selective laser melting (SLM), and electron beam melting (EBM). SLS is appropriate for complicated parts with intricate features. For example, aluminum alloy $AlSi_{10}Mg$, having a near-eutectic composition in the Al–Si phase diagram with $T_{melt} = 570°C$, is the most used aluminum alloy by laser melting techniques. EBM uses an electron beam to fuse raw materials in a vacuum and get them completely melted and formed into a 3D object.

3. Laminated Object Manufacturing (LOM): It is a sheet lamination technique. For example, laser cutting is used to make sheet metals, which can be bound together using USW. Material properties of a part made from sheet lamination depend on the bonding strength between laminae. The in-plane material properties of each lamina are typically the same as the feedstock used to make the part.

4. Photopolymerization: This is to use a vat of liquid photopolymer resin, out of which the model is constructed layer by layer.

In light of reducing cost and time to market in product development, this manufacturing technology utilizes on-demand digital design tools and offers tremendous manufacturing flexibility across most production industries.

Starting with technical prototyping, preproduction series, and short production series, the additive manufacturing based on 3D printing process is getting into more realistic manufacturing capacity. Materials used in additive 3D printing ranges from plastics, rubber, metals, ceramics, and composites. Advantages of 3D printing over traditional production processes include the following:

(a) No specialized facility needed, thus reducing tooling time and cost

(b) Economical to have a small production batch

(c) Versatile for most materials

(d) Quick to have a design change

(e) Rapid prototyping (RP)

(f) Waste reduction

(g) Short lead times for easy supply chain management and low inventories

These additive manufacturing techniques have raised interesting awareness from the production society as state-of-the-art technologies, by which engineers can build and adjust 3D physical objects quickly. One major challenge to 3D printing and additive manufacturing is how to tune up physicochemical properties of materials applied and how to quantify the design and operating parameters, aiming at reaching or exceeding the product quality of the counterparts produced with traditional manufacturing processes. Because of the inherent complexity, DOE turns out to be an excellent tool for exploiting the holistic reliability of 3D-printed products. Furthermore, porosities (not voids) can be used as a critical factor in the product design space with 3D printing. Note that porosities are defined as undesired empty spaces, while voids are designed empty spaces.

15.1.1. Additive Manufacturing of Plastic Parts

Plastics are the preferred materials for building parts with 3D printing techniques, especially for RP. Plastics that are frequently applied to mechanical parts using fused deposition modeling include ABS, PC, polycaprolactone (PCL), polyetheretherketone (PEEK), PEI, polylactic acid (PLA), polyphenylsulfone (PPSF), PS, PVA, and some thermoplastic elastomers (TPEs). In practice, automotive plastic parts made from additive manufacturing include side scuttles, interior trims, illuminated door sills, and light-emitting diode (LED) door projectors.

15.1.2. Additive Manufacturing of Metal Parts

With a single-layered 3D-printed pattern, screen printing technology has been widely used to form electronic parts such as conductors, resistors, and dielectrics since the mid-1960s, as a practice of 3D printing using metals. With a growing demand of denser packaging and a drive for higher pin count, SMT has evolved from standard SMT to ultrafine pitch SMT. For example, PCBs with 125-μm (5 mils) lines/spaces are cost-effective for a plastic BGA (PBGA), while PCBs with 100-μm (4 mils) lines/spaces or less are more cost-effective for the CSP of 1.0- and 0.8-mm pitch. Statistical DOE has been exploited to discover the effects of process parameters, including printing speed, squeegee hardness, squeegee pressure, and snap-off distance, on the refinement of screen printing [Pan et al. 1998]. The substrate used in their experiments is 50.8×50.8 mm ($2 \times 2''$) alumina (96%), and the paste is Ag/Pd conductor.

Additive metal 3D printing is also available to make 3D metallic parts that are hard to fabricate with material-subtracting technologies. These include aluminum-based alloys such as automotive transmission plates ($AlSi_9Cu_3$), automotive front-axle-differential housings and brackets, tire tread molds, tool steels, stainless steels, copper alloys, nickel alloys, and titanium alloys. Residual stresses cause major problems in 3D-printed metal parts such as warping, distortions, decreasing fatigue strength, oxidation, and even delamination. A higher material melt temperature for materials like titanium and stainless steel generally requires more energy, creating a process that rapidly heats up in a local area. A substantial temperature variation in either heating or cooling may generate local thermal stresses (strains) in areas lacking thermal dissipation (e.g., sharp edges), when the product is built layer by layer.

Instead of making metal parts directly from 3D printing, one may also cast metal parts using the plastic "lost foam" built with 3D printing patterns, in conjunction with the traditional investment casting process as an example. Lost-foam patterns are herein designed with 3D printing instead of being machined from a wax block.

15.1.3. Additive Manufacturing of Ceramics

There are five major processes for parts made of 3D printing ceramics, including 3D printing, binder jetting, powder sintering, nanoparticle jetting, and photopolymerization [digital light processing (DLP)]. 3D printing is able to produce ceramic parts with complex geometries that could not be fabricated through traditional injection molding techniques, and it can be also scaled up to a mass production volume required for automotive parts.

15.1.4. Additive Manufacturing of Composites

Composites are generally made of a matrix and reinforcements. 3D printing enables the fabrication of near-net-shaped complex 3D parts made of composite materials in a short time frame without expensive molds and machine tools, while dramatically improving the mechanical properties of parts. Additive manufacturing of composites can be divided into the following five categories:

1. Particulate Composites: Reinforcing particulates (particles) mixed with resin (matrix) can be used for making parts as is usually using resin only.
2. Laminates with Short Fibers: In light of shearing of resin in polymeric composites, alignment of short fibers and concentration of carbon fiber fraction in a resin solution can be controlled following a 3D printing procedure. For example, a direct ink writing (DIW) system is able to build complex composite parts with a raster width as low as 250 μm.
3. Making Preforms with Continuous Fibers: Preforms or fabrics can be built with 3D printing like a laminate.
4. Deposition onto Preforms (Woven Fabrics): 3D printing is performed directly onto a preform (textile) to make a fibrous-composite part.

5. Continuous Fiber-Reinforced Composites: Thermoplastic filaments and strings of continuous fibers are separately fed into the 3D printer. Fibers are impregnated with molten filaments prior to printing [Matsuzak et al. 2016]. The tensile strength along the fibers can be utilized significantly when reinforced with continuous fibers as shown in Figure 15.1.4.

Extending the applicability of 3D printing of polymers (plastics) to polymeric composites facilitates the structural performance required for both automotive and aerospace engineering components.

FIGURE 15.1.4   Tensile stress–strain curves of composites reinforced with continuous fibers loaded along the primary material axis: carbon-fiber-reinforced thermoplastics (CFRTPs) and jute-fiber-reinforced thermoplastics (JFRTPs) versus basic thermoplastic PLA [Matsuzak et al. 2016].

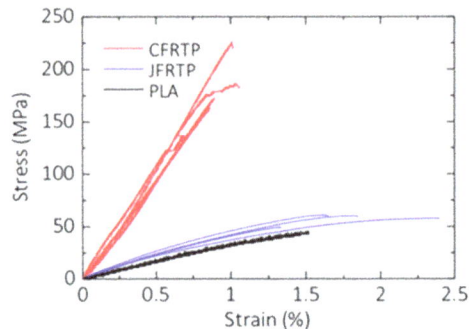

15.1.5. Design and Operating Factors of Fused Deposition Modeling

When parts are built with 3D printing, the microstructure of a part differs from that made using traditional manufacturing techniques and requires different standards to determine specific material properties. Because of its complexity involving both design and manufacturing in a 3D printing process, statistical DOE has gained its momentum in assessing the structural integrity [Galantucci et al. 2008, Nguyen et al. 2019, Renteria et al. 2019, Eguren et al. 2020] and off-line dimensional control [Sood et al. 2009, Galantucci et al. 2015] of parts. Major parameters of concern in the fused deposition modeling process are given as follows:

1. Layer Thickness (Layer Height): Building of a part begins with deposition of molten filament material that is called a "road," and the assembly of in-plane roads shall form a lamina layer. The layer thickness is the through-thickness dimension of deposited molten filament lamina, along the z-axis. The 3D printer of an additive manufacturing machine starts and stops the flow of the material while scanning to complete the layer. Another layer will then be deposited and bonded onto the previous layer. This process continues until the part is completed.

2. Air Gap (Air Space): It is the gap (spacing) between two deposited roads in the same layer. In general, a deposited road is used for defining the perimeter of the part to form a closed boundary, called shell, that will later be filled using a fill pattern. The density associated with a filling process is called fill density, and the fill style is called fill pattern. The fill density is 100% if the air gap is zero. A negative air gap is feasible but hard to control.

3. Raster Width (Extrusion Width): It is the width of deposited filament, i.e., width of a road.

4. Raster Angle (Infill Angle): It refers to the inclination of the infill roads (raster) with respect to the x-axis of the build plate. It is also called infill angle or internal fill angle that defines the orientation of a deposited layer.

5. Test Specimen Position: It identifies the part orientation on the build platform, in which geometries are manufactured horizontally or vertically. If the build direction is vertical, the interlayer thickness is stacked up out of planes 1–2 (Figure 15.1.5), i.e., along the 3-axis.

6. Nozzle Diameter: This is an operating factor, which is involved with the major process complexity subjected to rheological properties of the material that regulates the flow rate through process monitoring and control algorithms for improving printing resolution.

7. Nozzle Temperature (Extrusion Temperature): Input filaments are heated up to a certain temperature above the glass transition point (temperature) of the material applied. This is a major factor that controls the rheological properties of filament material [Turner and Gold 2015].

8. Printing Speed (Deposition Rate or Nozzle Speed): The moving speed of a nozzle is the linear speed of movement in the lamina x–y plane, referring to the attached build plate. A high printing speed may lead to more porosity and thus a loss of material strength.

9. Interlayer Cooling Time: It is defined as the time duration it takes from the moment when the nozzle begins the deposition of a layer up to the moment when it initiates the deposition of the next layer. It may have an influence on the interlayer bonding quality. The higher the diffusion temperature relative to the glass transition temperature (T_g), the faster the bonding process [Karimi 2011]. It is viable to keep the layer's temperature higher than its T_g. It means whether the interlayer cooling time is short enough for promoting the diffusion [Faes et al. 2016].

Major performance and quality concerns about the parts built with 3D printing are the dimensional control and stability, surface roughness, shrinkage, residual stress, material anisotropy, and interlayer stress (strain) intensities. Tensile and shear test specimens for identifying material properties and strengths are regulated by ASTM as shown in Figure 15.1.5. One major concern about occupational hazard with 3D printing is how to avoid operator's contact with volatile organic compounds and ultrafine particles.

FIGURE 15.1.5   Schematic drawings of tensile and shear test specimens with relevant dimensions (mm).

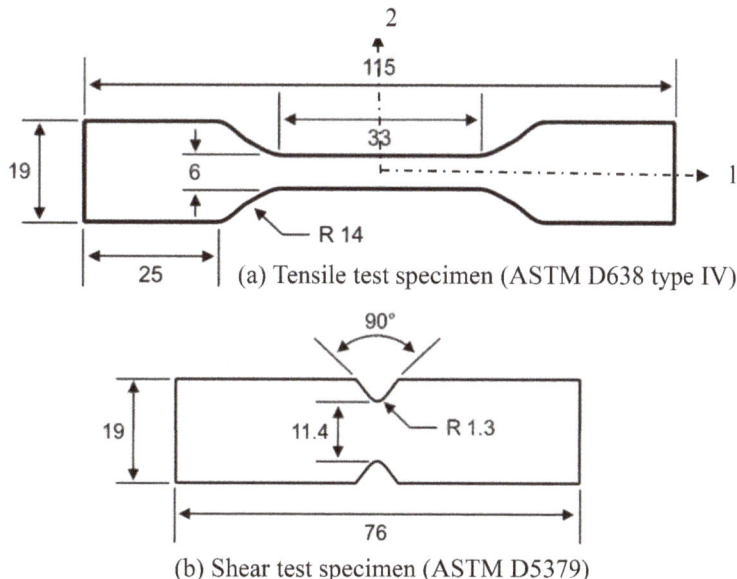

(a) Tensile test specimen (ASTM D638 type IV)

(b) Shear test specimen (ASTM D5379)

15.2. Rheology of 3D Printing

Rheological properties of each applied material in the 3D printing process include shear rate, dynamic viscosity, and after treatment. Initial operating settings for each submitted job carried out by a specific 3D printing machine are unique for meeting the rheological needs. For example, the following are default settings of a collected "MakerBot" replicator under the extruder panel category for making Lego-type polylactic (PLA) parts [Novoa and Flores 2019]:

Operating parameters	Setting
Filament diameter	1.77 mm
Filament retraction distance	1.0 mm
Filament retraction speed	50 mm/s
Filament restart speed	30 mm/s
Filament extra restart distance	0.1 mm
Extra restart speed	30 mm/s
Ooze distance	0.1 mm
Minimum ooze path length	0.1 mm

When the nozzle of a 3D printer moves across a gap without printing anything resulting in the filament being left in the hot nozzle, oozing occurs. Since the nozzle is set to extrude constantly, the filament melts enough to be printed during the retraction period. The part of the filament keeps getting hotter and runnier and may ooze out of the nozzle by itself. Ooze distance is a corresponding measure of the amount of oozed plastics used after the extrusion stops before the end of one move, while minimum ooze path length is to turn off the oozing on each short movement.

15.2.1. Shear Rate in Response to Applied Voltage

When the electric motor is on, the applied electric energy generates pressure to push the heated 3D printing material in the heating chamber in rheological state into the nozzle. The engineering performance of the rheological state is usually represented as a function of time and shear rate, as shown in Figure 15.2.1. The dynamic viscosity is consequently derived. The printed layer thickness depends on the pressure exerted and rheological properties of the melt. For example, when printing with a layer of small thickness, an elevated pressure is observed because of the pressure drop experienced between the nozzle and the previous layer [Coogan and Kazmer 2019].

FIGURE 15.2.1 Resultant pressures and shear rates at nozzle and heating chamber as voltage applied in a 3D printing process [Coogan and Kazmer 2019].

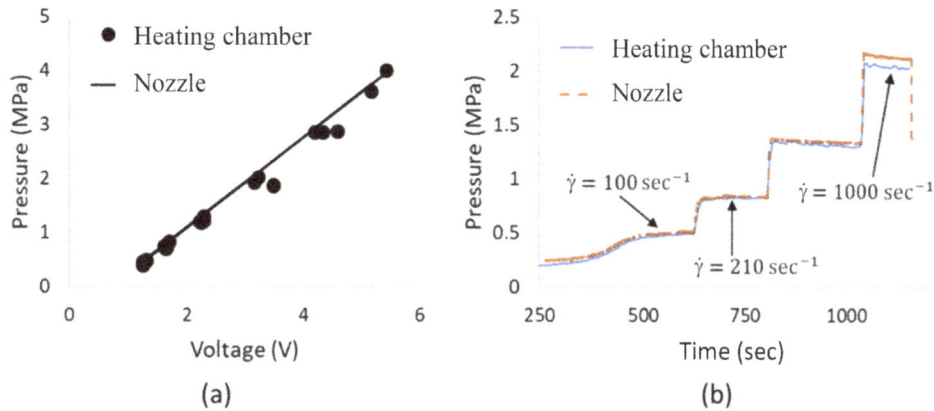

(a)   (b)

15.2.2. DOE for Dynamic Viscosity

Two different filament materials, i.e., PC with glass transition point $T_g = 112°C$ and high-impact PS with glass transition point $T_g = 98°C$, are used for a further understanding of the rheological behavior of a 3D printing process via an online measurement. Two operating factors, i.e., factor A: temperature (°C) and factor B: shear rate (sec^{-1}), of which their design levels are given in Table 15.2.1, are examined. Nine experimental tests based on a full factorial design 3^2 are conducted for each material, and their test results are shown in Table 15.2.2, including dynamic viscosities (Pa·s) at two locations—one at the nozzle and the other at the heating chamber.

TABLE 15.2.1 Process factors setup for 3D printing rheology [Coogan and Kazmer 2019].

Block (material)	Factor	Level (−)	Level (0)	Level (+1)
1 (PC)	A (°C): Temperature	250	270	290
1 (PC)	B (sec^{-1}): Shear rate	46	140	460
2 (PS, high impact)	A (°C): Temperature	220	250	275
2 (PS, high impact)	B (sec^{-1}): Shear rate	46	140	460

TABLE 15.2.2 Factorial design 2^3 for exploring rheology of a 3D printing process [Coogan and Kazmer 2019].

Run	Material	A (°C)	B (sec⁻¹)	μ_N (Pa·s)	μ_H (Pa·s)
1.1	PC	250	46	459	328
1.2	PC	250	140	428	311
1.3	PC	250	460	337	259
1.4	PC	270	46	282	249
1.5	PC	270	140	265	224
1.6	PC	270	460	216	157
1.7	PC	290	46	182	134
1.8	PC	290	140	154	118
1.9	PC	290	460	129	103
2.1	PS, H. impact	225	46	844	828
2.2	PS, H. impact	225	140	373	408
2.3	PS, H. impact	225	460	156	191
2.4	PS, H. impact	250	46	493	538
2.5	PS, H. impact	250	140	242	271
2.6	PS, H. impact	250	460	113	130
2.7	PS, H. impact	275	46	337	363
2.8	PS, H. impact	275	140	179	203
2.9	PS, H. impact	275	460	91	109

where
μ_N (Pa·s): Dynamic viscosity at the nozzle
μ_H (Pa·s): Dynamic viscosity at the heating chamber, right located above nozzle

Based on the ANOVA with the F-distribution, the overall contributions of individual factors and their interactions to the dynamic viscosities of the PC melt, with the given experimental conditions given in Table 15.2.1, are presented in Table 15.2.3. The analysis is summarized in the following two predictive equations:

$$\mu_{N,p} = 7211 - 43.95 \, A - 1.382 \, B + 0.0679 \, A^2 + 0.004420 \, A \, B$$

$$(R = 99.9\%, R_{adj} = 99.7\%, \text{ and } R_{pred} = 98.2\%), \tag{15.2.1}$$

and $$\mu_{H,p} = 1463.4 - 4.525 \, A - 0.1507 \, B$$

$$(R = 99.0\%, R_{adj} = 98.6\%, \text{ and } R_{pred} = 97.1\%). \tag{15.2.2}$$

Both equations are effective with high correlations, including unadjusted model correlation, adjusted model correlation, and predictive correlation, while the model adequacy is evidenced by the insignificance of lack of fit. Response surfaces corresponding to the above two equations are presented in Figure 15.2.2. The dynamic viscosity of PC material at the nozzle (μ_N) varies nonlinearly with respect to the temperature but linearly with respect to the shear rate. The two-factor interaction has also an impact on the dynamic viscosity at the nozzle. Nevertheless, the dynamic viscosity of PC material at the heating chamber varies linearly with respect to both temperature and shear rate.

TABLE 15.2.3 ANOVA for rheology of PC parts built with 3D printing.

Source	SS$_{adj}$	DOF	MS$_{adj}$	F$_{u,v}$	p-Value (%)
Regression	21.8397	4	5.45992	154.14	0.00
A	0.2865	1	0.28648	8.09	0.6
B	1.3416	1	1.34156	37.87	0.0
A^2	0.3897	1	0.38975	11.00	0.1
AB	0.7791	1	0.77910	22.00	0.0
Error	2.6920	76	0.03542		
Lack of fit	0.1045	4	0.02612	0.73	57.7%
Pure error	2.5875	72	0.03594		
Subtotal	24.5317	80			

FIGURE 15.2.2 Dynamic viscosities of PC as functions of temperature and shear rate.

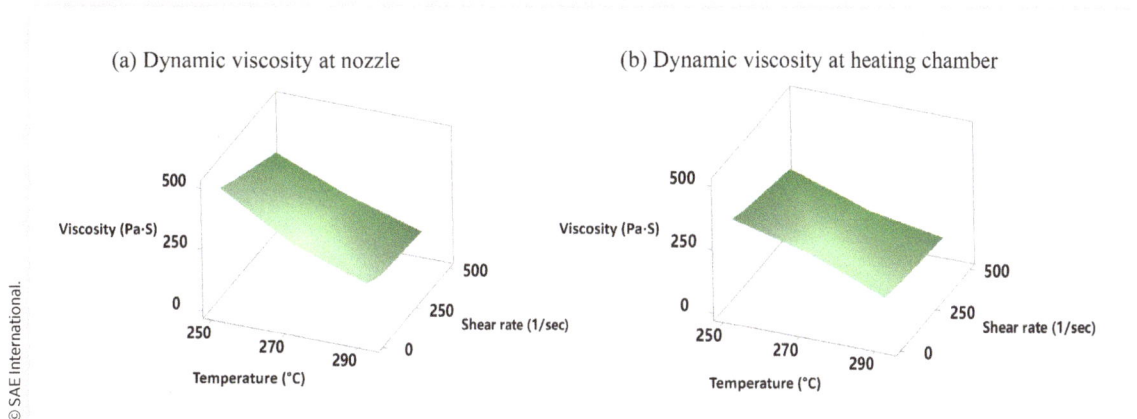

(a) Dynamic viscosity at nozzle

(b) Dynamic viscosity at heating chamber

By the same token, the overall contribution of individual factors and their interactions to the dynamic viscosities of high-impact PS, with the given experimental conditions given in Table 15.2.1, can be summarized in the following two predictive equations:

$$\mu_{N,p} = 2986 - 8.97\,A - 8.81\,B + 0.00645\,B^2 + 0.01795\,A\,B$$

$$(R = 97.7\%, R_{adj} = 95.3\%, \text{ and } R_{pred} = 85.6\%), \qquad (15.2.3)$$

and

$$\mu_{H,p} = 2854 - 8.40\,A - 8.08\,B + 0.00612\,B^2 + 0.01574\,A\,B$$

$$(R = 98.5\%, R_{adj} = 97.1\%, \text{ and } R_{pred} = 90.7\%). \qquad (15.2.4)$$

Both equations are effective with high correlations, including unadjusted model correlation, adjusted model correlation, and predictive correlation, while the model adequacy is evidenced by the insignificance of lack of fit. The dynamic viscosities of high-impact PS material at the nozzle (μ_N) and at the heating chamber (μ_H) resemble each other, as shown in Figure 15.2.3. Both respond quadratically with respect to the shear rate (factor B) but linearly with respect to temperature (factor A). The two-factor interaction also has a significant impact on the dynamic viscosity.

FIGURE 15.2.3   Dynamic viscosities of high-impact PS as functions of temperature and shear rate.

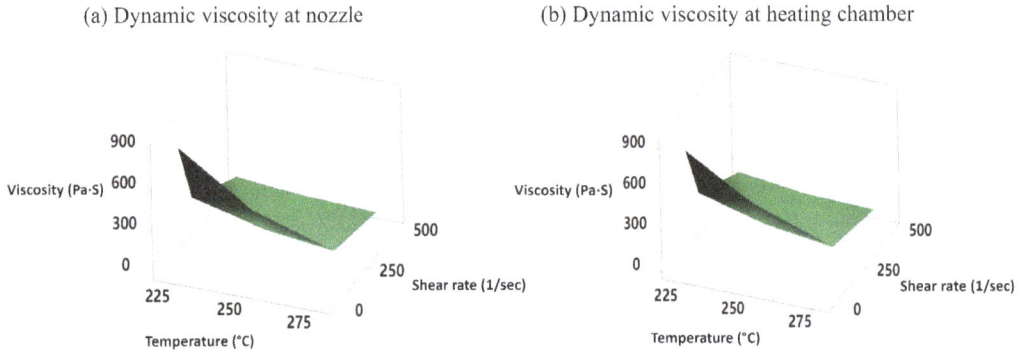

(a) Dynamic viscosity at nozzle

(b) Dynamic viscosity at heating chamber

15.2.3. Voids in 3D-Printed Parts

Voids are inherent in 3D-printed structures, and they are a key reliability factor to various applications, as evidenced in Table 15.2.4 [Wang et al. 2019]. Generation of voids may go beyond the theory of existing rheological theory that is exploited via continuum mechanics. With the applied PLA material, it shows that more than 99% of the voids are of the size less than 200 µm (0.2 mm). Both modulus of elasticity in tension E (GPa) and ultimate tensile strength σ_{uts} (MPa) of the laminates constructed increase with decreasing volume fraction of voids. A narrower raster width (e.g., 240 µm versus 480 µm) based on the given operating conditions tends to yield more voids and thus lowers both the modulus of elasticity and ultimate tensile strength.

TABLE 15.2.4 Influences of voids on mechanical properties of 3D-printed parts [Wang et al. 2019].

Run	Laminate	Raster width (µm)	Voids (%)	E (GPa)	σ_{uts}, ε_{uts} (MPa, %)
1	0°	480	4.26	E_{11} = 3.17	$\sigma_{11, uts}$ = 60.3, $\varepsilon_{11, uts}$ = 1.8%
2	90°	480	4.81	E_{22} = 2.97	$\sigma_{22, uts}$ = 55.9, $\varepsilon_{22, uts}$ = 1.9%
3	±45°	480	4.88	$E_{\pm45°}$ = 3.086	$\sigma_{\pm45°, uts}$ = 59.2, $\varepsilon_{\pm45°, uts}$ = 2.2%
4	0°/90°	480	5.84	$E_{0°/90°}$ = 2.809	$\sigma_{0°/90°, uts}$ = 56.1, $\varepsilon_{0°/90°, uts}$ = 2.1%
5	0°/90°	240	6.32	$E_{0°/90°}$ = 2.697	$\sigma_{0°/90°, uts}$ = 54.4, $\varepsilon_{0°/90°, uts}$ = 2.3%

Notes:
E (GPa): Young's modulus
σ_{uts} (MPa): Ultimate tensile strength
ε_{uts}: Ultimate tensile strain

15.3. Dimensional Integrity

3D printing imposes inherent manufacturing challenges of tolerance control that one may achieve on an "as built" part [Sood et al. 2009, Turner and Gold 2015]. The optimization of process parameters against dimensional accuracy is a major work task. Dimensional accuracy for both outer edges and inner slots is to be examined using the DOE.

15.3.1. Dimensional Accuracy of Outer Edges

An experimental design was done by Galantucci et al. [2015] to study ABS using an industrial Stratasys fused deposition modeling 3000 3D printer (Stratasys Inc., US). Raw ABS sticks, having a diameter of 3 mm, are fused in an extruder at 240°C. Pressure is applied to the feeding system to control the extruded filament diameter, which may range from 0.5 to 3 mm. The filament is ejected through the nozzle and deposited in layers onto a simple platform, which is covered by a cardboard moistened with dimethyl ketone (acetone) to allow a good adhesion of the first layer of ABS to the cardboard. The goal is to accurately form an 18 mm × 18 mm × 8 mm (length × width × height) block. The impact of three operating parameters on the dimensional accuracy is of great interest, and their design levels are listed in Table 15.3.1. Test results are given in Table 15.3.2.

TABLE 15.3.1 Design factors for dimensional control of 3D printing [Galantucci et al. 2015].

Factor (μm)	Level (−)	Level (+)	Variate (−1, 1)
A: Slice height	178	254	a = (A − 216)/38
B: Raster height	304	729	b = (B − 516.5)/212.5
C: Nozzle diameter	254	305	c = (C − 229.5)/25.5

Reprinted with permission. © Elsevier.

TABLE 15.3.2 Factorial design 2^3 for dimensional control of 3D printing [Galantucci et al. 2015].

Run	A	B	C	H (μm)	L (μm)	W (μm)
1	178	304	254	8150	17,900	17,920
2	254	304	254	8250	17,990	17,970
3	178	729	254	8070	18,330	18,050
4	254	729	254	8100	17,900	18,610
5	178	304	305	8050	17,870	17,920
6	254	304	305	8090	17,960	17,870
7	178	729	305	8270	17,910	17,910
8	254	729	305	8150	17,890	17,840

Based on the DOE with t-distribution, as shown in Table 15.3.3(i), the contribution of individual factors and their interactions to the inaccuracy of height (δL), along the z-axis, can be summarized using the following predictive equation:

$$\delta H_p = 141.25 + 6.25\,a + 6.25\,b - 1.25\,c$$

$$- 28.75\,ab - 26.25\,ac + 63.75\,bc - 11.25\,abc. \qquad (15.3.1)$$

It can be concluded that all the main effects of individual factors A (slice height), B (raster height), and C (nozzle diameter) are weak, but each of the two-factor interactions, i.e., AB, AC, or BC, has a practical effect on the height ("stackup" of deposited filaments). Interestingly, their three-factor interaction is also a significant effect.

TABLE 15.3.3 Factorial analysis for dimensional control in 3D printing.

(i) δH:

Run	a	b	c	ab	ac	bc	abc	δH (μm)
1	−1	−1	−1	1	1	1	−1	150 (=8150 − 8000)
2	1	−1	−1	−1	−1	1	1	250 (=8250 − 8000)
3	−1	1	−1	−1	1	−1	1	070 (=8070 − 8000)
4	1	1	−1	1	−1	−1	−1	100 (=8100 − 8000)
5	−1	−1	1	1	−1	−1	1	050 (=8050 − 8000)
6	1	−1	1	−1	1	−1	−1	090 (=8090 − 8000)
7	−1	1	1	−1	−1	1	−1	270 (=8270 − 8000)
8	1	1	1	1	1	1	1	150 (=8150 − 8000)
Contrast	12.5	12.5	−2.5	−57.5	−52.5	127.5	−22.5	
Effect	6.25	6.25	−1.25	−28.75	−26.25	63.75	−11.25	141.25 (average)

(ii) δL:

Run	a	b	c	ab	ac	bc	abc	δL (μm)
1	−1	−1	−1	1	1	1	−1	−100 (=17,900 − 18,000)
2	1	−1	−1	−1	−1	1	1	−10 (=17,990 − 18,000)
3	−1	1	−1	−1	1	−1	1	330 (=18,330 − 18,000)
4	1	1	−1	1	−1	−1	−1	−100 (=17,900 − 18,000)
5	−1	−1	1	1	−1	−1	1	−130 (=17,870 − 18,000)
6	1	−1	1	−1	1	−1	−1	−40 (=17,960 − 18,000)
7	−1	1	1	−1	−1	1	−1	−90 (=17,910 − 18,000)
8	1	1	1	1	1	1	1	−110 (=17,890 − 18,000)
Contrast	−42.5	52.5	−147.5	−132.5	127.5	−117.5	127.5	
Effect	−21.25	26.25	−73.75	−66.25	63.75	−58.75	63.75	−43.75 (average)

(iii) δW:

Run	a	b	c	ab	ac	bc	abc	δW (μm)
1	−1	−1	−1	1	1	1	−1	−80 (=17,920 − 18,000)
2	1	−1	−1	−1	−1	1	1	−30 (=17,970 − 18,000)
3	−1	1	−1	−1	1	−1	1	50 (=18,050 − 18,000)
4	1	1	−1	1	−1	−1	−1	610 (=18,610 − 18,000)
5	−1	−1	1	1	−1	−1	1	−80 (=17,920 − 18,000)
6	1	−1	1	−1	1	−1	−1	−130 (=17,870 − 18,000)
7	−1	1	1	−1	−1	1	−1	−90 (=17,910 − 18,000)
8	1	1	1	1	1	1	1	−160 (=17,840 − 18,000)
Contrast	122.5	182.5	−252.5	122.5	−182.5	−202.5	−132.5	
Effect	61.25	91.25	−126.3	61.25	−91.25	−101.3	−66.25	11.25 (average)

By the same token, one may formulate a predictive equation for the inaccuracy of length (δL) in the longitudinal direction according to Table 15.3.3(ii) as follows:

$$\delta L_p = -43.75 - 21.25 \, a + 26.25 \, b - 73.75 \, c$$

$$- 66.25 \, a \, b - 63.75 \, a \, c - 58.75 \, b \, c - 63.75 \, a \, b \, c. \tag{15.3.2}$$

It can be concluded that all main effects of individual factors A (slice height), B (raster height), and C (nozzle diameter), two-factor interactions (i.e., AB, AC, and BC), and the three-factor interaction (ABC) are statistically significant, having influences on the inaccuracy of length.

By the same token once again, one may also formulate a predictive equation for the inaccuracy of width (δW) in the transverse direction according to Table 15.3.3(iii) as follows:

$$\delta W_p = 11.25 + 61.25 \, a + 91.25 \, b - 126.25 \, c$$

$$+ 61.25 \, a \, b - 91.25 \, a \, c - 101.25 \, b \, c - 66.25 \, a \, b \, c. \tag{15.3.3}$$

Again, all main effects of individual factors A (slice height), B (raster height), and C (nozzle diameter), two-factor interactions (i.e., AB, AC, and BC), and the three-factor interaction (ABC) are statistically significant, having influences on the inaccuracy of width.

15.3.2. Dimensional Accuracy of Slots

A DOE is utilized to identify process parameters that may potentially affect the width of a rectangular slot (clearance) in a printed part for assembly [Luthria 2016]. Process parameters and their design levels are given in Table 15.3.4. Note that a slot angle is the raster (road) angle relative to the longitudinal axis of the slot. The experimental design matrix and test results are listed in Table 15.3.5. Slot width Y is denoted by the average of measured widths at four different locations along the longitudinal axis, i.e., Y_1, Y_2, Y_3, and Y_4 listed in the table, while factor B is the desired (target) value of the slot width as designated in the CAD model. The objective of the study is how to reduce the error, i.e., $E = Y - B$.

TABLE 15.3.4 Design factors for assembly of 3D-printed parts [Luthria 2016].

Factor	Level (1)	Level (2)	Level (3)	Level (4)
A: Slot angle (deg)	0	30	60	90
B: Slot width (mm)	0.5	0.75	1.0	—
C: Layer thickness (µm)	100	200	300	—
D: Slot length, fixed (mm)	35	—	—	—

TABLE 15.3.5 Design matrix for assembly of 3D-printed parts [Luthria 2016].

Run	Block	A	B	C	E = Y − B	Y	Y_1	Y_2	Y_3	Y_4
1	8	0	0.5	100	−0.470	0.03	0.120	0.000	0.000	0.000
2	8	0	0.5	100	−0.4625	0.0375	0.150	0.000	0.000	0.000
3	8	30	0.5	100	0.122	0.622	0.728	0.642	0.578	0.540
4	8	90	0.5	100	0.137	0.637	0.642	0.652	0.652	0.602
5	2	0	0.75	100	−0.6455	0.1045	0.278	0.140	0.000	0.000
6	2	30	0.75	100	0.02925	0.7793	0.711	0.825	0.759	0.822
7	2	60	0.75	100	0.005	0.755	0.767	0.748	0.707	0.798
8	2	90	0.75	100	0.1335	0.8835	0.872	0.908	0.872	0.882
9	4	0	1	100	−0.723	0.177	0.477	0.000	0.000	0.631
10	4	60	1	100	0.09975	1.0998	1.083	1.062	1.136	1.118
11	4	90	1	100	0.11075	1.1108	1.073	1.120	1.120	1.130
12	4	90	1	100	0.135	1.135	1.166	1.145	1.120	1.109
13	5	0	1	100	−0.739	0.261	0.615	0.114	0.000	0.315
14	5	30	1	100	−0.047	0.953	0.931	0.984	1.015	0.882
15	5	60	1	100	0.0595	0.0595	1.061	1.018	1.071	1.088
16	5	90	1	100	0.1145	1.1145	1.130	1.068	1.130	1.130
17	9	0	0.5	200	−0.475	0.025	0.100	0.000	0.000	0.000
18	9	30	0.5	200	−0.0555	0.4445	0.459	0.455	0.476	0.388
19	9	60	0.5	200	0.03825	0.5383	0.561	0.521	0.554	0.517
20	9	90	0.5	200	0.00525	0.5053	0.485	0.530	0.495	0.511
21	6	0	1	200	−0.5415	0.4585	0.615	0.159	0.418	0.642
22	6	0	1	200	−0.5435	0.4565	0.755	0.112	0.371	0.588
23	6	30	1	200	−0.08275	0.9173	0.915	0.900	0.977	0.877
24	6	90	1	200	−0.0415	0.9585	0.940	0.952	0.967	0.975
25	3	0	0.5	300	−0.271	0.229	0.293	0.000	0.245	0.378
26	3	0	0.5	300	−0.29025	0.2098	0.375	0.235	0.000	0.229
27	3	90	0.5	300	−1.0775	0.3923	0.384	0.397	0.386	0.402
28	3	90	0.5	300	−0.087	0.413	0.423	0.397	0.432	0.400
29	7	0	0.75	300	−0.541	0.209	0.410	0.240	0.000	0.186
30	7	0	0.75	300	−0.563	0.187	0.395	0.147	0.000	0.206
31	7	60	0.75	300	−0.00475	0.7453	0.740	0.721	0.767	0.753
32	7	90	0.75	300	−0.07975	0.6703	0.619	0.650	0.650	0.762
33	1	0	1	300	−0.6685	0.3315	0.634	0.224	0.000	0.468
34	1	0	1	300	−0.6825	0.3175	0.573	0.191	0.000	0.506
35	1	90	1	300	−0.069	0.931	0.869	0.904	0.940	1.011
36	1	90	1	300	−0.102	0.898	0.832	0.849	0.867	1.044

According to the ANOVA with F-distribution as exhibited in Table 15.3.6, the overall contribution of individual factors and their interactions to the dimensional error of the slot width as built with the given experimental conditions given in Table 15.3.4 can be summarized using the following predictive equation:

$$E_p = -0.1681 + 0.01699\ A - 0.4795\ B - 0.000144\ A^2$$

$$+ 0.00549\ A\ B - 0.000011\ A\ C$$

$$(R = 97.2\%,\ R_{adj} = 96.7\%,\ \text{and}\ R_{pred} = 96.2\%). \tag{15.3.4}$$

The error equation given above is validated by the three different correlations as given above with a statistical significance level of at least 5% for each individual component (Table 15.3.6).

TABLE 15.3.6 ANOVA for dimensional accuracy of a part slot built with 3D printing.

Source	SS_{adj}	DOF	MS_{adj}	$F_{u,v}$	p-Value
Regression	2.92665	5	0.585331	102.23	0.000
A	0.47168	1	0.471682	82.38	0.000
AB	0.07883	1	0.078832	13.77	0.001
A²	0.46482	1	0.464823	81.18	0.000
B	0.18437	1	0.184366	32.20	0.000
AC	0.11552	1	0.115520	20.18	0.000
Error	0.17177	30	0.005726		
Lack of fit	0.08311	14	0.005936	1.07	0.443
Pure error	0.08867	16	0.005542		
Subtotal	3.09843	35			
Grand average	—	1			
Total	—	36			

In light of Equation (15.3.4), the dimensional error of 3D printing varies nonlinearly with respect to the slot angle (factor A) but linearly with respect to the desired slot width (factor B). Nevertheless, two two-factor interactions, i.e., AB and AC, are also significant. The main effect of layer thickness (factor C) is not statistically significant. Let layer thickness C = 100 μm, and the response surface of dimensional error as a function of slot angle and desired slot width is depicted in Figure 15.3.2. It shows that the effect of slot orientation is statistically significant and has a dominant influence on the width of the 3D-printed slot.

FIGURE 15.3.2   Dimensional error as a function of slot angle and width.

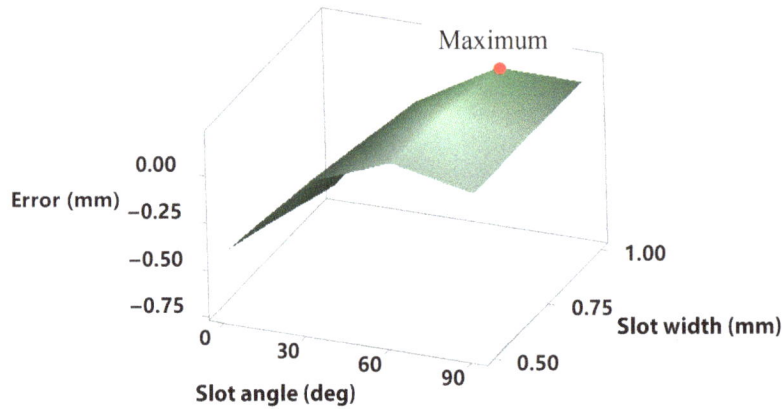

In search of the minimum error, one may take a differentiation of E_p against factor A:

$$dE_p / dA = 0.01699 - 0.000288 \, A + 0.00549 \, B - 0.000011 \, C. \tag{15.3.5}$$

Since $d^2E_p/dA^2 = -0.000288$, setting $dE_p/dA = 0$ will yield the slot angle that maximizes the error,

$$A = 58.99306 + 19.0625 \, B - 0.038194 \, C. \tag{15.3.6}$$

For example, B = 1 mm (slot width) and C = 100 μm (layer thickness), and then, A = 74.23°. For an expected "zero" error, one may set $E_p = 0$, and then, Equation (15.3.4) reduces to

$$- 0.1681 + 0.01699 \, A - 0.4795 \, B - 0.000144 \, A^2 + 0.00549 \, A \, B - 0.000011 \, A \, C = 0.$$

With B = 1 mm and C = 100 μm, A = 42.39° leads to a zero error.

15.4. **Surface Roughness**

3D-printed surfaces of parts made using 3D printing techniques tend to be rougher compared to conventional processes. The surface roughness of parts built with 3D printing techniques may vary significantly with layer orientation [Taufik and Jain 2017]. The surface roughness of parts built with 3D printing has also been of concern [Raol et al. 2014], as a rough surface shortens the part strength. As exhibited in Figure 15.4.1, it shows how a selective laser melted (SLM) Inconel 778 (a nickel-based alloy) part built with fused deposition modeling exhibits a lower fatigue life, as compared with that made out of wrought bar stock. Although a reasonable aftertreatment process such as electropolishing, chemical milling, and machining may improve the fatigue life of this aerospace alloy, it is not an inexpensive operation.

| FIGURE 15.4.1 | Influence of surface roughness on the fatigue life of SLA Inconel 778 Bar [MMPDS-08: 2013]. |

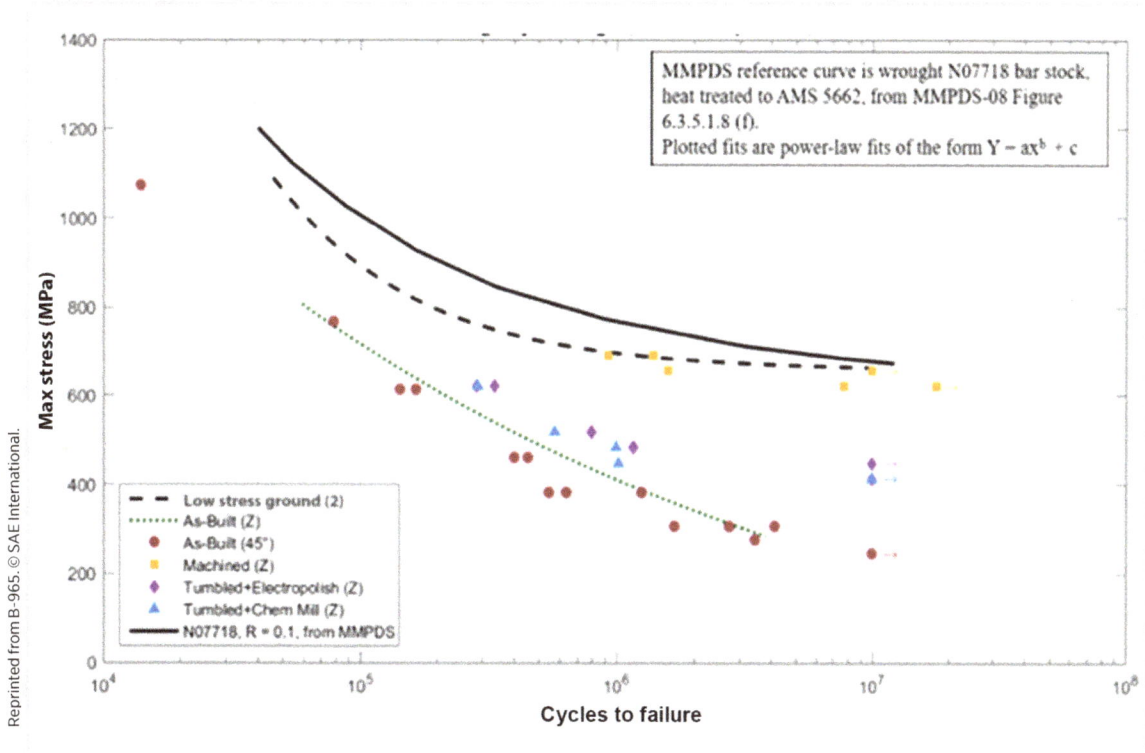

15.4.1. Improving Surface Roughness by Adjusting Process Parameters

The relationship of three process parameters, as listed in Table 15.4.1, to the surface roughness of plastic parts built with fused deposition modeling is investigated by Wu et al. [2019]. Full factorial design 3^3 is used to explore the influence of these parameters on the surface roughness as shown in Table 15.4.2, and three replications are taken for each treatment for examining the lack of fit.

TABLE 15.4.1 Design factors for dimensional control of 3D printing [Wu et al. 2019].

Factor	Level −1	Level 0	Level 1	Coded variable
A: Print speed, relative	0.85	1	1.15	c = (C − 1)/0.15
B: Layer thickness (μm)	200	250	300	a = (A − 250)/25
C: Nozzle temperature (°C)	210	220	230	b = (B − 220)/10

Reprinted with permission. © Taylor & Francis Group.

TABLE 15.4.2 Factorial design 3^3 for surface roughness of parts by 3D printing [Wu et al. 2019].

Treatment (run)	a	b	c	A	B	C	Y (μm)
1 (16)	−1	−1	−1	0.85	200	210	4.414, 4.476, 4.589
2 (18)	−1	−1	0	0.85	200	220	4.750, 4.888, 5.038
3 (17)	−1	−1	1	0.85	200	230	5.619, 5.342, 5.907
4 (13)	−1	0	−1	0.85	250	210	5.184, 5.258, 5.238
5 (15)	−1	0	0	0.85	250	220	5.498, 5.421, 5.427
6 (14)	−1	0	1	0.85	250	230	5.359, 5.453, 5.582
7 (10)	−1	1	−1	0.85	300	210	5.989, 5.791, 6.172
8 (12)	−1	1	0	0.85	300	220	5.828, 6.299, 6.006
9 (11)	−1	1	1	0.85	300	230	6.124, 6.240, 6.155
10 (7)	0	−1	−1	1	200	210	4.146, 4.569, 4.820
11 (9)	0	−1	0	1	200	220	4.925, 4.724, 5.314
12 (8)	0	−1	1	1	200	230	5.565, 4.758, 4.987
13 (4)	0	0	−1	1	250	210	4.995, 5.032, 5.080
14 (6)	0	0	0	1	250	220	5.664, 5.436, 5.142
15 (5)	0	0	1	1	250	230	5.613, 5.715, 5.709
16 (1)	0	1	−1	1	300	210	5.972, 6.056, 5.898
17 (3)	0	1	0	1	300	220	6.296, 6.319, 6.234
18 (2)	0	1	1	1	300	230	6.183, 6.152, 6.062
19 (25)	1	−1	−1	1.15	200	210	4.626, 4.623, 4.445
20 (27)	1	−1	0	1.15	200	220	5.151, 4.953, 4.936
21 (26)	1	−1	1	1.15	200	230	5.354, 5.306, 5.357
22 (22)	1	0	−1	1.15	250	210	5.142, 5.086, 5.489
23 (24)	1	0	0	1.15	250	220	5.267, 5.311, 5.390
24 (23)	1	0	1	1.15	250	230	5.359, 5.749, 5.379
25 (19)	1	1	−1	1.15	300	210	5.918, 5.749, 5.975
26 (21)	1	1	0	1.15	300	220	5.729, 6.155, 6.223
27 (20)	1	1	1	1.15	300	230	6.217, 6.327, 6.251

Based on the ANOVA with F-distribution as shown in Table 15.4.3, the overall contribution of individual factors and their interactions to the surface roughness of the plane perpendicular to the z-axis can be summarized using the following predictive equation:

$$Y = -15.31 + 0.0467\ B + 0.0978\ C + 0.000059\ B^2 - 0.000294\ B\ C$$

$$(R = 94.4\%,\ R_{adj} = 94.0\%,\ \text{and}\ R_{pred} = 93.5\%). \tag{15.4.1}$$

The model accuracy is assured by the three correlation functions given above, and its adequacy is guarded by the insignificance of the lack of fit (57.7%).

The main effects of factor B (layer thickness) and factor C (nozzle temperature) are statistically significant, so is their two-factor interaction (i.e., BC). Nevertheless, the variation of factor A (printing speed) within the ±15% range has no impact on the surface roughness. A surface plot of the surface roughness (Y) versus layer thickness and nozzle temperature is presented in Figure 15.4.2. A decrease in layer thickness and/or the nozzle temperature within the applied level ranges of the parameters given in Table 15.4.2 will reduce the surface roughness.

TABLE 15.4.3 ANOVA for surface roughness of parts built with 3D printing.

Source	SS_{adj}	DOF	MS_{adj}	$F_{u,v}$	p-Value (%)
Regression	21.8397	4	5.45992	154.14	0.00
B	0.2865	1	0.28648	8.09	0.6
C	1.3416	1	1.34156	37.87	0.0
BB	0.3897	1	0.38975	11.00	0.1
BC	0.7791	1	0.77910	22.00	0.0
Error	2.6920	76	0.03542		
Lack of fit	0.1045	4	0.02612	0.73	57.7%
Pure error	2.5875	72	0.03594		
Subtotal	24.5317	80			

© SAE International.

FIGURE 15.4.2 Plot of surface roughness (Y) versus layer thickness and nozzle temperature.

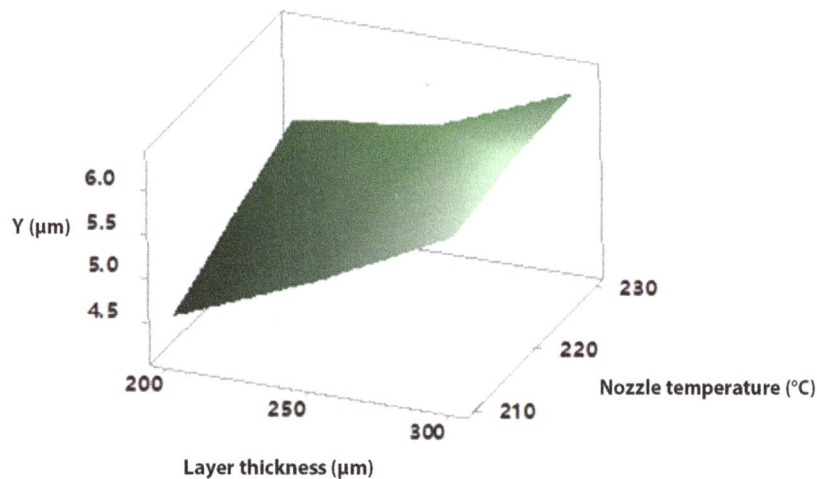

© SAE International.

15.4.2. Plating of Parts Built with 3D Printing

From chrome-plated door handles and grills of an automobile, electroplated parts are everywhere. Parts built with 3D printing techniques can be electroplated with chromium and silver, or other high-quality metals by electrolysis, to improve the appearance and even the material properties of 3D-printed parts. There are two main ways of plating 3D-printed parts:

1. Electroplating: The part is immersed in a solution of water and metal salts, in which an electrical current is applied through it causing metal cations to form a thin coating around the part.

2. Electroless Plating (i.e., autocatalytic plating): The part is immersed in a solution of water and metal salts, in which chemical reactions cause the metal to bond onto the part.

A resultant plated part is quite different from its original plastic state, such as improved surface roughness, wear resistance, UV resistance, chemical inertness, moisture resistance, and even electrical conductivity.

15.4.3. Machining 3D-Printed Parts

How to remove 3D-printed parts from build plates is another technical concern. Wire electronic discharge machining (EDM) may be used for separating a 3D-printed part from its build plate. Wire EDM involves a thin wire used to cut the workpiece, acting as an electrode. The wire is fed through an automatic feed, whereby cuts are made around the workpiece using electrical discharges as shaped. The wire may be held with diamond guides, and the wire itself is generally made from brass or copper. Parts cut with EDM are under no pressure, damage, or material degradation, and there is little material waste.

Post-machining, using a CNC machine, for example, offers another opportunity to create superior dimensional control, as well as great surface finishes and neat features, to supplement 3D printing.

Laser cutting is another option, in addition to EDM, for controlling the dimensions and surface roughness of 3D-printed parts. The feasibility of laser cutting of PLA parts built with 3D printing techniques was investigated using the DOE by Moradi et al. [2020].

15.5. Elastoplasticities of 3D-Printed Laminae

A lamina is a stackup of printed layers of filaments (fibers) that are aligned in the same direction, while a laminate is a stackup of laminae with different filament (fibrous) directions. Each individual lamina built with 3D printing exhibits quasi-orthotropic material properties. There are nine independent material properties associated with an orthotropic material, such as a lamina rubber reinforced with unidirectionally aligned fibers (e.g., steel or polyester). A lamina with unidirectional fibers follows orthotropic elastoplasticities. Orthotropic materials are a subset of anisotropic materials; their properties depend on the direction in which they are measured. Orthotropic materials possess three orthogonal planes of symmetry. If the material properties along the 1-axis, 2-axis, and 3-axis are mutually different in the orthogonal material coordinate system, it is an orthotropic material.

FIGURE 15.5.1 Schematic drawing of "unidirectional laminae" built with 3D printing.

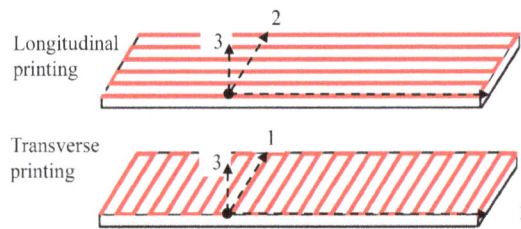

© SAE International.

15.5.1. Elasticity of Orthotropic Material

By convention, the elastic constants in orthotropic constitutive equations are comprised of Young's moduli E_{11}, E_{22}, and E_{33}, Poisson's ratios ν_{12}, ν_{23}, ν_{31}, ν_{21}, ν_{32}, and ν_{13}, and shear moduli G_{12}, G_{23}, G_{31}, G_{21}, G_{32}, and G_{13}, based on the primary material coordinate system denoted as (1, 2, 3). A layer of unidirectional fiber-reinforced lamina, shown in Figure 15.5.1, can be described using the following orthotropic stress–strain relationships in the primary material coordinate system:

$$\varepsilon_{11} = \frac{1}{E_{11}}\sigma_{11} + \frac{-\nu_{21}}{E_{22}}\sigma_{22} + \frac{-\nu_{31}}{E_{33}}\sigma_{33} + \alpha_1 \Delta T + \beta_1 \Delta C_m, \tag{15.5.1}$$

$$\varepsilon_{22} = \frac{-v_{12}}{E_{11}} \sigma_{11} + \frac{1}{E_{22}} \sigma_{22} + \frac{-v_{32}}{E_{33}} \sigma_{33} + \alpha_2 \Delta T + \beta_2 \Delta C_m, \qquad (15.5.2)$$

$$\varepsilon_{33} = \frac{-v_{13}}{E_{11}} \sigma_{11} + \frac{-v_{23}}{E_{22}} \sigma_{22} + \frac{1}{E_{33}} \sigma_{33} + \alpha_3 \Delta T + \beta_3 \Delta C_m, \qquad (15.5.3)$$

$$\gamma_{23} = \tau_{23} / G_{23}, \qquad (15.5.4)$$

$$\gamma_{31} = \tau_{31} / G_{31}, \qquad (15.5.5)$$

and $\quad \gamma_{12} = \tau_{12} / G_{12}, \qquad (15.5.6)$

where

    1-axis is along the longitudinal direction of fibers in the unidirectional lamina
    2-axis is transverse to the longitudinal direction of fibers in the lamina plane
    3-axis is perpendicular to the lamina plane
    E_{11}, E_{22}, and E_{33} (GPa) are Young's moduli in axial, transverse, and out-of-plane directions
    G_{12}, G_{23}, G_{31}, G_{21}, G_{12}, and G_{23} (GPa) are the shear moduli of a unidirectional lamina
    v_{12}, v_{23}, v_{31}, v_{21}, v_{13}, and v_{32} are Poisson's ratios of a unidirectional lamina
    α_1, α_2, and α_3 (μm/m/°C) are the coefficients of linear thermal expansion in 1, 2, and 3 directions
    β_1, β_2, and β_3 are the coefficients of linear moisture expansion in 1, 2, and 3 directions
    ΔT is the temperature variation
    ΔC_m is the variation in concentration of moisture

Note that the tensor shear strains ε_{ij} and the engineering shear strains γ_{ij} are related by the following three equations:

$$\gamma_{12} = 2 \varepsilon_{12}, \qquad (15.5.7)$$

$$\gamma_{23} = 2 \varepsilon_{23}, \qquad (15.5.8)$$

and $\quad \gamma_{31} = 2 \varepsilon_{31}. \qquad (15.5.9)$

The elastic constants in Equations (15.5.1)–(15.5.6) can be regrouped into a compliance matrix, $[s_{ij}]$, based on tensor shear strain without taking the hydrothermal (both temperature and moisture) effects into consideration, as follows:

$$\{\varepsilon\} = [s_{ij}] \{\sigma\}, \qquad (15.5.10)$$

i.e. $\quad \begin{Bmatrix} \varepsilon_{11} \\ \varepsilon_{22} \\ \varepsilon_{33} \\ \varepsilon_{23} \\ \varepsilon_{31} \\ \varepsilon_{12} \end{Bmatrix} = \begin{bmatrix} s_{11} & s_{12} & s_{13} & 0 & 0 & 0 \\ s_{12} & s_{22} & s_{23} & 0 & 0 & 0 \\ s_{13} & s_{23} & s_{33} & 0 & 0 & 0 \\ 0 & 0 & 0 & s_{44} & 0 & 0 \\ 0 & 0 & 0 & 0 & s_{55} & 0 \\ 0 & 0 & 0 & 0 & 0 & s_{66} \end{bmatrix} \begin{Bmatrix} \sigma_{11} \\ \sigma_{22} \\ \sigma_{33} \\ \sigma_{23} \\ \sigma_{31} \\ \sigma_{12} \end{Bmatrix}, \qquad (15.5.11)$

where $\quad s_{11} = 1 / E_{11}, \qquad (15.5.12)$

$$s_{22} = 1 / E_{22}, \tag{15.5.13}$$

$$s_{33} = 1 / E_{33}, \tag{15.5.14}$$

$$s_{12} = - \nu_{21} / E_{22} = s_{21} = - \nu_{12} / E_{11}, \tag{15.5.15}$$

$$s_{23} = - \nu_{32} / E_{33} = s_{32} = - \nu_{23} / E_{22}, \tag{15.5.16}$$

$$s_{31} = - \nu_{13} / E_{11} = s_{13} = - \nu_{31} / E_{33}, \tag{15.5.17}$$

$$s_{44} = 1 / (2G_{23}), \tag{15.5.18}$$

$$s_{55} = 1 / (2G_{31}), \tag{15.5.19}$$

$$\text{and} \quad s_{66} = 1 / (2G_{12}), \tag{15.5.20}$$

The compliance matrix is symmetric and must be positive definite for the strain energy density to be positive. The factor 1/2 multiplying the shear moduli in the compliance matrix results from the difference between tensor shear strain and engineering shear strain, as described by Equations (15.5.7)–(15.5.9).

15.5.2. Orthotropic Elastic Constants

As given in Equations (15.5.1)–(15.5.6), there are 15 elastic constants for an orthotropic material. However, there are six special relationships for the elastic constants of an orthotropic material. They are given here without detailed derivations. The reciprocal theory relating Young's moduli to Poisson's ratios holds for an orthotropic material:

$$\nu_{ij} E_{jj} = \nu_{ji} E_{ii} \qquad (i = 1, 2, 3 \ \& \ j = 1, 2, 3; \ i \neq j), \tag{15.5.21}$$

$$\text{i.e.} \quad \nu_{21} = \nu_{12} E_{22} / E_{11}, \tag{15.5.22}$$

$$\nu_{32} = \nu_{23} E_{33} / E_{22}, \tag{15.5.23}$$

$$\text{and} \quad \nu_{13} = \nu_{31} E_{11} / E_{33}. \tag{15.5.24}$$

and so does the reciprocal relation hold for shear moduli:

$$G_{ij} = G_{ji} \qquad (i = 1, 2, 3 \ \& \ j = 1, 2, 3; i \neq j), \qquad (15.5.25)$$

i.e. $\quad G_{21} = G_{12},$ $\qquad\qquad\qquad\qquad\qquad\qquad (15.5.26)$

$$G_{31} = G_{13}, \qquad\qquad\qquad\qquad\qquad\qquad (15.5.27)$$

and $\quad G_{32} = G_{23}.$ $\qquad\qquad\qquad\qquad\qquad\qquad (15.5.28)$

Enlightened by the reciprocal equations given above, there are only nine independent elastic constants. By convention, the nine independent elastic constants in orthotropic constitutive equations are the three Young's moduli, three Poisson's ratios, and the three shear moduli given as follows:

(a) Young's moduli: E_{11}, E_{22}, and E_{33}
(b) Poisson's ratios: ν_{12}, ν_{23}, and ν_{31}
(c) Shear moduli: G_{12}, G_{23}, and G_{31}

All these material properties have to be characterized using physical tests. Industry standards such as ASTM D638 and ASTM D5379 are available for carrying out these tests. Anisotropy is defined as

$$\text{Anisotropy} = E_{11} / E_{22}, \qquad\qquad\qquad\qquad (15.5.29a)$$

or $\quad \text{Anisotropy} = E_{11} / E_{33}, \qquad\qquad\qquad\qquad (15.5.30a)$

15.5.3. Elastoplastic Stress–Strain Relationship of 3D-Printed Laminae

Experimental tests are conducted to find out the mechanical behavior of different laminae with various orientations as shown in Figure 15.5.2(a). Elastoplastic stress–strain curves of PLA laminae loaded in various directions are presented in Figure 15.5.2(b), which shows representative strain fields captured in physical tests [Song et al. 2017] corresponding to the laminae built with 3D printing or injection molding. The dog-bone specimens experience the lowest strength when the laminae built with 3D printing are loaded in tension, while the compressive strength of laminae built with 3D printing is compatible with that of injection-molded laminae. This sounds like a structural performance instead of material properties, for example, the effect of voids. In the development of new materials for 3D printing processes, understanding the effects of 3D printer build orientation and raster pattern on physical properties and failure modes is paramount. Design engineers would like to align the loading direction with the strongest structural orientation of a part (Table 15.5.1).

FIGURE 15.5.2 Tensile strengths of plastic parts built with 3D printing.

(a) Specimen orientations [Zhao, Y. et al.]

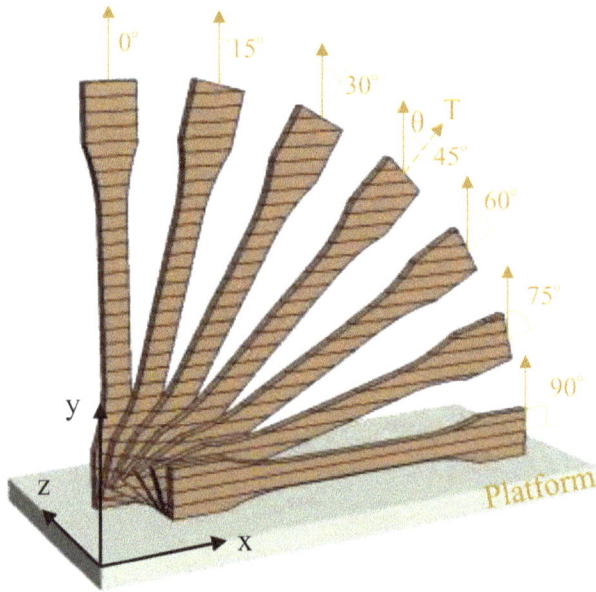

(b) True stress–strain curves of 3D-printed PLA laminae [Song et al.]

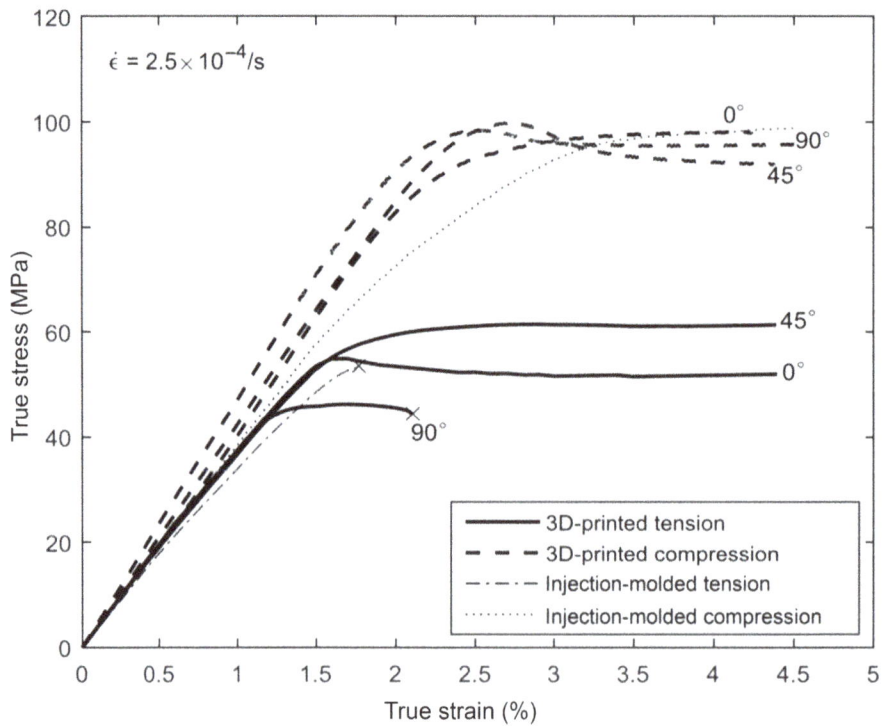

TABLE 15.5.1 Quasi-orthotropic elasticities of plastics used for and parts built with 3D printing [ASTM D638, ASTM D3039, Torrado and Roberson 2016, Ligon et al. 2017, Goo et al. 2020, Materialise Manufacturing 2020].

Material-DAM	Temp.	ρ	E_{11}	E_{22}	E_{33}	G_{12}	G_{13}	G_{23}	ν_{12}	ν_{13}	ν_{23}
ABS (homogeneous)	−25°C (low service temperature)	—				—	—	—			—
	−18°C	—	2.8	2.8	2.8	—	—	—	—	—	—
	23°C	1.07	2.3	2.3	2.3	—	—	—	0.35	0.35	0.35
	98°C (high service temperature)										
	107°C (T_g)										
	240°C (T_m)										
ABS (3D printing)	23°C	—	2.23	—	—	—	—	—	—	—	—
	107°C (T_g)										
	240°C (T_m)										
ABS/short carbon fiber (SCF) (3D printing; ASTM D3039; 158 μm thick) [Somireddy et al. 2020]	23°C	—	4.12	1.65	—	0.77	—	—	0.32	—	—
ABS/SCF (3D printing; ASTM D3039; 317 μm thick) [Somireddy et al. 2020]	23°C	—	2.68	1.55	—	0.625	—	—	0.34	—	—
Onyx (PA6/carbon fiber)	23°C	1.2	1.4	1.4	1.4	—	—	—	0.35	0.35	0.35
PA6	23°C	1.7	—	—	—	—	—	—	—	—	—
PC (homogeneous)	23°C	1.21	2.2	2.2	2.2	—	—	—	0.37	0.37	0.37
	121°C (high service temperature)										
	147°C (T_g)										
PEEK (homogeneous)	23°C	1.31	3.8	3.8	3.8	—	—	—	0.38	0.38	0.38
	143°C (T_g)										
	292°C (T_c)										
	343°C (T_m)										
PEEK (3D printing)	23°C	—	—	—	—	—	—	—	—	—	—
PET (homogeneous)	23°C	1.38	2.95	2.95	2.95	—	—	—	0.39	0.39	0.39
	74°C (T_g)										
PET (3D printing)	23°C	—	—	—	—	—	—	—	—	—	—
	74°C (T_g)										
PETG (homogeneous)	23°C	1.27	2.1	2.1	2.1	—	—	—	—	—	—
	81°C (T_g)										
	140°C (T_m)										

(Continued)

TABLE 15.5.1 (Continued) Quasi-orthotropic elasticities of plastics used for and parts built with 3D printing [ASTM D638, ASTM D3039, Torrado and Roberson 2016, Ligon et al. 2017, Goo et al. 2020, Materialise Manufacturing 2020].

Material-DAM	Temp.	ρ	E_{11}	E_{22}	E_{33}	G_{12}	G_{13}	G_{23}	ν_{12}	ν_{13}	ν_{23}
PETG (3D printing; 5 mm thick) [Szykiedans et al. 2017]	23°C	1.27	0.91	0.594	0.485	—	—	—	—	—	—
pNIPAm (3D Laser Lithography) [Hippler et al. 2019]	23°C	1.1	5×10^{-5}	5×10^{-5}	5×10^{-5}	—	—	—	0.5	0.5	0.5
	31°C	—	2×10^{-5}	2×10^{-5}	2×10^{-5}	—	—	—	—	—	—
	34°C	—	4×10^{-5}	4×10^{-5}	4×10^{-5}	—	—	—	—	—	—
	37°C	—	1×10^{-4}	1×10^{-4}	1×10^{-4}	—	—	—	—	—	—
	40°C	—	2×10^{-4}	2×10^{-4}	2×10^{-4}	—	—	—	—	—	—
	96°C (T_m)										
PLA (amorphous) [Farah et al. 2016]	23°C	1.24	3.5	3.5	3.5	—	—	—	0.36	0.36	0.36
	52°C (high service temperature)										
	63°C (T_g)										
	155°C (T_m)										
PLA (3D printing) [Wang et al. 2020]	23°C	1.2	3.17	2.97	—	—	—	—	0.331	0.333	0.33
PP (homopolymer)	<0°C (brittle)	0.906	1.3	1.3	1.3	—	—	—	0.42	0.42	0.42
	23°C										
	163°C (T_m)										
PS, high impact (homogeneous)	23°C	1.04	1.9	1.9	1.9	—	—	—	0.41	0.41	0.41
	98°C (T_g)										
PVA, pure (hydrogel)	23°C	1.25	1×10^{-4}	1×10^{-4}	1×10^{-4}	—	—	—	0.45	0.45	0.45
	62.4°C (T_g)										
	186°C (T_c)										
	221°C (T_m)										
Thermoplastic polyurethane (TPU) [Qi and Boyce 2005]	−25°C (T_g)	0.02	Hyperelastic						0.5	0.5	0.5
	23°C										

Note:
Data are listed for reference only, as material properties of laminae built with 3D printing depend on porosity and assigned operating parameters.

15.5.4. DOE to Characterize the Orthotropic Elasticities

Mechanical models of orthotropic elasticities and tensile strengths, such as E_{11}, E_{33}, $\sigma_{11,uts}$, and $\sigma_{33,uts}$, of PLA parts built with 3D printing techniques are examined by Zhao et al. [2019b]. Results from their physical tests are summarized in Table 15.5.2. Besides these four orthotropic material properties, off-axis tensile tests are also conducted.

TABLE 15.5.2 Design factors for laminar tensile properties due to 3D printing [Zhao et al. 2019b].

Factor	Level (−1)	Level (0)	Level (+1)
A: Layer thickness (μm)	100	200	300
B: Raster angle (deg)	0	45	90
C: Extrusion temperature (°C)	−	220	−
D: Softening temperature (°C)	−	140	−
E: Build plate temperature (°C)	−	60	−
F: Diameter of input filament (mm)	−	1.75	−

TABLE 15.5.3 Factorial design for tensile strength of laminae built with 3D printing [Zhao et al. 2019b].

Run	A (μm)	B (°)	E: Ave/SD/sample	σ_{uts}: Ave/SD/sample	E_{11}	E_{33}	$\sigma_{11,uts}$	$\sigma_{33,uts}$
1	100	0	2.444/0.122/5	23.40/1.58/5	−	2.444	−	23.4
2	100	15	2.507/0.087/5	25.69/1.23/5	−	−	−	−
3	100	30	2.549/0.041/5	29.24/1.56/4	−	−	−	−
4	100	45	2.659/0.090/5	30.17/1.47/5	−	−	−	−
5	100	60	2.726/0.111/5	34.59/0.97/4	−	−	−	−
6	100	75	2.788/0.077/5	43.65/1.91/5	−	−	−	−
7	100	90	2.864/0.122/5	49.66/1.12/4	2.864	−	49.66	−
8	200	0	2.108/0.076/5	21.35/0.75/5	−	2.108	−	21.35
9	200	15	2.187/0.100/5	24.14/0.38/5	−	−	−	−
10	200	30	2.224/0.077/5	25.24/1.37/4	−	−	−	−
11	200	45	2.273/0.032/5	28.12/1.86/5	−	−	−	−
12	200	60	2.548/0.015/5	35.85/1.04/4	−	−	−	−
13	200	75	2.705/0.049/5	43.44/0.63/5	−	−	−	−
14	200	90	2.816/0.087/5	48.28/0.96/5	2.816	−	48.28	−
15	300	0	1.825/0.058/5	19.16/0.54/4	−	1.825	−	19.16
16	300	15	1.856/0.058/5	23.02/0.28/5	−	−	−	−
17	300	30	1.968/0.106/5	23.97/1.15/5	−	−	−	−
18	300	45	2.037/0.042/5	26.36/0.25/5	−	−	−	−
19	300	60	2.320/0.048/5	31.57/1.59/5	−	−	−	−
20	300	75	2.516/0.039/5	36.16/0.15/4	−	−	−	−
21	300	90	2.683/0.019/5	44.57/1.25/5	2.683	−	44.57	−

Notes:
E (GPa): Young's modulus; E ➜ E_{33} for B = 0°; E ➜ E_{11} for B = 90°
σ_{uts} (MPa): Ultimate tensile strength; σ_{uts} ➜ $\sigma_{33,uts}$ for B = 0°; σ_{uts} ➜ $\sigma_{33,uts}$ for B = 90°
Ave: Averaged value for the available sample size
SD: Standard deviation

Assume that the classical two-dimensional (2D) in-plane lamination theory is correct for the 3D-printed parts, and then, the Young's modulus in an arbitrary direction can be obtained as

$$\frac{1}{E_{xx}} = \left[\frac{\cos^4\theta}{E_{11}} + \left(\frac{1}{G_{13}} - \frac{2v_{31}}{E_{11}}\right)\sin^2\theta\cos^2\theta + \frac{\sin^4\theta}{E_{33}}\right]. \tag{15.5.29b}$$

The above equation can be rewritten as

$$\frac{1}{G_{13}} - \frac{2v_{31}}{E_{11}} = \left[\frac{1}{E_{xx}} - \frac{\cos^4\theta}{E_{11}} - \frac{\sin^4\theta}{E_{33}}\right] / (\sin^2\theta\cos^2\theta). \tag{15.5.30b}$$

It means that for the case study layer thickness of 100 μm (factor A = 100 μm), substituting the data of E_{xx}, E_{11}, and E_{33} corresponding to θ = 15, 30, 45, 60, and 75° one should have a "constant" value for ($1/G_{13}$ – $2v_{31}/E_{11}$) since E_{11}, G_{13}, and v_{31} are constants. This is not true for the test data given in Table 15.5.3. Therefore, the classical 2D in-plane lamination theory borrowed from "mechanics of fibrous composites" as stated in Casavola et al. [2016] is not applicable to a lamina (layer) built with 3D printing without considering the out-of-lamina effect (along the axis 3). In other words, 3D orthotropic constitutive equations, i.e., Equations (15.5.1)–(15.5.6), must be applied simultaneously when resolving the structural performance of a product built with 3D printing techniques. The infill effect with air gap along axis 2, the interlayer effect along axis 3, and the performance of filaments along axis 1 are mutually coupled.

Furthermore, fused deposition modeling parts are pretty complicated as they have a lot more process parameters that affect their constitutive equations and structural strength than just layer adhesion. Since any "solid" fused deposition modeling part has infill, each layer cannot really be modeled just as a simple lamina.

15.5.5. Influences of Nozzle Temperature and Printing Speed on 3D-Printed Part Strength

Effects of 3D printing parameters on mechanical properties of laminae fabricated with open-source 3D printers in PLA by fused deposition modeling were investigated by Ouhsti et al. [2018]. Three process parameters at three different levels per factor, as shown in Table 15.5.4, are examined. Again, off-axis Young's moduli are not fundamental "material" properties. Only test results, related to E_{11}, are listed in Table 15.5.5 for furthering the statistical DOE based on ANOVA.

TABLE 15.5.4 Design factors for tensile elasticity (E_{11}) of 3D-printed laminae [Ouhsti et al. 2018].

Factor	Level (−)	Level (0)	Level (1)
A: Raster orientation (deg)	0° (axial)	30°	60°
B: Nozzle temperature (°C)	190	200	210
C: Printing speed (mm/s)	30	50	70
D: Infill	100%	—	—
E: Nozzle diameter (μm)	—	400	—
F: Layer height (μm)	—	200	—

© SAE International.

TABLE 15.5.5 Experimental design for tensile elasticity E_{11} of 3D-printed laminae [Ouhsti et al. 2018].

Run	a	b	c	A	B	C	E_{11} (GPa)
1	−1	−1	−1	0	190	30	3.942
2	−1	1	−1	0	210	30	3.5292
3	−1	−1	1	0	190	70	3.7144
4	−1	1	1	0	210	70	2.5682
5	−1	0	0	0	200	50	3.4856

Based on the ANOVA with F-distribution as shown in Table 15.5.6, the overall contribution of individual factors and their interactions to the in-plane Young's modulus along the x-axis can be summarized using the following predictive equation:

$$E_{11} = 4190.8 + 144.83 \ C - 0.7984 \ B \ C$$

$$(R = 99.8\%, \ R_{adj} = 99.6\%, \ and \ R_{pred} = 97.8\%), \tag{15.5.31}$$

The main effect of factor C (printing speed) is statistically significant, so is its two-factor interaction with factor B (nozzle temperature), i.e., BC. A surface plot of E_{11} (Young's modulus) versus printing speed and nozzle temperature is presented in Figure 15.5.3.

TABLE 15.5.6 ANOVA for tensile elasticity E_{11} of 3D-printed laminae.

Source	SS_{adj}	DOF	MS_{adj}	$F_{u,v}$	p-Value (%)
Regression	1,092,683	2	546,342	249.65	0.4
C	597,456	1	597,456	273.01	0.4
BC	739,491	1	739,491	337.92	0.3
Error	4377	2	2188		
Subtotal	1,097,060	4			
Grand average	—	1			
Total	—	5			

FIGURE 15.5.3 Response surface plot of Young's modulus versus nozzle temperature and printing speed.

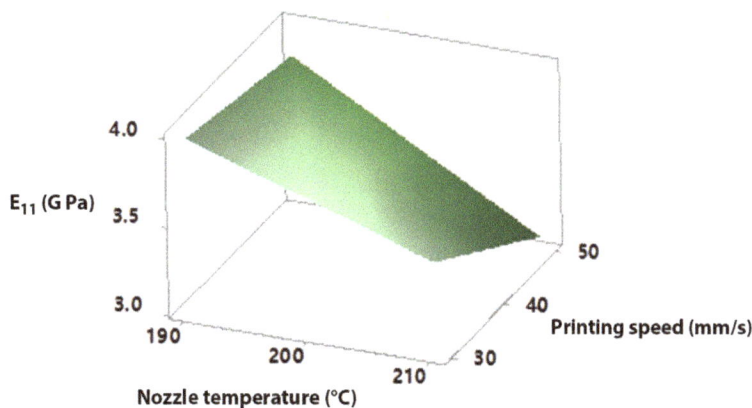

15.6. Hydrothermal Properties of 3D-Printed Laminae

Another design concern is the hydrothermal property of the applied materials for 3D-printed parts. Hydrothermal effects include both thermal and hygroscopic effects. A change in temperature and moisture of a part may cause variations in both physical and chemical properties. For example, as PLA is poor in heat resistance, its 3D-printed parts are not suitable for applications that need to be exposed to temperatures beyond a nominal range of 45°C. On the other hand, 3D-printed PLA parts may be used as the "lost foam" in an investment casting of metal parts because of this inherence.

Nevertheless, hygroscopic strains can be induced as the material reacts to the presence of moisture in the air by attracting and holding water molecules from the environment.

15.6.1. Quasi-Orthotropic Coefficients of Thermal Expansion of 3D-Printed Laminae

Directional thermal deformations and the resulting coefficients of linear thermal expansion of "unidirectional" specimens (e.g., laminae and bars) built with 3D printing techniques, along the roads, are conventionally used to characterize the "quasi-orthotropic" properties. Coefficients of linear thermal expansion of some popular plastics used for 3D printing are listed in Table 15.6.1.

TABLE 15.6.1 Hydrothermal thermomechanical properties of plastics used for and parts built with 3D printing [Torrado and Roberson 2016, Ligon et al. 2017, Goo et al. 2020, Materialise Manufacturing 2020].

Material	Temp.	α_1	α_2	α_3	k_1	k_2	k_3	γ	β_1	β_2	β_3
ABS (homogeneous)	23°C	80	80	80	0.17	0.17	0.17	2050	—	—	—
	107.8°C (T_g)	—	—	0.18	0.18	0.18	—	—	—	—	—
ABS (3D printing)	23°C	~1530	950	2222	—	—	—	—	—	—	—
	80–100°C (print-bed temperature)										
	80–120°C (slumping temperature)										
	220–235°C (printing temperature)										
Onyx	23°C	—	—	—	—	—	—	—	—	—	—
	27°C (T_g)										
	161°C (T_c)										
	197°C (T_m)										
PA6 (homogeneous)	23°C	95	95	95	0.25	0.25	0.25	1700	—	—	—
	22°C (T_g)										
	152°C (T_c)										
	197°C (T_m)										
PA (3D printing)	23°C	—	—	—	—	—	—	—	—	—	—
	100°C (print-bed temperature)										
	245°C (printing temperature)										
PA6,6 (homogeneous)	23°C	90	90	90	0.25	0.25	0.25	1670	—	—	—
PA12 (homogeneous)	23°C	110	110	110	—	—	—	—	—	—	—
PC (homogeneous)	23°C	68	68	68	0.2	0.2	0.2	1250	—	—	—
	147°C (T_g)										
PC (3D printing)	23°C	—	—	—	—	—	—	—	—	—	—
	80–120°C (print-bed temperature)										
	147°C (T_g)										
	260–310°C (printing temperature)										
PE (homogeneous)	23°C	150	150	150	0.33	0.33	0.33	2100	—	—	—
PE, HD (homogeneous)	23°C	150	150	150	0.5	0.5	0.5	2100	—	—	—

(Continued)

TABLE 15.6.1 (Continued) Hydrothermal thermomechanical properties of plastics used for and parts built with 3D printing [Torrado and Roberson 2016, Ligon et al. 2017, Goo et al. 2020, Materialise Manufacturing 2020].

Material	Temp.	α_1	α_2	α_3	k_1	k_2	k_3	γ	β_1	β_2	β_3
PEEK	23°C	47/108	47/108	47/108	0.25	0.25	0.25	1340	—	—	—
	143°C (T_g)										
	292°C (T_c)										
	343°C (T_m)										
PEEK (3D printing)	23°C	—	—	—	—	—	—	—	—	—	—
PET (homogeneous)	23°C	70	70	70	0.2	0.2	0.2	1075	—	—	—
	74°C (T_g)										
	260°C (T_m)										
PET (3D printing)	23°C	—	—	—	—	—	—	—	—	—	—
	55–70°C (print-bed temperature)										
	230–255°C (printing temperature)										
PETG (homogeneous)	23°C	68	68	68	0.29	0.29	0.29	1200	—	—	—
	81°C (T_g)										
	75–90°C (print-bed temperature)										
	140°C (T_m)										
PETG (3D printing)	23°C	—	—	—	—	—	—	—	—	—	—
	70°C (print-bed temperature)										
	230°C (printing temperature)										
PLA (amorphous) [Farah et al. 2016]	23°C	41	41	41	0.1	0.1	0.1	1800	—	—	—
	48°C				0.111	0.111	0.111	—	—	—	—
	55°C (T_g)						1590	—	—	—	—
	100°C						—	1955	—	—	—
	109°C				0.197	0.197	0.197	—	—	—	—
	155°C (T_m)										
	190°C				0.195	0.195	0.195	2060	—	—	—
PLA (3D printing) [Qahtani et al. 2019]	23°C	—	—	—	—	—	—	—	—	—	—
	61°C (T_g)										
	40–70°C (print-bed temperature)										
	70–80°C (slumping temperature)										
	151°C (T_m)										
	190–220°C (printing temperature)										

(Continued)

TABLE 15.6.1 (Continued) Hydrothermal thermomechanical properties of plastics used for and parts built with 3D printing [Torrado and Roberson 2016, Ligon et al. 2017, Goo et al. 2020, Materialise Manufacturing 2020].

Material	Temp.	α_1	α_2	α_3	k_1	k_2	k_3	γ	β_1	β_2	β_3
pNIPAm (3D printing) [Hippler et al. 2019]	23°C	325	325	325	0.51	0.51	0.51	1950	—	—	—
	32°C [lower critical solution temperature (LCST)]										
	45°C	—	—	—	0.35	0.35	0.35	—	—	—	—
	96°C (T_m)	—	—	—							
PP (homogeneous)	23°C	150	150	150	0.2	0.2	0.2	2500	—	—	—
	100°C (high service temperature)										
	163°C (T_m)	—	—	0.16	0.16	0.16	—	—	—	—	—
PP (3D printing)	23°C	—	—	—	—	—	—	—	—	—	—
	85–100°C (print-bed temperature)										
	220–250°C (printing temperature)										
PS, high impact (homogeneous)	23°C	80	80	80	0.22	0.22	0.22	1400	—	—	—
	100–115°C (print-bed temperature)										
	230–245°C (printing temperature)										
PVA (hydrogel)	23°C	85	85	85	0.34	0.34	0.34	1010	—	—	—
	45–60°C (print-bed temperature)										
	62.4°C (T_g)										
	180–200°C (printing temperature)										
	186°C (T_c)										
	221°C (T_m)										
TPU—[Bajsic et al. 2005]	−25°C (T_g)	—	—	—	—	—	—		—	—	—
	23°C	—	—	—	—	—	—		—	—	—
	40–60°C (print-bed temperature)										
	230–260°C (printing temperature)										

Note:
Data are listed for reference only, as material properties of laminae built with 3D printing depend on porosity and assigned operating parameters.

15.6.2. Warping and Delamination of 3D-Printed Parts

Warping occurs in composites built with 3D printing, especially for a large part of which the first layer is deformed and detached from the printing surface, frequently starting from the corners as a boundary (edge) effect. Delamination, i.e., interlaminar separation, may also occur between two adjacent layers because of uneven cooling rates and/or rough contact surfaces between them as the temperature varies.

Because of its relatively low coefficient of thermal expansion as exhibited in Table 15.6.1, PLA is one of the most suitable thermoplastics to avoid warping and delamination for printing large parts. Other methods include having a large contact surface with the build bed and avoiding sharp corners. Control of the relative temperature between the build plate and environment is another key factor in a 3D printing process to reduce potential warping. Of course, a lamina with thin first several layers is helpful in warpage reduction.

15.6.3. Degree of Porosity as Built with 3D Printing

Mechanical properties of 3D-printed parts depend on the degree of porosity that is an inherent exhibition born with 3D printing. Porosities are found to be sensitive to the printing parameters as exhibited in Table 15.6.2.

TABLE 15.6.2 Degree of porosity varying with operating parameters [Chaturvedi 2009].

Run	A (mm)	B (°C)	C (mm/s)	Porosity (%)
1	200	200	45	5.66
2	200	200	45	7.53
3	200	210	45	4.71
4	200	230	45	5.83
5	200	240	45	7.28
6	100	220	45	3.43
7	300	220	45	13.54
8	400	220	45	7.01
9	200	220	30	4.49
10	200	220	60	1.46
11	200	230	75	5.45
12	200	230	100	6.67
13	200	240	100	5.79
14	200	240	125	6.32
15	200	240	150	7.70

© SAE International.

15.7. Mechanical Strength of 3D-Printed Laminae

One drawback of parts built with 3D printing is the inherent structural problems, including tensile fracture, compressive buckling, low fracture toughness, free boundary effect (free-edge effect), and short fatigue life [Montero et al. 2001, Galantucci et al. 2008, Zaman et al. 2019]. Just like fibrous composites, the structural integrity of the accumulated laminar layers (parts) built with 3D printing, as one of the RP technologies, is a major challenge [Galantucci et al. 2008, Zaman et al. 2019] because of the nonuniform road paths and material anisotropy.

An fused deposition modeling-formed part may advance its potential material strength in a cure environment such as a UV curing chamber. Curing helps initiate and keep molecular chain growth (e.g., tire rubber) but does not take part in it. However, the build and post-processing parameters applied in an additive manufacturing process need to be closely controlled to eliminate defects. Up to failure, stress–strain curves of frequently used 3D-printed parts under a quasi-static loading condition are listed in Table 15.7.1. Just like homogeneous materials, the ultimate tensile strength of a 3D-printed cross-ply laminate or 3D-printed hatched ply laminate increases with an increasing strain rate [Hibbert et al. 2019].

TABLE 15.7.1 Failure profiles of plastic parts built with 3D printing [Torrado and Roberson 2016, Ligon et al. 2017, Materialise Manufacturing 2020].

Material	Temp.	$(\sigma_{11u}, \varepsilon_{11u})$	$(\sigma_{22u}, \varepsilon_{22u})$	$(\sigma_{33u}, \varepsilon_{33u})$	$(\sigma_{12u}, \varepsilon_{12u})/(\sigma_{23u}, \varepsilon_{23u})/(\sigma_{13u}, \varepsilon_{13u})$
	−25°C	(low service temperature)			
	−18°C	(−64, −3.4%)	—	—	—
		(36, 15%)			
ABS (homogeneous)	23°C	(−63, −20%)	—	—	—
		(43, 3.5%)			
		(39, 25%)			
	107°C (T_g)				
	240°C (T_m)				
ABS (fused deposition modeling 3D printing; ASTM D638 −10: type V) [Torrado and Roberson 2016]	23°C	(34.5, 2.7%)	(28.6, 2.4%)	(29.5, 2.2%)	—
	107°C (T_g)				
	240°C (T_m)				
ABS/SCF (3D printing; ASTM D3039; 158 µm thick) [Somireddy et al. 2020]	23°C	(40.7, 2.1%)	(14.9, 2.0%)	—	$\sigma_{12,us}$ = 14.6
ABS/SCF (3D printing; ASTM D3039; 317 µm thick) [Somireddy et al. 2020]	23°C	(26.1, 1.6%)	(14.6, 1.7%)	—	$\sigma_{12,us}$ = 11.8
Onyx (PA/CF)	23°C	(14, 1%)	(14, 1%)	(14, 1%)	—
		(30, 2%)	(30, 2%)	(30, 2%)	—
		(36, 25%)	(36, 25%)	(36, 25%)	—
		(53, 38%)	(53, 38%)	(53, 38%)	—
PA6 (homogeneous) [Pascual-González et al. 2020]	23°C	(17, 1%)	(17, 1%)	(17, 1%)	—
		(29, 25%)	(29, 25%)	(29, 25%)	—
		(27, 75%)	(27, 75%)	(27, 75%)	—
		(69, 311%)	(69, 311%)	(69, 311%)	—
PA6 (3D printing, SLS)	23°C	(35, 6%)	—	—	—
PC (homogeneous)	23°C	(68, 4.8%)	(68, 4.8%)	(68, 4.8%)	—
		$\sigma_{11,ucs}$ = −80	$\sigma_{11,ucs}$ = −80	$\sigma_{11,ucs}$ = −80	
PC (3D printing)	23°C	$\sigma_{11,uts}$ = 59.7	$\sigma_{22,uts}$ = 19	—	—
PEI (3D printing)	23°C	$\sigma_{11,uts}$ = 59	$\sigma_{22,uts}$ = 40	—	—
PEEK (pressed/homogeneous) [Muhsin et al. 2019]	23°C	(95, 50%)	—	—	—
	100°C	$\sigma_{11,uts}$ = 91	—	—	—
	143°C (T_g)				
	150°C	$\sigma_{11,uts}$ = 96	—	—	—
	175°C	$\sigma_{11,uts}$ = 93	—	—	—
	200°C	$\sigma_{11,uts}$ = 97	—	—	—

(Continued)

TABLE 15.7.1 (Continued) Failure profiles of plastic parts built with 3D printing [Torrado and Roberson 2016, Ligon et al. 2017, Materialise Manufacturing 2020].

Material	Temp.	$(\sigma_{11u}, \varepsilon_{11u})$	$(\sigma_{22u}, \varepsilon_{22u})$	$(\sigma_{33u}, \varepsilon_{33u})$	$(\sigma_{12u}, \varepsilon_{12u})/(\sigma_{23u}, \varepsilon_{23u})/(\sigma_{13u}, \varepsilon_{13u})$
PET (homogeneous)	23°C	—	—	—	—
	74°C (T_g)				
PETG (homogeneous)	23°C	$\sigma_{11,ucs} = -55$ (50, 0.2%) (63, 110%)	$\sigma_{11,ucs} = -55$ (50, 0.2%) (63, 110%)	$\sigma_{11,ucs} = -55$ (50, 0.2%) (63, 110%)	—
	81°C (T_g)				
	140°C (T_m)				
PLA (amorphous) [Farah et al. 2016]	23°C	(70, 2%) (65, 6%)	(70, 2%) (70, 2%)	(70, 2%) (70, 2%)	—
	98°C (T_g)		—	—	—
	155°C (T_m)				
PLA (3D printing) [Wang et al. 2020]	23°C	(31.7, 1%) (60.3, 1.9%) (61, 10.7%)	(29.7, 1%) (55.9, 2%) (56, 3.5%)	(24, 2%)	—
pNIPAm (3D printing) [Hippler et al. 2019]	23°C	$\sigma_{11,uts} = 0.001$	$\sigma_{11,uts} = 0.001$	$\sigma_{11,uts} = 0.001$	—
	32°C (LCST)				
	96°C (T_m)				
PP (homogeneous)	23°C	(21.4, 529%)	(21.4, 529%)	(21.4, 529%)	—
PP (3D printing)	23°C	$\sigma_{11,uts} = 36$	$\sigma_{22,uts} = 32$	—	—
PS, high impact (homogeneous)	23°C	(32, 40%)	(32, 40%)	(32, 40%)	—
PVA (hydrogel) [Nazouri 2020]	23°C	(5, 18%) (10, 50.5%)	(5, 18%) (10, 50.5%)	(5, 18%) (10, 50.5%)	—
TPU [Bajsic et al. 2005, Qi and Boyce 2005]	−25°C (T_g)		—	—	—
	23°C	(1.5, 20%) (3, 2500%)	—	—	—

© SAE International.

Note:
Data are listed for reference only, as material properties of laminae built with 3D printing depend on porosity and assigned operating parameters [Wang et al. 2019].

15.7.1. Failure of 3D-Printed ABS Laminae Loaded in Primary Material Directions

An experimental design with five process parameters, as shown in Table 15.7.2, is proposed by Montero et al. [2001] to improve the tensile strength. Fractional factorial design 2_V^{5-1}, with alias equation E = ABCD, was employed to explore the relationships between these five factors and the tensile strength of ABS laminae built with 3D printing techniques [Montero et al. 2001]. As listed in Table 15.7.3, each experimental treatment is run with two replications and the error (i.e., sample standard deviation) was calculated from replicated runs.

TABLE 15.7.2 Design factors for laminar tensile strength of ABS laminae due to 3D printing [Montero et al. 2001].

Factor	Level (−)	Level (+)	Variate (−1, 1)
A: Air gap (µm)	0	−50.8	a = [A − (−25.4)]/(−25.4)
B: Layer thickness (µm)	508	1005.84	b = (B − 756.92)/248.92
C: Nozzle temperature (°C)	270	280	c = (C − 275)/5
D: ABS color	Blue	White	d = D
E: Raster orientation (deg)	90°, Trans.	0°, Axial	e = (e − 45)/(−45)

In light of the DOE based on the t-distribution as shown in Table 15.7.3, the contribution of individual factors and their interactions to the tensile strength along the principal axes can be summarized using the following predictive equation:

$$Y_p = 14.065 + 2.714\ a + 7.583\ e - 1.569\ a\ e. \tag{15.7.1}$$

It can be concluded that the main effects of factors A and E are statistically significant with a significance level less than 10% ($t_{16,\ 10\%} = 1.337$), and so is their two-factor interaction. Factor E (raster angle) has the greatest impact on the tensile strength. Thus, specimens with zero-degree raster (layer) orientation, in which printing direction is parallel to the tensile direction, yield enhanced mechanical strength over those with other raster (layer) orientations.

The predicted tensile strengths in the axial and transverse directions become, respectively,

$$Y_p = 21.548 + 1.145\ a \qquad \text{(Axial Direction, } e = 1\text{),} \tag{15.7.2}$$

$$\text{and} \quad Y_p = 6.482 + 4.283\ a \qquad \text{(Transverse Direction, } e = -1\text{).} \tag{15.7.3}$$

It means that both tensile strengths increase with increasing interference between roads, while the transverse tensile strength gains more. Note that a negative air gap ($a > -1$) is hard to control. The minima of both tensile strengths occur at no interference, i.e., $a = -1$. With $a = -1$, Equations (15.7.2) and (15.7.3) reduce to

$$Y_p = 20.403 \qquad \text{(Axial direction with zero air gap)}, \tag{15.7.4}$$

$$\text{and} \quad Y_p = 2.199 \qquad \text{(Transverse direction with zero air gap).} \tag{15.7.5}$$

The anisotropy of the ABS lamina built with 3D printing with zero air gap in tensile strength, as loaded uniaxially in the laminar plane, is 9.28 (=20.403/2.199). This is a strong disparity. Z-pinning approach may be used to reduce the mechanical anisotropy of 3D-printed parts, in which continuous materials (e.g., filaments) are deposited across multiple layers within the volume of the part [Duty et al. 2019].

TABLE 15.7.3 Fractional factorial design 2_V^{5-1} (E = ABCD) for tensile strength of 3D-printed laminae [Montero et al. 2001].

(i) Main effect							
Treatment	A	B	C	D	E	Y (MPa)	Y_{ave} (MPa)
1	−1	−1	−1	−1	1	19.17, 20.02	19.60
2	1	−1	−1	−1	−1	10.00, 14.08	12.04
3	−1	1	−1	−1	−1	1.14, 2.96	2.05
4	1	1	−1	−1	1	22.77, 23.20	22.98
5	−1	−1	1	−1	−1	3.40, 1.54	2.47
6	1	−1	1	−1	1	21.57, 21.42	21.50
7	−1	1	1	−1	1	20.93, 20.48	20.71
8	1	1	1	−1	−1	11.38, 10.98	11.18
9	−1	−1	−1	1	−1	1.42, 3.17	2.30
10	1	−1	−1	1	1	22.99, 22.50	22.74
11	−1	1	−1	1	1	20.77, 21.39	21.08
12	1	1	−1	1	−1	5.42, 8.18	6.80
13	−1	−1	1	1	1	20.27, 19.90	20.08
14	1	−1	1	1	−1	12.07, 12.92	12.50
15	−1	1	1	1	−1	0.90, 1.98	1.44
16	1	1	1	1	1	23.50, 23.30	23.40
Contrast	5.428	−0.448	0.460	−0.272	15.165		
Effect	2.714	−0.224	0.230	−0.136	7.583	—	— 14.065 (average)
Error	1.091	1.091	1.091	1.091	1.091		
t-Ratio	2.488	0.205	0.211	0.125	6.951		
$t_{16, 10\%}$	1.337	1.337	1.337	1.337	1.337		
Significant?	Yes	No	No	No	Yes		

(ii) 2-factor interactive effect										
Treatment	AB	AC	AD	AE	BC	BD	BE	CD	CE	DE
1	1	1	1	−1	1	1	−1	1	−1	−1
2	−1	−1	−1	−1	1	1	1	1	1	1
3	−1	1	1	1	−1	−1	−1	1	1	1
4	1	−1	−1	1	−1	−1	1	1	−1	−1
5	1	−1	1	1	−1	1	1	−1	−1	1
6	−1	1	−1	1	−1	1	−1	−1	1	−1
7	−1	−1	1	−1	1	−1	1	−1	1	−1
8	1	1	−1	−1	1	−1	−1	−1	−1	1
9	1	1	−1	1	1	−1	1	−1	1	−1
10	−1	−1	1	1	1	−1	−1	−1	−1	1
11	−1	1	−1	−1	−1	1	1	−1	−1	1
12	1	−1	1	−1	−1	1	−1	−1	1	−1
13	1	−1	−1	−1	−1	−1	−1	1	1	1
14	−1	1	1	−1	−1	−1	1	1	−1	−1
15	−1	−1	−1	1	1	1	−1	1	−1	−1
16	1	1	1	1	1	1	1	1	1	1
Contrast	−0.655	0.540	−0.293	−3.138	0.492	−0.778	1.510	0.665	−0.640	0.905
Effect	−0.328	0.270	−0.146	−1.569	0.246	−0.389	0.755	0.333	−0.320	0.452
Error	1.091	1.091	1.091	1.091	1.091	1.091	1.091	1.091	1.091	1.091
t-Ratio	0.300	0.248	0.134	1.438	0.226	0.356	0.692	0.305	0.293	0.415
$t_{16, 10\%}$	1.337	1.337	1.337	1.337	1.337	1.337	1.337	1.337	1.337	1.337
Significant?	No	No	No	Yes	No	No	No	No	No	No

Notes:
Y (MPa): Ultimate tensile strength
Y_{ave} (MPa): Ultimate tensile strength, the average of two replicated runs

15.7.2. In-Plane Stress–Strain Transformations for 3D-Printed Laminae

In stress analysis of an orthotropic lamina with fiber orientation angle θ, measured from 1-axis to x-axis, the in-plane stress transformations can be done using the following three equations [Chiang 2019]:

$$\sigma_{xx} = \sigma_{11} \cos^2\theta + \sigma_{22} \sin^2\theta + 2\,\tau_{12} \cos\theta\,\sin\theta, \qquad (15.7.6)$$

$$\sigma_{yy} = \sigma_{11} \sin^2\theta + \sigma_{22} \cos^2\theta - 2\,\tau_{12} \cos\theta\,\sin\theta, \qquad (15.7.7)$$

$$\text{and} \quad \tau_{xy} = -(\sigma_{11} - \sigma_{22}) \sin\theta\,\cos\theta + \tau_{12} (\cos^2\theta - \sin^2\theta). \qquad (15.7.8)$$

Similarly, the in-plane strain transformations can be accomplished as

$$\varepsilon_{xx} = \varepsilon_{11} \cos^2\theta + \varepsilon_{22} \sin^2\theta + \gamma_{12} \sin\theta\,\cos\theta, \qquad (15.7.9)$$

$$\varepsilon_{yy} = \varepsilon_{11} \sin^2\theta + \varepsilon_{22} \cos^2\theta - \gamma_{12} \sin\theta\,\cos\theta, \qquad (15.7.10)$$

$$\text{and} \quad \gamma_{xy} = -2(\varepsilon_{11} - \varepsilon_{22}) \sin\theta\,\cos\theta + \gamma_{12} (\cos^2\theta - \sin^2\theta). \qquad (15.7.11)$$

15.8. 3D-Printed Composites

Composite materials used for 3D-printed parts may improve their mechanical, thermal, electrical, and chemical properties. For example, the superiority of 3D-printed ABS/SCF laminae, i.e., ABS reinforced with SCFs, to 3D-printed ABS laminae is demonstrated in terms of their mechanical strengths that are exhibited in Tables 15.5.1, 15.6.1, and 15.8.1, respectively.

The 3D printing technology can be also used to build multifunctional composite structures. For example, continuous copper wire can be printed to power electronics for electric conduction, continuous nichrome wire can be printed to embed heat for anti-icing applications, or continuous fiber optics can be printed for a real-time structural health monitoring (SHM) system [Gardiner 2018]. All in all, these versatile functionalities facilitate the performance optimization of composite structures for specific applications.

15.8.1. Multifactorial Analysis of Simply Laminated Composites [±45°] and [0°/90°]

Attempting to detect the influence of the applied strain rate on the failure of simple composites that were built with ABS using 3D printing machine, Cantrell et al. [2017] and Hibbert et al. [2019] set up a physical experiment based on factorial design $2^3 \times 4$, of which design factors and their design levels are identified in Table 15.8.2. Resulting from physical tests, the ultimate tensile strengths corresponding to these 32 treatments under uniaxial loading are also listed in the table. Physical tests are done using the dog-bone specimens of type I per ASTM D638 specifications.

Here, a further analysis based on the t-distribution with the DOE is exploited to investigate potential influences of individual factors and their interactions. Findings from the study, as extended from the work done by Hibbert et al. [2019], reveal more advanced design information for understanding the mechanical behaviors of 3D-printed ABS composites [±45°] and [0°/90°], subjected to uniaxial loading in tension.

TABLE 15.8.1 Design factors and their design levels for failure analysis of composites [±45°] and [90°/0°] [Hibbert et al. 2019].

Factor	Level (−1)	Level (+1)	Coded variable
A: Raster angle (deg)	[±45°]	[0°/90°]	a = A
B: Layer thickness (μm)	254	330.2	b = (B − 292.1)/38.1
C: Interior fill style	Solid	High density	c = C
D: Strain rate (cm/s)	0	10	d = (D − 5)/5

© SAE International.

TABLE 15.8.2 Factorial design $2^3 \times 4$ for failure analysis of composites [±45°] and [90°/0°] [Hibbert et al. 2019].

Run	a	b	c	d	A	B	C	D	Y	Y_p
1	−1	−1	−1	−0.9746	[±45°]	254	Solid	0.127	20.89	21.87
2	1	−1	−1	−0.9746	[0°/90°]	254	Solid	0.127	20.82	22.11
3	−1	1	−1	−0.9746	[±45°]	330.2	Solid	0.127	20.62	21.71
4	1	1	−1	−0.9746	[0°/90°]	330.2	Solid	0.127	19.06	19.19
5	−1	−1	1	−0.9746	[±45°]	254	HD	0.127	18.00	17.94
6	1	−1	1	−0.9746	[0°/90°]	254	HD	0.127	17.17	18.18
7	−1	1	1	−0.9746	[±45°]	330.2	HD	0.127	16.62	17.77
8	1	1	1	−0.9746	[0°/90°]	330.2	HD	0.127	15.31	15.25
9	−1	−1	−1	−0.9	[±45°]	254	Solid	0.5	22.96	22.31
10	1	−1	−1	−0.9	[0°/90°]	254	Solid	0.5	22.89	22.55
11	−1	1	−1	−0.9	[±45°]	330.2	Solid	0.5	23.99	22.14
12	1	1	−1	−0.9	[0°/90°]	330.2	Solid	0.5	19.79	19.62
13	−1	−1	1	−0.9	[±45°]	254	HD	0.5	19.03	18.38
14	1	−1	1	−0.9	[0°/90°]	254	HD	0.5	19.37	18.62
15	−1	1	1	−0.9	[±45°]	330.2	HD	0.5	18.48	18.21
16	1	1	1	−0.9	[0°/90°]	330.2	HD	0.5	17.24	15.69
17	−1	−1	−1	0	[±45°]	254	Solid	5	26.06	26.05
18	1	−1	−1	0	[0°/90°]	254	Solid	5	26.24	26.29
19	−1	1	−1	0	[±45°]	330.2	Solid	5	25.86	25.88
20	1	1	−1	0	[0°/90°]	330.2	Solid	5	22.61	23.36
21	−1	−1	1	0	[±45°]	254	HD	5	21.64	22.12
22	1	−1	1	0	[0°/90°]	254	HD	5	22.75	21.64
23	−1	1	1	0	[±45°]	330.2	HD	5	21.99	21.95
24	1	1	1	0	[0°/90°]	330.2	HD	5	19.37	19.43
25	−1	−1	−1	1	[±45°]	254	Solid	10	27.03	26.91
26	1	−1	−1	1	[0°/90°]	254	Solid	10	27.30	27.15
27	−1	1	−1	1	[±45°]	330.2	Solid	10	27.44	26.74
28	1	1	−1	1	[0°/90°]	330.2	Solid	10	24.06	24.22
29	−1	−1	1	1	[±45°]	254	HD	10	22.68	22.97
30	1	−1	1	1	[0°/90°]	254	HD	10	23.72	23.21
31	−1	1	1	1	[±45°]	330.2	HD	10	22.41	22.81
32	1	1	1	1	[0°/90°]	330.2	HD	10	19.86	20.29

© SAE International.

Notes:
Y (MPa): Ultimate strength under uniaxial loading by physical tests
Y_p (MPa): Ultimate strength under uniaxial tension by predictive equation based on DOE

TABLE 15.8.3 Coded coefficients based on t-distribution for failure analysis of composites [±45°] and [90°/0°].

Term	Effect	Coefficient	SE coeff.	T-value	p-Value (%)
Constant	—	23.428	0.291	80.48	0.0
A	−1.139	−0.570	0.152	−3.75	0.1
B	−1.549	−0.774	0.152	−5.10	0.0
C	−3.934	−1.967	0.152	−12.96	0.0
D	5.185	2.593	0.183	14.18	0.0
D^2	−3.470	−1.735	0.362	−4.79	0.0
A B	−1.380	−0.690	0.146	−4.71	0.0
A C	0.376	0.188	0.146	1.28	21.4
A D	−0.025	−0.013	0.183	−0.07	94.6
B C	−0.145	−0.073	0.146	−0.50	62.6
B D	−0.269	−0.134	0.183	−0.74	47.1
C D	−0.275	−0.138	0.183	−0.75	46.0

The predictive equation for the ultimate strength under uniaxial loading is formulated using "Minitab → Stat → DOE." After the insignificant terms are dropped one by one, the predictive equations are derived with the statistical significance level of 5% (α = 5%) by means of the F-distribution as in Table 15.8.3 as follows:

$$Y_p = 23.428 - 0.570\ a - 0.774\ b - 0.690\ a\ b - 1.967\ c + 2.593\ d - 1.735\ d^2$$

$$(R = 97.9\%, R_{adj} = 96.8\%, \text{ and } R_{pred} = 95.2\%) \qquad . \qquad (15.8.1)$$

Effectiveness of the above equation is validated by the three different correlations given above. Everyone one of them is above 95%. The predicted values of ultimate strength using Equation (15.8.1), i.e., column Y_p in Table 15.8.2, correlate very well with the data obtained from physical tests.

The main effect of every factor is statistically significant: Hatched ply [±45°] is stronger than cross-ply [0°/90°], a thin layer thickness (factor A) yields a stronger composite, and a solid fill behaves better than a HD fill (factor C). Furthermore, the impact of strain rate on the ultimate strength under uniaxial loading is quite nonlinear. In the meanwhile, the interaction between the raster angle (factor A) and layer thickness (factor B) is also statistically significant. By taking partial differentiations of Y_p with respect to individual factors and setting them to zeros, one is able to solve them for maximizing the tensile strength under uniaxial loading as follows:

$$\partial Y_p / \partial a = -0.57 - 0.69\ b = 0 \quad \rightarrow \quad b = -0.8261 \quad \rightarrow \quad B = 260.63\ \mu m$$

$$\partial Y_p / \partial b = -0.744 - 0.69\ a = 0 \quad \rightarrow \quad a = -1.0783 \quad \rightarrow \quad A = [\pm 45°]$$

$$\partial Y_p / \partial c = -1.967 \quad \rightarrow \quad c = -1 \quad \rightarrow \quad C = \text{Solid}$$

$$\text{and} \quad \partial Y_p / \partial d = 2.593 - 3.47\ d = 0 \quad \rightarrow \quad d = 0.7473 \quad \rightarrow \quad D = 8.7363\ cm/sec.$$

Therefore, the optimal design configuration for reaching the ultimate strength in tension under axial loading is to use cross-ply composite [±45°] (factor A; a = −1) with the layer thickness of 260.63 μm (factor B; b = −0.8261) and solid fill (factor C; c = −1). With these conditions, Equation (15.8.1) reduces to

$$Y_p = 23.428 - 0.570\ (-1) - 0.774\ (-0.8261) - 0.690\ (-1)\ (-0.8261) - 1.967\ (-1)$$

$$+ 2.593\ [(D - 5) / 5] - 1.735\ [(D - 5) / 5]^2$$

$$= 21.7064 + 1.216\ D - 0.0994\ D^2. \qquad (15.8.2)$$

A plot of Y_p (predicted ultimate strength subjected to uniaxial loading) versus the applied strain rate (factor D) is exhibited in Figure 15.8.2. It shows that the resultant strength of the composite [±45°] built with ABS is a parabolic function of the applied strain rate with the peak strength ($Y_p = 27.0$ MPa) at D = 8.7363 cm/s.

FIGURE 15.8.2   Ultimate strength of composite [±45°] built with 3D printing of ABS under uniaxial tension at various strain rates.

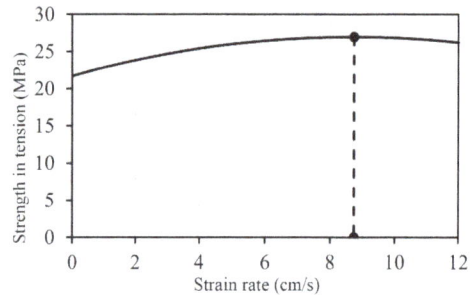

© SAE International.

15.8.2. Fracture Toughness by Double Cantilever Beam (DCB) Test

Delamination may also occur between two adjacent layers of dissimilar materials, including 3D-printed laminates with different filament (fibrous) directions in the boundary surface of two adjacent laminae, when subjected to applied mechanical and/or thermal loads [Taylor et al. 2019].

A major drawback of parts built with 3D printing is the interlaminar (interlayer) strength, just like fiber-reinforced composites, predominantly characterized by weak stress intensity factors that are responsible for delamination (separation) between laminae (layers), as demonstrated in Figure 15.8.3(a). Microstructural features such as reinforcements and porosities, including the size, shape, orientation, distribution of the reinforcements, and discontinuities in the material of the printed parts, have influences on the overall material properties.

Reprinted with permission. © Elsevier.

FIGURE 15.8.3 Mode I interlaminar fracture toughness tests [Somireddy et al. 2020].

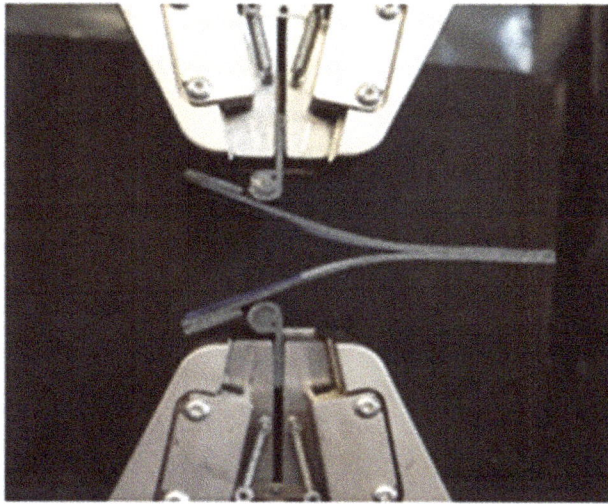

(a)

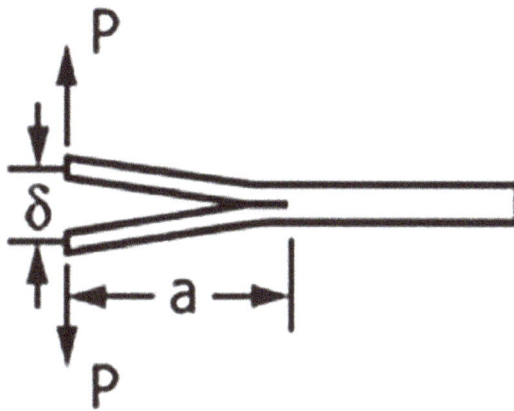

(b)

FIGURE 15.8.4   Mode I interlaminar fracture toughness varying with layer thickness [Somireddy et al. 2020].

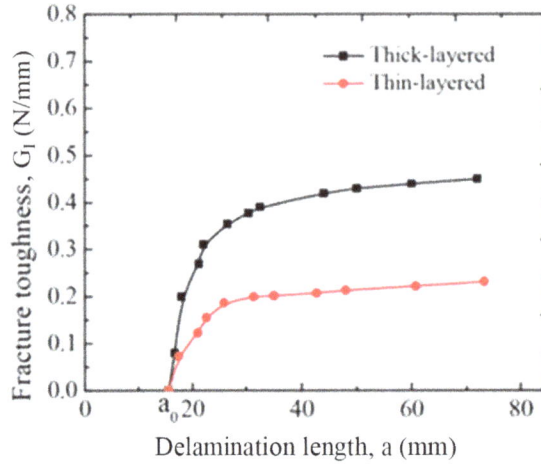

Interlaminar fracture toughness is frequently used to represent the stress intensity factors [Chiang 2019]. DCB test specimens of ABS/SCF made in accordance with ASTM D5528, as shown in Figure 15.8.3, can be built with 3D printing to investigate the interlaminar fracture toughness of the part subjected to mode I crack opening mode. Note that the fracture toughness in mode I failure under tensile loading is assessed as follows:

$$G_{IC} = \frac{P^2}{2B}\left(\frac{dC}{da}\right) \approx \frac{3P\delta}{2Ba_0},$$
(15.8.3)

where
G_{IC} (K/m^2) is the fracture toughness of mode I, i.e., crack opening mode
B (m or mm) is the specimen width
P (N) is the critical force applied
C (m/N) is the compliance, i.e., reciprocal of the initial slope of the load–displacement curve
a (m or mm) is the crack length
a_0 (m or mm) is the initial crack length
δ (m or mm) is the opening

The approximating formula given above is valid only if there is no "side" twisting or shear effects involved in the loading function. It is shown in Figure 15.8.4 that the interlaminar fracture toughness of the thin-layered laminae (thickness = 158 μm) is not so robust as the thick-layered parts (thickness = 317 μm). On the other hand, the in-plane tensile strengths in both the primary (printing) direction and transverse direction of the thin-layered laminae are stronger than those of the thick-layered laminae [Somireddy et al. 2020].

Fracture toughness of 3D-printed ABS laminae studied by Sheth [2019] is further explained here. Given that specimen width B = 20 mm, the data of fracture toughness at the room temperature can be calculated using Equation (15.8.3) and the data are listed in Table 15.8.4. Again, the idealized data of fracture toughness of specific 3D-printed laminae given here are for reference only, since they depend on many operating parameters of 3D printing and physicochemical properties of materials.

TABLE 15.8.4 Fracture toughness of ABS laminae estimated by DCB tests per ASTM D5528 at 23°C [Sheth 2019].

Run	Force P (N)	Opening δ (mm)	Crack a_o (mm)	G_{IC} (K/m²)
1	64.44	37.37	71	2.544
2	64.08	38.04	72	2.539
3	63.39	38.70	73	2.520
4	59.99	39.31	74	2.390
5	57.71	39.67	75	2.289
6	57.28	40.80	76	2.307
7	58.35	43.27	77	2.459
8	57.33	44.85	78	2.472
9	56.46	45.74	79	2.451

Printer Settings: Nozzle temperature: 230°C; bed temperature = 110°C; infill = 100%; infill angle = 0°; printing speed = 50 mm/s; layer height = 0.2 mm; extrusion width: auto; no. of perimeter shells: 0; generate supports: on

Fracture toughness of 3D-printed parts can be improved by means of heat treatment right after laminae are printed. The effect of heat treatment can be examined directly by bare eyes as shown in Figure 15.8.5. After a heat treatment under 160°C, adjacent layers soundly fuse to each other creating a cross section nearing a solid as the volume of voids tremendously reduces.

FIGURE 15.8.5   Effect of heat treatment on reduction in voids in 3D-printed laminae [Sheth 2019].

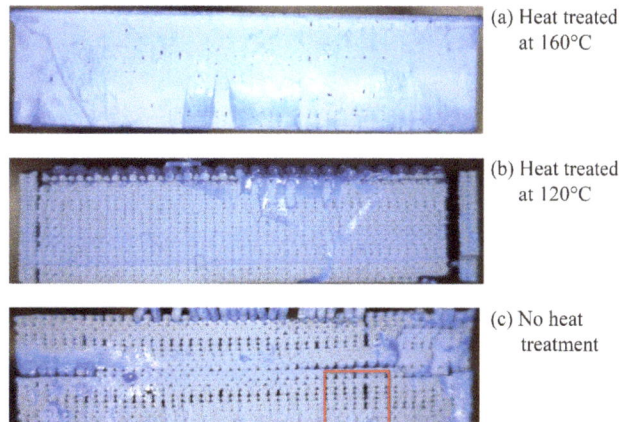

(a) Heat treated at 160°C

(b) Heat treated at 120°C

(c) No heat treatment

15.8.3. Fracture Toughness by Charpy Impact Test

An experimental design focused on the mechanical properties of ABS parts built with 3D-printing was carried out by Chaturvedi [2009]. Five process parameters (factors) of concern and their corresponding application levels are identified in Table 15.8.5. The objective is to identify viable process parameters for producing parts of high-impact strength (kJ/m²) based on a spherical composite design. Note that two independent angular transformations have to be taken into consideration, i.e., the layer orientation and raster angle given in Table 15.8.5.

TABLE 15.8.5 Design factors and their design levels for mechanical performance of 3D-printed parts [Chaturvedi 2009].

Factor	Levels				
	(−2)	(−1)	(0)	(1)	(2)
A: Layer thickness (μm)	127	158	190	222	254
B: Sample orientation (deg)	0	15	30	45	60
C: Raster angle (deg)	0	15	30	45	60
D: Raster width (μm)	406.4	426.4	446.4	466.4	486.4
E: Air gap (μm)	0	2	4	6	8

© SAE International.

TABLE 15.8.6 Fracture toughness of 3D-printed laminae using spherical composite design based on 2_V^{5-1} design (with alias equation E = ABCD) [Chaturvedi 2009].

Run	a	b	c	d	e	A	B	C	D	E	Y	Y_p
1	−1	−1	−1	−1	1	158	15	15	426.4	6	11.80	11.34
2	1	−1	−1	−1	−1	222	15	15	426.4	2	11.30	11.41
3	−1	1	−1	−1	−1	158	45	15	426.4	2	10.50	11.16
4	1	1	−1	−1	1	222	45	15	426.4	6	14.00	13.80
5	−1	−1	1	−1	−1	158	15	45	426.4	2	10.85	10.64
6	1	−1	1	−1	1	222	15	45	426.4	6	11.35	11.40
7	−1	1	1	−1	1	158	45	45	426.4	6	12.85	13.03
8	1	1	1	−1	−1	222	45	45	426.4	2	11.60	11.22
9	−1	−1	−1	1	−1	158	15	15	466.4	2	10.79	10.52
10	1	−1	−1	1	1	222	15	15	466.4	6	13.80	13.90
11	−1	1	−1	1	1	158	45	15	466.4	6	10.90	11.51
12	1	1	−1	1	−1	222	45	15	466.4	2	11.90	12.33
13	−1	−1	1	1	1	158	15	45	466.4	6	12.80	13.13
14	1	−1	1	1	−1	222	15	45	466.4	2	10.40	10.58
15	−1	1	1	1	−1	158	45	45	466.4	2	11.70	11.56
16	1	1	1	1	1	222	45	45	466.4	6	11.37	11.57
17	−2	0	0	0	0	127	30	30	446.4	4	11.50	11.74
18	2	0	0	0	0	254	30	30	446.4	4	12.14	12.47
19	0	−2	0	0	0	190	0	30	446.4	4	10.80	11.43
20	0	2	0	0	0	190	60	30	446.4	4	12.30	12.24
21	0	0	−2	0	0	190	30	0	446.4	4	13.40	13.49
22	0	0	2	0	0	190	30	60	446.4	4	12.30	12.78
23	0	0	0	−2	0	190	30	30	406.4	4	11.40	12.04
24	0	0	0	2	0	190	30	30	486.4	4	12.40	12.32
25	0	0	0	0	−2	190	30	30	446.4	0	11.50	11.90
26	0	0	0	0	2	190	30	30	446.4	8	14.30	14.47
27	0	0	0	0	0	190	30	30	446.4	4	14.90	15.11
28	0	0	0	0	0	190	30	30	446.4	4	15.50	15.11
29	0	0	0	0	0	190	30	30	446.4	4	14.70	15.11
30	0	0	0	0	0	190	30	30	446.4	4	15.00	15.11
31	0	0	0	0	0	190	30	30	446.4	4	15.10	15.11
32	0	0	0	0	0	190	30	30	446.4	4	14.90	15.11

© SAE International.

Experimental treatments at 23°C and relative humidity of 50% are listed in Table 15.8.6, so are the data resulting from the Charpy impact test (Wolpert pendulum impact test machine by Instron) of 3D-printed ABS parts at a speed of 3.8 m/s. Column Y is used to store the data of impact energy (Joules), i.e., force * displacement, resulting from impact. The predictive equation for the impact energy is formulated using "Minitab ➔ Stat ➔ Regression." After insignificant terms are removed one by one, the predictive equation is obtained with the statistical significance level of 5%, except two-factor interaction BE, by means of ANOVA based on F-statistics exhibited in Table 15.8.7:

$$Y_p = -412.1 + 0.3313 \, A + 0.775 \, B + 0.3797 \, C + 1.673 \, D + 1.473 \, E$$

$$- 0.000747 \, A^2 - 0.003645 \, B^2 - 0.002200 \, C^2 - 0.001831 \, D^2 - 0.1206 \, E^2$$

$$- 0.001366 \, A \, C - 0.001160 \, B \, D - 0.00623 \, B \, E$$

$$(R = 98.0\%, R_{adj} = 96.6\%, \text{ and } R_{pred} = 92.3\%). \tag{15.8.4}$$

Predicted values of fracture toughness corresponding to the 32 treatments (runs) using the above equation are listed as column Y_p in Table 15.8.6. Taking partial differentiations of Y_p with respect to factors A, B, C, D, and E, and setting them to zeros,

$$\partial Y_p / \partial A = 0.3313 - 0.001494 \, A - 0.001366 \, C = 0$$

$$\partial Y_p / \partial B = 0.775 - 0.00729 \, B - 0.00116 \, C = 0$$

$$\partial Y_p / \partial C = 0.3797 - 0.0044 \, C - 0.001366 \, A = 0$$

$$\partial Y_p / \partial D = 1.673 - 0.003662 \, D - 0.00116 \, B = 0$$

and $\quad \partial Y_p / \partial E = 1.473 - 0.2412 \, E - 0.00623 \, B = 0.$

Since $\partial^2 Y_p/\partial A^2 < 0$, $\partial^2 Y_p/\partial B^2 < 0$, $\partial^2 Y_p/\partial C^2 < 0$, $\partial^2 Y_p/\partial D^2 < 0$, and $\partial^2 Y_p/\partial E^2 < 0$, the solution to the above five equations attains the maximum value of fracture toughness when A = 207.21 μm, B = 103.58°, C = 17.15°, D = 424.04 μm, and E = 3.432 μm.

TABLE 15.8.7 ANOVA for impact strength of 3D-printed parts.

Source	SS_{adj}	DOF	MS_{adj}	F-value	p-Value (%)
Regression	72.9127	13	5.6087	34.39	0.0
A	21.4755	1	21.4755	131.69	0.0
B	4.2239	1	4.2239	25.90	0.0
C	11.6240	1	11.6240	71.28	0.0
D	16.4136	1	16.4136	100.65	0.0
E	10.3768	1	10.3768	63.63	0.0
A^2	16.7299	1	16.7299	102.59	0.0
B^2	19.7354	1	19.7354	21.02	0.0
C^2	7.1922	1	7.1922	44.10	0.0
D^2	15.7485	1	15.7485	96.57	0.0
EE	6.8336	1	6.8336	41.90	0.0
AC	6.8775	1	6.8775	42.17	0.0
BD	1.9391	1	1.9391	11.89	0.3
BE	0.5588	1	0.5588	3.43	
Error	2.9353	18	0.1631		
Lack of fit	2.5670	13	0.1975	2.68	14.2%
Pure error	0.3683	5	0.0737		
Subtotal	75.8480	31			
Average	—	1			
Total	Grand average	32			

15.8.4. Stress Intensity Factor

Interpretation of fracture toughness because of interlaminar delamination in 3D-printed laminae can be also represented using the concept of stress intensity factors per ASTM E1820-01 [Gardan et al. 2018, Chiang 2019].

15.9. Printed Acoustic Noise Attenuators

Materials with tunable porosity in 3D-printed parts, e.g., plastics and metals, can be used in the noise control of vehicles, aircraft, buildings, etc. A 3D-printed part with such materials has a "transversely placed bilayer medium with large degrees of contrast in the layers' acoustic properties [Liu et al. 2016]." It consequently releases an asymmetrical transmission, similar to the phenomenon of Fano interference, i.e., an interference between a background and a resonant scattering process producing the asymmetric line-shape transmission for noise reduction. Precise dimensions and specifications, the structure would need to have, could interfere with the transmitted sound waves that prevent sound flow while permitting airflow. Furthermore, a planar profile and subwavelength thickness ($\sim\lambda/8$) acoustic ventilation barrier prohibitive for sound in a broad range, such as hollowed-out horn-like helical metasurface [Sun et al. 2020a], allows for noise reduction and airflow passage simultaneously in some practical applications.

A series of open-cell porous materials with straight micro-tube-shaped holes has been built with 3D printing techniques by Jiang et al. [2017]. The design properties of the 3D-printed PC parts with varying geometries are given in Table 15.9.1. The objective is to attenuate the noise as measured using the peak coefficient of sound absorption, which occurs somewhere between 5400 and 6200 Hz, i.e., labial sounds giving presence to speech, for all nine run conditions. Note that the generally accepted standard hearing range for humans falls between 20 Hz and 20×10^3 Hz. The experimental design matrix and test results are listed in Table 15.9.2.

TABLE 15.9.1 Potential factors and their design levels for 3D-printed noise attenuators [Jiang et al. 2017].

Factor	(1)	(2)	(3)	(4)	(5)	(6)	(7)
A: Number of holes	45	69	97	146	269		
B: Hole diameter (mm)	0.6	0.8	1				
C: Degree of porosity (%)	5.35	8.2	11.11	11.53			
D: Thickness (mm)	5	6	7	8	9	10	
E: Aspect ratio (thickness/diameter)	0.06	0.08	0.1	0.111	0.125	0.143	0.167

TABLE 15.9.2 Design matrix for 3D-printed parts for acoustic noise reduction [Jiang et al. 2017].

Run	A	B	C	D	E	Y (%)	Y_p (%)	F (Hz)	F_p (Hz)
1	97	1	11.53	10	0.1	100	103.5	6200	6262
2	69	1	8.20	10	0.1	92	90.6	5900	5812
3	45	1	5.35	10	0.1	79	79.6	5380	5427
4	146	0.8	11.11	10	0.08	98	98.3	6200	6205
5	269	0.6	11.51	10	0.06	96.5	96.3	6200	6259
6	97	1	11.53	9	0.111	82	78.6	6400	6301
7	97	1	11.53	8	0.125	61	56.6	6400	6336
8	97	1	11.53	7	0.143	34	38.5	6300	6367
9	97	1	11.53	6	0.167	25	26.1	6400	6393
10	97	1	11.53	5	0.2	24	22.5	6400	6416
11*	0	0	0	5	5/29	22	—	6400	—

Y (%): Coefficient of noise attenuation
F (Hz): Frequency
* : Not included in regression models

First, consider the data of coefficients of noise attenuation in column Y of Table 15.9.2. Based on the ANOVA with F-distribution as shown in Table 15.9.3(a), the overall contribution of individual factors and their interactions to the peak noise attenuation (column Y) can be summarized using the following predictive equation:

$$Y (\%) = -314.2 - 78.7\, B + 3.868\, C + 35.52\, D + 966\, E$$

$$(R = 99.6\%,\ R_{adj} = 99.3\%,\ \text{and}\ R_{pred} = 97.9\%). \tag{15.9.1}$$

With a cutoff statistical significance level of 5%, the main effects of factors B (hole diameter), C (degree of porosity), D (thickness), and E (aspect ratio) are statistically significant. The predicted values of peak noise attenuation using Equation (15.9.1) are listed in column Y_p in Table 15.9.2, which correlate well with test data listed in column Y.

Next, consider the data of frequencies at peak noise attenuation in column F of Table 15.9.2. Based on the ANOVA with the F-distribution as shown in Table 15.9.3(b), the overall contribution of individual factors and their interactions to locating the peak noise attenuation (column F) can be summarized using the following predictive equation:

$$F (Hz) = 4908 + 135.2\, C - 2.047\, D^2$$

$$(R = 98.0\%,\ R_{adj} = 97.5\%,\ \text{and}\ R_{pred} = 94.3\%). \tag{15.9.2}$$

With a cutoff statistical significance level of 5.5%, the main effects of factor C (degree of porosity) and the quadratic term D^2 (thickness) are statistically significant. The predicted values of peak noise attenuation using Equation (15.9.2) are listed as column F_p in Table 15.9.2, which correlate well with test data listed in column F.

TABLE 15.9.3 ANOVA for 3D-printed parts for acoustic noise reduction.

(a) Coefficient of noise attenuation (Y)					
Source	**SS_{adj}**	**DOF**	**MS_{adj}**	**F-value**	**p-Value (%)**
Regression	8537.06	4	2134.26	154.73	0.0
B	201.56	1	201.56	14.61	1.2
C	341.81	1	341.81	24.78	0.4
D	950.75	1	950.75	68.93	0.0
E	272.75	1	272.75	19.77	0.7
Error	68.97	5	13.79		
Subtotal	8606.02	9			
Grand average	—	1			
Total	—	10			
(b) Frequency at peak noise attenuation (F)					
Source	**SS_{adj}**	**DOF**	**MS_{adj}**	**F-value**	**p-Value (%)**
Regression	891,577	2	445,789	86.72	0.0
C	586,994	1	586,994	114.19	0.0
DD	27,132	1	27,132	5.28	5.5
Error	35,983	7	5140		
Subtotal	927,560	9			
Grand average	—	1			
Total	—	10			

Besides what was studied above, the use of 3D-printed porous coatings in a single or tandem cylinder system is also one of the passive flow control methods that are effectively used to reduce turbulence and noise and vibrations generated as the fluid flows in a pipe. With a layer of coating with higher porosity, a more reduction in turbulence and mean wake velocity can be achieved [Bathla and Kennedy 2020].

15.10. Porous Structures and Triply Periodic Minimal Surfaces

Porosity in a 3D-printed product is a viable design factor in the design space, e.g., scaffolds applied in bone tissue engineering (BTE). An artificial scaffold has to be strong in material strength but soft enough in structural performance in order to be compatible with real bones with the extracellular matrix of cells.

Lattice structures demonstrate the fundamental concept of designing with porosity. A lattice structure starts with the choice of type of lattice cells. Three frequently used lattice cells are exhibited in Figure 15.10.1, including the primitive, gyroid, and diamond. Smooth interlinking solids of a lattice structure may enhance various areas of performance and often utilize less material without weakening the strength while compromising other functional requirements.

FIGURE 15.10.1 Lattice constructions with triply periodic minimal surfaces—unit cells and 3D-printed structures [Restrepo et al. 2017].

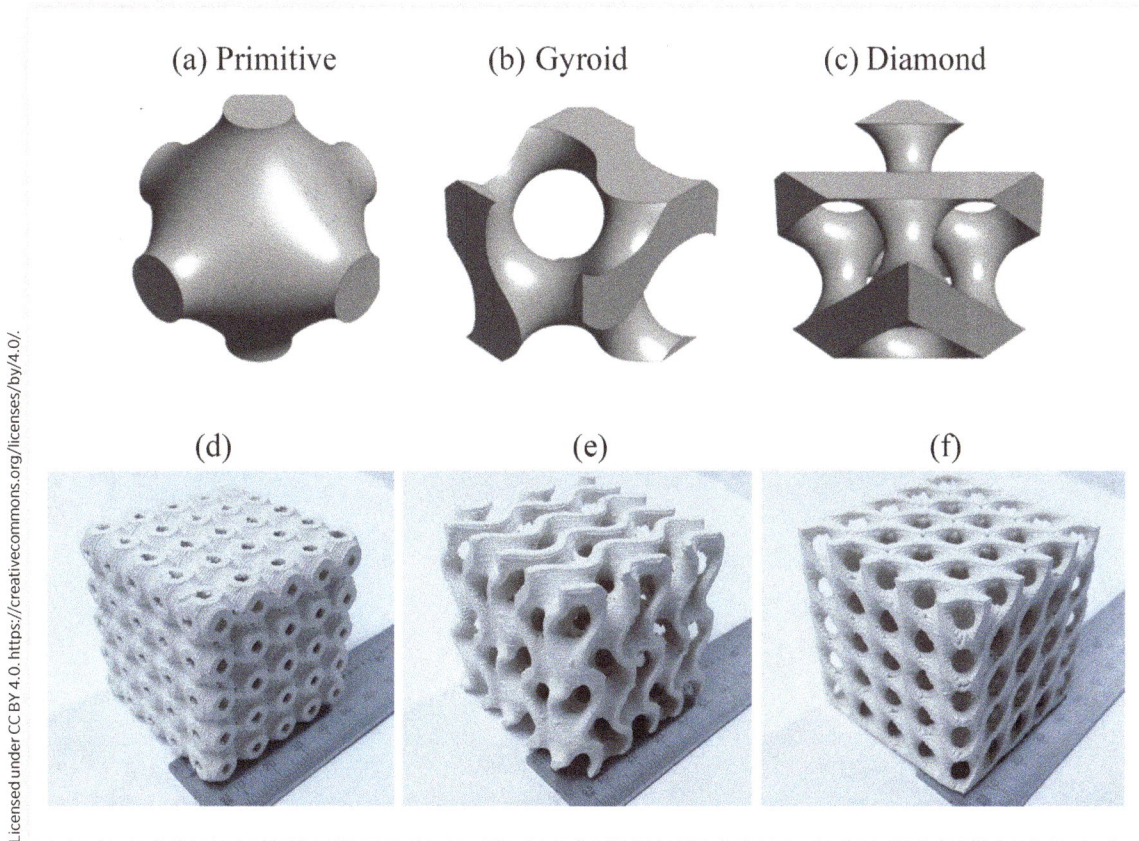

(a) Primitive (b) Gyroid (c) Diamond

(d) (e) (f)

15.10.1. Triply Periodic Minimal Surfaces

How to design 3D-printed porous structures with triply periodic minimal surfaces? It can be carried out using 3D printing techniques that are more convenient than subtractive manufacturing. Diamond, gyroid, I-WP graph, neovius, and primitive structures are members of cellular structures with the triply periodic minimal surfaces, featuring high surface-to-volume ratios with no apparent discontinuity in theory [Torquato and Donev 2004, Galusha 2008, Schroder-Turk et al. 2011, Abuiedda et al. 2019]. Three frequently used lattice structures, which come with gyroid-, diamond-, and primitive-shaped triply periodic minimal surfaces, are shown in Figure 15.10.1.

When a 3D-printed structure is designed with a triply periodic minimal surfaces, it is also expected to carry a higher load than those using the same amount of material as the effects of stress concentration and/or stress intensity are reduced [Maskely et al. 2018]. Functional equations for primitive (type P), gyroid (type G), and diamond (type D) lattice structures are, respectively, given as follows:

$$\Phi(x, y, z) = \cos(a\,x) + \cos(b\,y) + \cos(c\,z) + d \quad \text{(Primitive)}, \tag{15.10.1}$$

$$\Phi(x, y, z) = \cos(a\,x)\sin(b\,y) + \cos(b\,y)\sin(c\,z)$$

$$+ \cos(c\,z)\sin(a\,x) + d \qquad \text{(Gyroid)}, \tag{15.10.2}$$

and $\quad \Phi(x, y, z) = \sin(a\,x)\sin(b\,y)\sin(c\,z) + \sin(a\,x)\cos(b\,x)\cos(c\,z)$

$$+ \cos(a\,x)\sin(b\,y)\cos(c\,z)$$

$$+ \cos(a\,x)\cos(b\,y)\sin(c\,z) + d \qquad \text{(Diamond)}, \tag{15.10.3}$$

where

$\Phi = 0$ is the isosurface that represents the solid-void boundary of a scaffold

a, b, and c are the geometric parameters that control pore size and surface architectures

d is the parameter that controls the volume fraction of pores in the resulting lattice

That $\Phi < 0$ is to guarantee a porous design; otherwise ($\Phi \geq 0$), it is a solid. As for graded scaffolds, the offset value of parameter d in primitive type is defined as a function of scaffold height along the z-axis in order to obtain a graded structure. For a linear function, $d = k\,z$, Equations (15.10.1)–(15.10.3) become

$$\Phi(x, y, z) = \cos\left(\frac{n_x}{L_x}\,x\right) + \cos\left(\frac{n_y}{L_y}\,y\right) + \cos\left(\frac{n_z}{L_z}\,z\right) + k\,z \quad \text{(Primitive)}, \tag{15.10.4}$$

$$\Phi(x, y, z) = \cos\left(\frac{n_x}{L_x}\,x\right)\sin\left(\frac{n_y}{L_y}\,y\right) + \cos\left(\frac{n_y}{L_y}\,y\right)\sin\left(\frac{n_z}{L_z}\,z\right) + \cos\left(\frac{n_z}{L_z}\,z\right)\sin\left(\frac{n_x}{L_x}\,x\right), \tag{15.10.5}$$

$$+ k\,z \qquad \text{(Gyroid)}$$

and $\quad \Phi(x, y, z) = \sin\left(\frac{n_x}{L_x}\,x\right)\sin\left(\frac{n_y}{L_y}\,y\right)\sin\left(\frac{n_z}{L_z}\,z\right) + \sin\left(\frac{n_x}{L_x}\,x\right)\cos\left(\frac{n_y}{L_y}\,y\right)\cos\left(\frac{n_z}{L_z}\,z\right)$

$$+ \cos\left(\frac{n_x}{L_x}\,x\right)\sin\left(\frac{n_y}{L_y}\,y\right)\cos\left(\frac{n_z}{L_z}\,z\right) + \cos\left(\frac{n_x}{L_x}\,x\right)\cos\left(\frac{n_y}{L_y}\,y\right)\sin\left(\frac{n_z}{L_z}\,z\right), \tag{15.10.6}$$

$$+ k\,z \qquad \text{(Diamond)}$$

where

n_x, n_y, and n_z are the cell repetition along x-, y-, and z-axes, respectively

L_x, L_y, and L_z are the absolute sizes of the structure along x-, y-, and z-axes, respectively

Scaffolds based on the above triply periodic minimal surfaces equations yield the salient features of a high surface-to-volume ratio and a uniform stress distribution in compression. Automotive engineers utilize 3D printing techniques to build complex metal lattice structures for structural parts and battery components for EVs. With triply periodic minimal surfaces-shaped structural parts, the weight reduction in automobile components is made feasible by decreasing their apparent densities.

Another advantage of 3D-printed parts with triply periodic minimal surfaces is their usage for function-ally graded bone implants. For example, the porosity of a bone implant has to have porosities within the range of those of the trabecular bone that is to be replaced accordingly. The lower the porosity, the higher the compressive strength and mechanical stiffness (e.g., Young's modulus). The mechanical stiffness of bone implants has to be similar to the original bone to avoid the potential stress-shielding effect [Huiskes et al. 1992, Vijayavenkataraman et al. 2020]. Furthermore, bone implants must display relatively uniform stress distributions across all members under compression or tension, leading to stable collapse mechanisms and desired energy absorption performance, should it happen. Note that the ultimate compressive strength of trabecular bones ranges from −2 to −12 MPa [Velasco et al. 2015]. It is demonstrated in Table 15.10.3 that all three triply periodic minimal surfaces built with 3D-printed ABS scaffolds are applicable for generating bone implants [de Aquino et al. 2020].

TABLE 15.10.3 Factorial design for triply periodic minimal surfaces of 3D-printed ABS scaffolds [de Aquino et al. 2020].

Run	A	B	σ_{ucs} (MPa)	E_{ii} (GPa) in compression
1	Primitive	0°	$\sigma_{11,ucs} = -11.58$	$E_{11} = 0.17026$
2	Gyroid	0°	$\sigma_{11,ucs} = -4.13$	$E_{11} = 0.10534$
3	Diamond	0°	$\sigma_{11,ucs} = -3.59$	$E_{11} = 0.08521$
4	Primitive	90°	$\sigma_{22,ucs} = \sigma_{33,ucs} = -10.02$	$E_{22} = E_{33} = 0.17392$
5	Gyroid	90°	$\sigma_{22,ucs} = \sigma_{33,ucs} = -4.02$	$E_{22} = E_{33} = 0.08315$
6	Diamond	90°	$\sigma_{22,ucs} = \sigma_{33,ucs} = -3.08$	$E_{22} = E_{33} = 0.06720$

Notes:
Factor A: Type of triply periodic minimal surfaces
Factor B: Loading direction
σ_{ucs} (MPa): Ultimate compressive strength; for trabecular bones, $-2 > \sigma_{ucs} > -12$

15.10.2. 3D Porous Graphene Structure as a Significant Breakthrough

Graphene has been hailed as a revolutionary material with the potential to energize 3D printing, and some important product breakthroughs have been achieved recently. Graphene is a material that consists of a 2D sheet of three covalently bonded carbon atoms in a sp^2 hybridization resulting in a planar hexagonal lattice with excellent material properties:

(a) Intrinsic electronic mobility: 20 m²/(V·s)

(b) Current density tolerance: 10^{12} A/m²

(c) Thermal conductivity: 5×10^3 W/(m °K)

Note that various mechanical strengths of graphene under tensile and compressive and shearing loads are measured in GPa, instead of MPa that is used for bulky materials.

It is a game changer to engineer a 3D porous graphene assembly that has almost the full electric and mechanical properties of a 2D graphene membrane with an atomic lattice structure. Nevertheless, the advantage of a 3D porous graphene assembly over woven structures made of lightweight thermoplastics such as PS may only exist with a specific gravity greater than 0.01 (i.e., 0.01 g/cm³ = 10 mg/cm³) as illustrated in Figure 15.10.2 [Qin et al. 2017].

FIGURE 15.10.2   Mechanical properties of 3D-printed porous graphene structure based on gyroid geometry [Figure 15.10.1(b)] relative to thermoplastic PS [Qin et al. 2017]. (a) Tensile modulus of elasticity and (b) ultimate tensile strength.

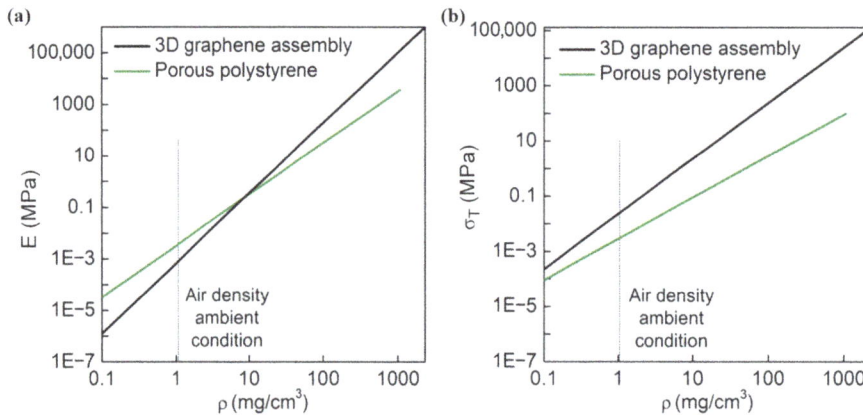

15.11. Printed Electrochemical Storage Devices

The capacity of electrochemical reactions in EV batteries can be enhanced using 3D-printed electrodes [Zhang et al. 2020]. Researchers also devise 3D structures for battery electrodes that exhibit a significant surface area for increasing the capacity of electrochemical reactions to power EVs and other electrical energy storage (EES) devices. This turns out to be a design with lattice-structured parts, having their interdigitated flow fields in 3D-printed scaffolds.

Li-air batteries are expected to revolutionize the automotive industry for use in EVs and electrochemical energy storage systems by surpassing the energy capacities of conventional Li-ion batteries [Yoo et al. 2017]. However, it is necessary for Li-air batteries to overcome low cyclic performance and high charging voltage, resulting from oxygen transport limitations, electrolyte degradation [Nomura et al. 2024], and the formation of irreversible reduction products [McCloskey et al. 2012].

For the oxygen transport in Li-air batteries, the cathode network can be streamlined for flow of air/oxygen (O_2) using 3D printing technique. The properties of 3D pore networks in battery electrodes are enhanced using effective transport capacity, including both the porous area in contact and the degree of streamlining (or constriction) in flow paths.

15.11.1. 3D-Printed EV Battery Electrodes

Micro- and mesostructured electrodes may be fabricated by combining 3D holographic lithography with conventional photolithography. It enables deterministic control of both the internal electrode mesostructured and the spatial distribution of the electrodes on the substrate, as demonstrated by Ning et al. [2015]. Battery electrodes of micro- or meso-lattice structures exhibit significantly greater battery capacity (denoted by mAh) carried per gram of weight [Lu et al. 2020] as made possible using higher-loaded active materials, larger surface area, and shorter ion transport path.

15.11.2. **3D-Printed Supercapacitors**

3D-printed electric double-layered capacitors (EDLCs) are a promising competitor for alternative energy storage because of their low cost, high power density, and long cycle life. A specific supercapacitor built with 3D printing is demonstrated in Figure 15.11.2 [Yao et al. 2019]. The performance of graphene/manganese dioxide (MnO_2) capacitor is charted against layer thickness in Figure 15.11.3 in terms of three different measures (areal, gravimetric, and volumetric capacitances). As reduced from graphene oxide, the 3D-printed lattice-structured graphene increases the contact area between graphene (electrode) and MnO_2-nano (electrolyte). Appropriation of MnO_2-nano-deposition to the highly conductive current collector is another significant contributing factor. The 3D printing process for supercapacitors is detailed in Yao et al. [2019].

FIGURE 15.11.2 MnO_2/graphene/MnO_2 supercapacitor [Yao et al. 2019].

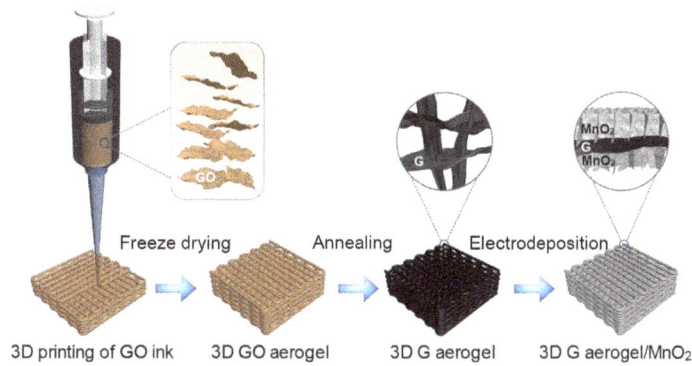

FIGURE 15.11.3   Performance of graphene/MnO_2 capacitor [Yao et al. 2019].

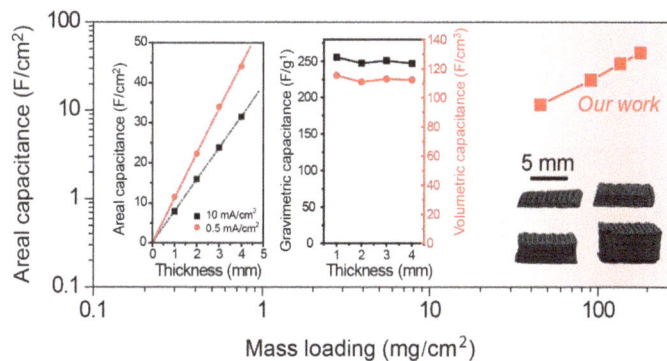

Furthermore, rechargeable batteries based on flexible double-layered capacitors that can be bent without affecting the battery capacity are desired for some applications such as wearable energy storage [Areir et al. 2017]. PVA has been used for flexible double-layered capacitors.

Material performances used for electric layered capacitors built with 3D printing techniques may be measured using gravimetric capacitance, volumetric capacitance, and areal capacitance, listed as follows [Yao et al. 2019]:

Electrode	Electrolyte	Gravimetric	Volumetric	Areal
Graphene aerogel	MnO_2	225 F/g	111 F/cm^3	—
Graphene aerogel	KOH	63.6 F/g	—	—
Reduced graphene oxide/Au	PVA-H_2SO	—	—	0.0038 F/cm^2
Reduced graphene oxide	PVA-H_2SO	—	41.8 F/cm^3	—
PI	PVA-H_2SO	—	0.0023 F/cm^3	—
Polypyrene (PPy)	PVA-H_3PO	—	2.4 F/cm^3	—
Graphene PLA	PVA-H_2SO_4	0.00049 F/g	—	—
Activated carbon particles	PVA-H_3PO	—	—	1.48 F/cm^2

15.12. Printed Sensors Such as Strain Gauges and Antennae

It is possible to control the dielectric permittivity by combining several filaments together or dynamically mixing inks that yield various design flexibility and functionality for 3D-printed sensors such as strain gauges and microwave antennae [Jeong et al. 2020].

The electromagnetic resonance frequency of remote antenna-type sensors varies according to the mechanical deformation of antennae. When these battery-free sensors are bonded onto a part surface, induced strains in daily operations and even potential cracks can be detected through convenient wireless interrogation as a SHM tool.

15.12.1. 3D-Printed Strain Gauges

3D-printed flexible and conformal-phased strain gauges have emerged as a great option as an appealing printed conductor for applications in sensors [Barši Palmić et al. 2020]. The top three common procedures in practice available for printing high-precision such metals are ink jet printing (IJP), DIW, and aerosol jet printing (AJP). Their technical capabilities are given as follows [Rosker et al. 2018]:

Printing	Ink viscosity (cp)	Nozzle diameter (µm)	Feature size (µm)	Line thickness (µm)
Ink jet	1–30	20–60	20–100	>0.6
Direct ink	1–10^6	0.1–1	1–1000	>0.5
Aerosol jet	1–10^3	150–300	10–200	>0.1

As a non-contact maskless method to print inks with tiny particles, AJP allows a higher print resolution and feasibility of deposition on non-planar surfaces, it is considered as an advanced alternative to screen printing and direct write methods like inkjet. Tiny particles can be silver (Ag), gold (Au), and CNTs.

A schematic drawing of AJP setup is exhibited in Figure 15.12.1. There will be a post-print curing, and UV irradiation is widely applied. One approach to reduce the surface roughness of a printed dielectric substrate is to follow with printing a more conformal dielectric material, which spreads over the rough substrate surface (a film like a build plate) before the UV cure is applied. The build plate is a 50 mm × 50 mm × 150-µm PI film

(E = 2.5 GPa and ν = 0.35), as specifically used by Rosker et al. [2018]. Connecting wires are attached to the contact pad of the strain sensor using conductive material (e.g., epoxy/silver) that has to be cured at an elevated temperature (e.g., 60°C for 30 min for epoxy/silver).

A porous film that contains many tiny holes is made possible using the 3D printing techniques. Because of the porosity in the film, the apparent Poisson ratio goes beyond the material property as the lateral contraction is assisted by pore contraction when stretched. For example, a possible 40% or more increase in the lateral contraction for a given deformation of the film makes the strain gauge much more sensitive to measurement [Rahman et al. 2018].

FIGURE 15.12.1   Schematic drawing of AJP facility [Agarwala et al. 2018].

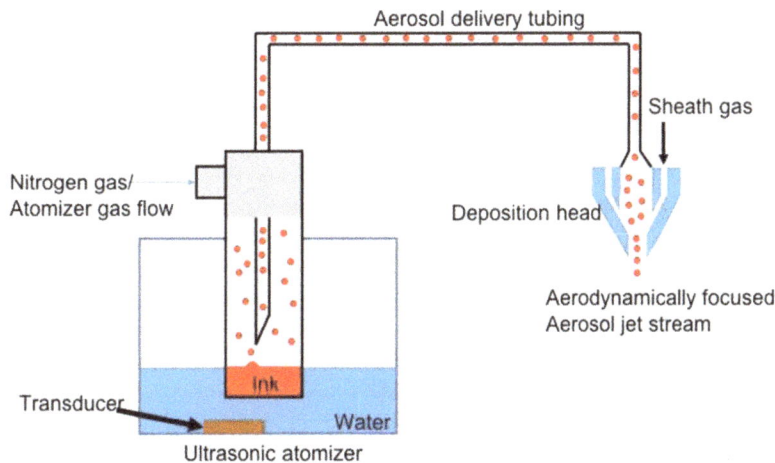

Factors that may have impacts on making strain gauges using the AJP method are listed in Table 15.12.1. Simple DOE involving two process parameters (factors)—ultrasonic current (Amp) and atomizer flow rate (sccm), i.e., factors D and E given in the table, were conducted by Agarwala et al. [2018]. Sixteen different experimental treatments (run conditions) pertaining to factorial design 4^2 are assigned. The objective is to generate good printed lines in terms of the line width and line quality (Figure 15.12.2), as denotably quantified by responses Y and Q in Table 15.12.2. Since there are no data available, the maximum-likelihood method has to be used for the analysis.

TABLE 15.12.1   Generic factors and their design levels for 3D-printed strain gauge [Agarwala et al. 2018].

Factor	Level
A: Ink fluid (PRELECT TPS)	Clariant
B: Ink viscosity (mPa·S)	10
C: Silver nanoparticles (E = 71 GPa, ν = 0.37)	50% (weight)
D: Ultrasonic current (Amp)	0.5
E: Atomizer flow rate (sccm)	30
F: Sheath gas flow rate (sccm)	30
G: Printing speed (mm/s)	7
H: Nozzle diameter (μm)	150 μm
I: Build plate (PI) temperature (°C)	40
J: Z-offset, i.e., gap—nozzle tip to printed surface (μm)	—
K: Aftertreatment—sintering temperature (°C)	200
L: Aftertreatment—sintering time (h)	2

TABLE 15.12.2 Factorial design for 3D-printed strain gauge [Agarwala et al. 2018].

Treatment	D	E	Y (μm)	Q (%)
1	0.3	14	Suspended	40
2	0.4	14	14.9	40
3	0.5	14	15.2	40
4	0.6	14	17	40
5	0.3	16	20.7	100
6	0.4	16	18.8	100
7	0.5	16	21.8	100
8	0.6	16	22.3	100
9	0.3	20	37.5	100
10	0.4	20	51.2	100
11	0.5	20	31.1	80
12	0.6	20	42.9	100
13	0.3	25	90.5	100
14	0.4	25	74.5	80
15	0.5	25	73.6	80
16	0.6	25	55.3	80

FIGURE 15.12.2 Qualification of strain gauge prototypes built with 3D printing using aerosol jet [Agarwala et al. 2018].

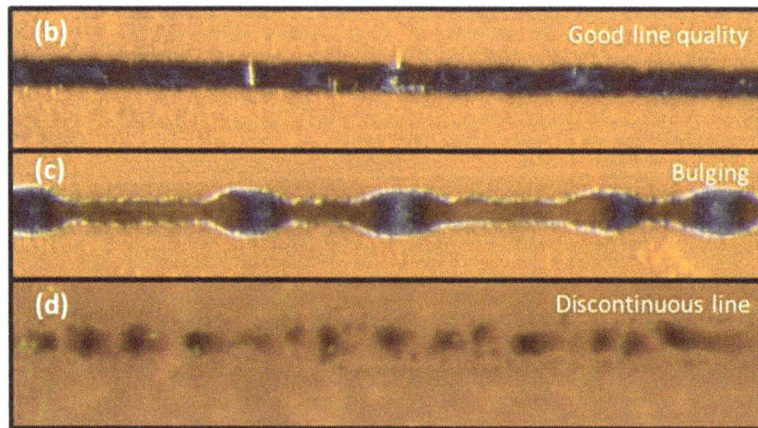

(b) Good line quality

(c) Bulging

(d) Discontinuous line

First, consider the data of line width in column Y of Table 15.12.2, without treatment 1. Based on the ANOVA with F-distribution as shown in Table 15.12.2, the overall contribution of individual factors and their interactions to the line width can be summarized using the following predictive equation:

$$Y = -49.5 + 178.7\,D + 0.2605\,E^2 - 10.64\,D\,E$$

$$(R = 97.9\%,\ R_{adj} = 97.3\%,\ \text{and}\ R_{pred} = 96.0\%). \qquad (15.12.1)$$

The main effect of factor D (ultrasonic current) is statistically significant, so is the quadratic term of factor E (atomizer flow rate). The two-factor interaction between D and E is also viable. A surface plot of Y (line width) versus ultrasonic current and atomizer flow rate is presented in Figure 15.12.3.

TABLE 15.12.3 ANOVA for line width of 3D-printed strain gauges.

Source	SS_{adj}	DOF	MS_{adj}	F-value	p-Value (%)
Regression	8195.9	3	2731.98	84.26	0.0
D	282.7	1	282.65	8.72	1.3
E^2	1615.5	1	1615.51	49.82	0.0
DE	392.8	1	392.75	12.11	0.5
Error	356.7	11	32.42		
Subtotal	8552.6	14			
Grand average	—	1			
Total	—	15			

© SAE International.

FIGURE 15.12.3 Response surface of printed line width as a function of atomizer flow rate (factor D) and atomizer flow rate (factor E).

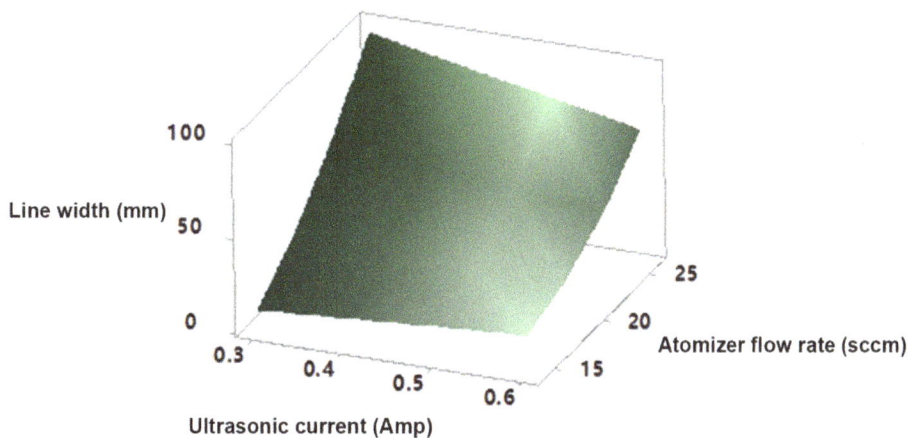

© SAE International.

15.12.2. 3D-Printed Antennae

A basic 3D-printed antenna pattern consists of three bonded layers, which include the top metal cladding (e.g., silver or gold) having an antenna pattern, the middle layer made of dielectric substrate (e.g., glass micro-fiber-reinforced PTFE material), and another metal cladding as the ground plane. Vias are used to connect the top copper cladding with the ground plane to achieve "folding," which may reduce the patch antenna size. After attaching an antenna sensor on a structure, the dimensions of the antenna change in response to strains developed in the structure. Consequently, resonance frequencies shift according to strains (i.e., dimensional variations per unit length) and can be wirelessly interrogated, e.g., utilizing RF identification (RFID) technology.

Nevertheless, as it is for the 3D-printed antenna sensor, the resonance frequency under strain ε for a thin substrate is related to the relative dielectric constant [Balanis 1997, Yi et al. 2015] as follows:

$$\omega = \frac{C}{4\,[L\,(1+\varepsilon)+L'\,(1-\nu\,\varepsilon)]\,(\kappa_r)^{1/2}} \approx \frac{C}{4\,(L+L')\,(1+\varepsilon)\,(\kappa_r)^{1/2}}, \tag{15.12.2}$$

where

C (m/s) is the speed of the light

ν is Poisson's ratio of the substrate

κ_r is the relative dielectric constant; κ_r = permittivity of substance/permittivity of free space

ω (Hz) is the resonance frequency at current strain ε

L (m) is the patch length

L' (m) is the additional length, compensating the effect of edge-fringing field [Finkenzeller 2003]

As strain ε is small, the above equation can be rewritten using Taylor's expansion series as

$$\omega \approx \frac{C}{4\,(L+L')\,(\kappa_r)^{1/2}}\,(1-\varepsilon). \tag{15.12.3}$$

An example output from the design configuration shown in Figure 15.12.3(a) and (b) is exhibited in Figure 15.12.3(c). With initial strain ε_0, Equation (15.2.3) reduces to

$$\omega_0 \approx \frac{C}{4\,(L+L')\,(\kappa_r)^{1/2}}\,(1-\varepsilon_0), \tag{15.12.4}$$

of which ω_0 is the initial resonance frequency at initial strain ε_0. Relating Equation (15.12.4) to Equation (15.12.3) leads to

$$\omega\,/\,\omega_0 = (1-\varepsilon)\ \ /\,(1-\varepsilon_0). \tag{15.12.5}$$

If initial strain $\varepsilon_0 = 0$ or small, the above equation reduces to

$$\omega \approx \omega_0\,(1-\varepsilon), \tag{15.12.6}$$

or

$$\omega - \omega_0 \approx -\,\omega_0\,\varepsilon. \tag{15.12.7}$$

If the die electric constant changes significantly, the above equation needs to be modified accordingly. As a generalized case, the strain sensitivity of a strain gauge is purposely defined as the frequency change over the strain change as

$$S = \frac{\omega - \omega_0}{\varepsilon - \varepsilon_0}, \tag{15.12.8}$$

i.e.

$$\Delta\omega = S\,\Delta\varepsilon, \tag{15.12.9}$$

where

S is the strain sensitivity, i.e., resonance frequency change in response to strain change

$\Delta\omega = \omega - \omega_0$ is the resonance frequency change

$\Delta\varepsilon = \varepsilon - \varepsilon_0$ is the strain change

For example, the ω–ε plot in Figure 15.12.3(c) reveals a constant strain sensitivity, S = −0.000703 MHz/ (μm/m). Because of the imperfection in material and geometric nonlinearities of the sensor structure, the strain sensitivity may not be a constant and should be updated with respect to the strain change accordingly [Yi et al. 2015].

Recently, 3D-printed flexible and conformal-phased array antennae have emerged as a popular choice for radar communications. These RF-based dielectric components in fully 3D forms outperform their planar counterparts and give more freedom for applications such as additionally having an extra function or being potentially embedded [Rosker et al. 2018]. For example, these antennae are printed on plastic films (carrier) that can be transferred to the plastic body parts.

Furthermore, wireless rechargeable sensor nodes can gather their energy from the transmission of energy sources such as RFID readers or harvesting from mechanical vibration rather than electric batteries. Given its small-form factors and universal sensing capabilities, it is expected that wireless rechargeable sensor will be a promising platform for different applications such as warehouse inventory management [Liu et al. 2006, Bijwaard et al. 2011], supply chain monitoring [Poon et al. 2009], and SHM (e.g., proper bolt-fastening force).

15.12.3. **3D-Printed Pressure Gauges**

3D-printed polymeric pressure gauges are so sensitive to the applied pressure that a small deformation to the gauge pad generates a change in the electrical resistance. This provides an analog signal to enable multi-functionalities, such as measurement, input, and control, possibly in a wide range of applications. 3D-printed pressure gauges can be directly integrated into a structural surface such as a PCB and skin of animals.

A simple wireless pressure sensor is also made possible by placing a thin layer of specially designed rubber between two strips of copper that act like radio antennae [Chen et al. 2014]. The rubber serves as an insulator. When the two strips of copper (antennae) squeeze the rubber (insulator) and move infinitesimally closer together, beaming radio waves go through this simple antenna-and-rubber sandwich as a wireless pressure sensor. As a simple example application, it allows a medical doctor to read a patient's pulse without touching him.

15.13. **Four-Dimensional (4D) Printing**

A part made of reactive materials (or structures) may be built with 4D printing techniques, i.e., 3D printing +1D reacting. The additional 1D is realized using special designs that exist in one form and behave otherwise in response to various levels of environmental stimuli, such as heat, moisture, pH value, chemical species, electric field, magnetic field, and even time (e.g., one constituent material degrading faster than the others) according to Tibbits [2014] and Rafiee et al. [2020]. Therefore, a 4D-printed part usually has two, or more, different physical appearances: The original shape before the stimulus is applied and the programmed shapes as the applied stimulus level changes. For example, it is demonstrated that electric heating of thermoplastics may form a flexible actuator built with 4D printing [Zhao et al. 2019a]. Nevertheless, potential material degradations including creep relaxation and ratcheting have to be avoided with proper design configurations and operating conditions.

4D printing techniques based on traditional smart materials, with 1D being designed to fit the future evolution in time domain and 3D printing as acute printing, can be applied to various fields in geometric domains ranging from simply shape changes to bio-printing for organisms. For example, one may create soft robots that move like jellyfish or even shape-changing medical implants that configure themselves to suit a human being's body.

15.13.1. DOE for Curvature Change due to 4D Printing

Laminates consisting of PLA and TPU plies, as shown in Figure 15.13.1(a), are built with 3D printing techniques. Note that PLA is more rigid than TPU. In the figure, circular dots are used to denote the "longitudinal direction of filaments," being coincident with raster direction, i.e., x-axis. The structure bends into U-shape when the environmental temperature goes higher than the glass transition point (T_g) of PLA (i.e., around 60°C), beyond which PLA laminae, working as the active material, get shortened in the raster direction and slightly expand in the other two directions. On the other hand, the glass transition point of TPU laminae is −25°C that is far lower than the room temperature and they deform little as long as the working temperature is above their glass transition point.

In an attempt to examine the influence of the top four operating parameters given in Table 15.13.1 on the bending of such laminated products built with 3D printing techniques, Kacergis et al. [2019] set up a physical experiment based on the factorial design given in Table 15.13.2. The strain induced by the curvature is employed to denote the induced deformation due to printing speed, temperature variation, and number of layers, as well as the test results, which are registered in column Y in the table. The temperature of the specimen is controlled by its immersion in the corresponding hot water that functions as the main stimulus.

> **FIGURE 15.13.1** Hinge bearing built with 3D printing reshaped by temperature variation [Kacergis et al. 2019]. (a) Multimaterial laminate built with 3D printing using PLA and TPU; (b) Reshaped by temperature variation.

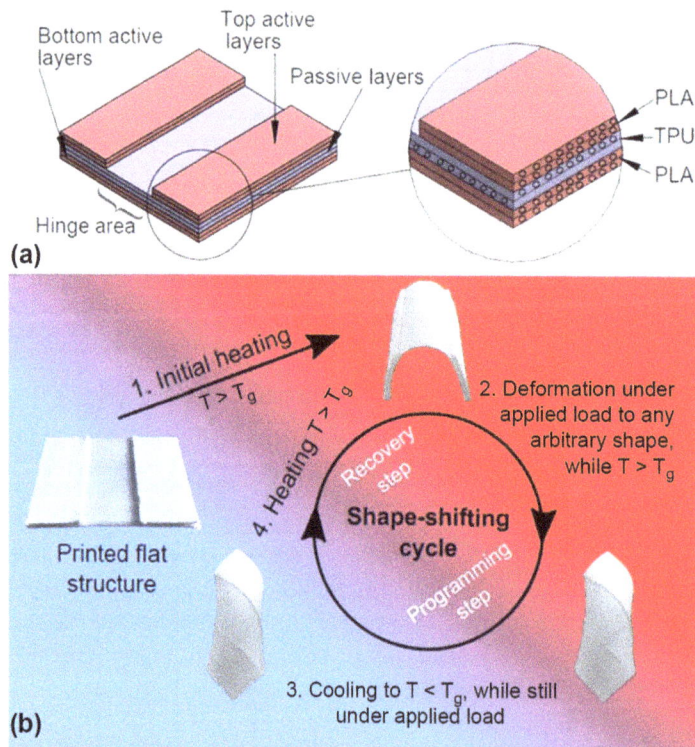

TABLE 15.13.1 Factors and their design levels for angular deformations and related strains [Kacergis et al. 2019].

Factor	(1)	(2)	(3)	(4)	(5)
A: Printing speed for PLA (mm/s)	35	50	65	80	95
B: Build plate temperature (°C)	25	40	60	—	—
C: Layers, total number of cycles	1	2	3	4	—
D: Reversibility (number of cycles)	1	3	10	—	—
E: Printing speed for TPU (mm/s)	30	—	—	—	—
F: Nozzle temperature for PLA (°C)	215	—	—	—	—
G: Nozzle temperature for TPU (°C)	230	—	—	—	—

© SAE International.

TABLE 15.13.2 Factorial design for angular deformations and related strains [Kacergis et al. 2019].

Run	A	B	C	D	Y (strain)	Y_p
1	50	60	1	1	0.047	0.045
2	80	60	1	1	0.061	0.070
3	50	25	2	1	0.157	0.159
4	80	25	2	1	0.186	0.184
5	50	40	2	1	0.130	0.137
6	80	40	2	1	0.168	0.162
7	50	60	2	1	0.072	0.077
8	80	60	2	1	0.102	0.102
9	35	60	2	1	0.062	0.065
10	65	60	2	1	0.088	0.090
11	95	60	2	1	0.112	0.115
12	50	60	3	1	0.089	0.101
13	80	60	3	1	0.125	0.126
14	50	60	4	1	0.120	0.117
15	80	60	4	1	0.139	0.142
16	35	60	2	3	0.064	0.065
17	65	60	2	3	0.088	0.090
18	95	60	2	3	0.113	0.115
19	35	60	2	10	0.071	0.065
20	65	60	2	10	0.092	0.090
21	95	60	2	10	0.114	0.115

© SAE International.

The predictive equation for the induced strain is formulated using "Minitab → Stat → Regression." As exhibited in Table 15.13.3, after insignificant terms are dropped one by one, the final predictive equation that is confirmed with the statistical significance level of 5% ($\alpha = 5\%$) by means of ANOVA on F-statistics turns out to be

$$Y_p = 0.12488 + 0.000833\ A - 0.000045\ B^2 - 0.00410\ C^2 + 0.000742\ B\ C$$

$$(R = 99.3\%,\ R_{adj} = 99.1\%,\ \text{and}\ R_{pred} = 98.6\%) \qquad . \qquad (15.13.1)$$

Effectiveness and adequacy of the above equation are validated by the three different correlations given above. Every one of them goes above 95%. The predicted values of strain using Equation (15.13.1), i.e., column Y_p in Table 15.13.2, correlate very well with the data obtained from physical tests, having $R_{pred} = 98.6\%$.

It can be seen that the strain varies linearly with the temperature (factor A) but nonlinearly with build plate temperature (factor B) and the total number of layers (factor C), respectively. When the printing speed (factor A) is fixed at 65 mm/s, Equation (15.13.1) reduces to

$$Y_p = 0.17903 - 0.000045\ B^2 - 0.00410\ C^2 + 0.000742\ B\ C \qquad .(15.13.2)$$

The strain is a quadratic function of the build plate temperature (factor B) and total number of layers (factor C), as demonstrated in Figure 15.13.2, while complicated with their significant two-factor interactive effect (interaction BC).

TABLE 15.13.3 ANOVA for angular deformations and related strains.

Source	SS_{adj}	DOF	MS_{adj}	F-value	p-Value (%)
Regression	0.027527	4	0.006882	280.59	0.0
A	0.005625	1	0.005625	229.34	0.0
BB	0.006481	1	0.006481	264.24	0.0
CC	0.000212	1	0.000212	8.64	1.0
BC	0.000891	1	0.000891	36.33	0.0
Error	0.000392	16	0.000025		
Lack of fit	0.000335	10	0.000034	3.51	0.069
Pure error	0.000057	6	0.000010		
Subtotal	0.02792	20			
Grand average		1			
Total		21			

FIGURE 15.13.2 Response surface of strain as a function of build plate temperature (factor B) and number of layers (factor C) at the printing speed of 65 mm/s for PLA.

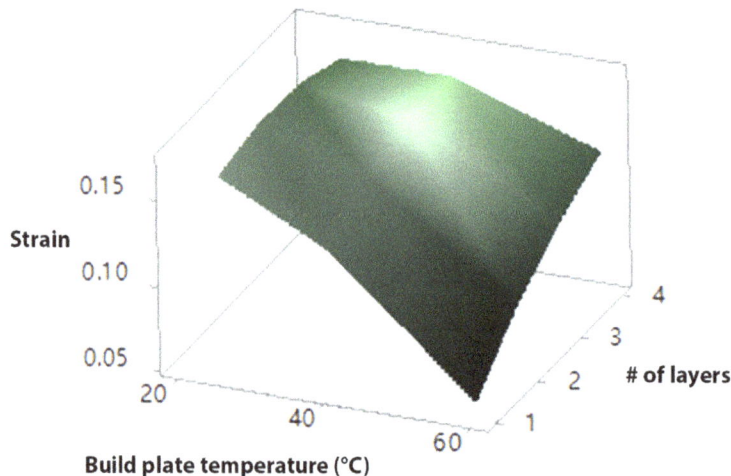

15.13.2. **Hydrogel**

Hydrogel such as pNIPAm, i.e., poly(N-isopropylacrylamide) with a chemical formula as $(C_6H_{11}NO)_n$, is a well-established polymer family [Schild 1992], which has previously been used in the field of 3D printing [Han et al. 2018]. When the temperature rises above its lower critical solution temperature, approximately 32°C, pNIPAm becomes hydrophobic. Consequently, it shrinks and stiffens considerably as the phase transition from a swollen hydrated state to a shrunken dehydrated state occurs, losing approximately 90% of its volume. Introduction of additives of pNIPA, or copolymerization, may vary the lower critical solution temperature. For example, pNIPAm/CNF hydrogels lower the lower critical solution temperature by 8°C and possess reversible optical, bio-adhesion, and thermal performance, making them suitable to be used as durable temperature-sensitive sensors and functional biomedical devices [Sun et al. 2020b]. Besides bio-applications, pNIPAm has been used in gel actuators, which convert external stimuli into mechanical motions.

A straight cantilever beam (20 µm in length) with hetero materials, e.g., (pNIPAm+additives)/pentae-rythritol triacrylate (PETA), becomes curved at 45°C as displayed in Figure 15.13.3. Upon decreasing the temperature, such as falling below 20°C (temperature < lower critical solution temperature), the less cross-linked material starts to swell and the beam reverts back to its initial position. The temperature variation can be controlled by varying the local exposure in 3D two-photon laser lithography [Hippler et al. 2019].

FIGURE 15.13.3   Cyclic actuation of bi-material (pNIPAm+PEAT) heated by laser lithography [Hippler et al. 2019].

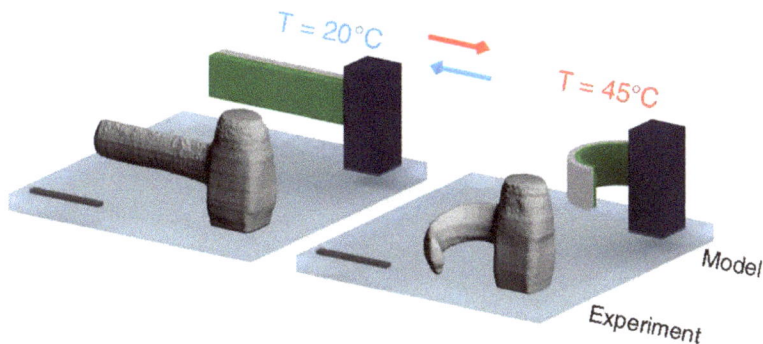

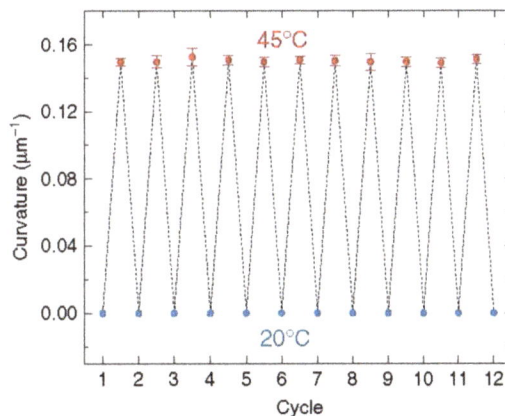

15.14. Finite Element Methods for 3D-Printed Structures

On top of geometric and material nonlinearities, finite element modeling of a 3D-printed structure involves operation-dependent geometry with material properties:

1. Geometry: How to model the design configuration in a layered structure, especially inter-layered connecting configurations/mechanisms, and soothe the explicit geometry of voids within the "road" will add a serious level of complexity to the FEA, making the problem-solving process much more complex and time-consuming. For example, should shells and looped interconnections between roads be modeled in detail?

2. Material Properties: Besides material anisotropy, the material properties of each 3D-printed part depend on the specific 3D printing process, including layer height, layer (raster) width, raster angle, nozzle diameter, nozzle temperature, printing speed, and intercooling time. Furthermore, the material behavior in compression differs from that in tension.

How to do FEA of structures built with the emergent 3D printing technology? Savvy FEA engineers are looking forward to blazing new trails and developing novel approaches to these problems. Parametric optimization techniques based on finite element methods in combination with DOE can be used for exploring the desired setting of design parameters to reduce the part weight and manufacturing cost [Hernández et al. 2017].

15.14.1. Finite Element Formulations Using Strain Energy

Taking initial stresses $\{\sigma_0\}$ and initial strains $\{\varepsilon_0\}$ into consideration, one can obtain the relationship between current stresses $\{\sigma\}$ and strains $\{\varepsilon\}$ [Chiang and Tang 1995] as

$$\{\sigma\} = [c] (\{\varepsilon\} - \{\varepsilon_0\}) + \{\sigma_0\}. \tag{15.14.1}$$

The initial strain $\{\varepsilon_0\}$ in the above equation may include the thermal, moisture, plastic, and creeping strains in a nonlinear large deformation. The corresponding strain energy per unit volume is

$$U_0 = \iiint_V dU_0 = \iiint_V \{\sigma\}^T \{d\varepsilon\}$$

$$= \tfrac{1}{2} \{\varepsilon\}^T [c] \{\varepsilon\} - \{\varepsilon\}^T [c] \{\varepsilon_0\} + \{\varepsilon\}^T \{\sigma_0\}. \tag{15.14.2}$$

Based on the formula for strains $\{\varepsilon\}$ expressed in terms of nodal displacements $\{\phi\}$ as described by Equation (3.2.25), the strain energy per unit volume can be further extended into

$$U_0 = \tfrac{1}{2} \{\phi\}^T [B]^T [c] [B] \{\phi\} - \{\phi\}^T [B]^T [c] \{\varepsilon_0\} + \{\phi\}^T [B]^T \{\sigma_0\}. \tag{15.14.3}$$

Furthermore, the total potential of a continuum with volume domain V and surface domain S consists of four parts, i.e., strain energy, and work done by body forces $\{b\}$, surface tractions $\{S\}$, and concentrated forces $\{Q\}$:

$$\Pi = \iiint_V U_0 \, dV - \iiint_V \{g\}^T \{b\} \, dV - \iint_S \{g\}^T \{S\} \, dS - \{\Phi\}^T \{Q\}. \tag{15.14.4}$$

Since the concentrated forces in {Q} are usually expressed in the global Cartesian coordinate system and independent of integration, the total displacement vector for the entire structure {Φ} is used as its corresponding displacement. Substitution of Equations (15.14.1) and (15.14.3) into the above equation yields

$$\Pi = \iiint_V (\tfrac{1}{2} \{\phi\}^T [B]^T [c] [B] \{\phi\} - \{\phi\}^T [B]^T [c] \{\varepsilon_o\} + \{\phi\}^T [B]^T \{\sigma_o\}) \, dV$$

$$- \iiint_V \{\phi\}^T [N]^T \{b\} \, dV - \iint_S \{\phi\}^T [N]^T \{S\} \, dS - \{\Phi\}^T \{Q\}, \qquad (15.14.5)$$

in which the summation means the contribution from all elements assembled. Then, the total potential of the entire structure is

$$\Pi = \tfrac{1}{2} \Sigma \{\phi\}^T [k] \{\phi\} - \Sigma \{\phi\}^T \{f\} - \Sigma \{\Phi\}^T \{Q\}, \qquad (15.14.6)$$

$$\text{where} \quad [k] = \iiint_V [B]^T [c] [B] \, dV, \qquad (15.14.7)$$

$$\text{and} \quad \{f\} = \iiint_V [B]^T [c] \{\varepsilon_o\} \, dV$$

$$- \iiint_V [B] \{\sigma_o\} \, dV + \iiint_V [N]^T \{b\} \, dV + \iint_S [N]^T \{S\}. \qquad (15.14.8)$$

By stacking the elemental displacements {ϕ} into a vector column to form the total displacement vector {Φ} according to the sequence of element numbers, the above equation can be rewritten as

$$\Pi = \tfrac{1}{2} \{\Phi\}^T [K] \{\Phi\} - \{\Phi\}^T \{F\}, \qquad (15.14.9)$$

and taking the first derivative with respect to the total displacement vector {Φ} to obtain the equation of equilibrium:

$$[K] \{\Phi\} = \{F\}. \qquad (15.14.10)$$

It means that static equilibrium prevails with the variation principle, when {Φ} satisfies the above set of simultaneous equations. The total stiffness [K] and total load {F} are assembled from individual elements:

$$[K] = \text{Assembled} \, [k], \qquad (15.14.11)$$

$$\text{and} \quad \{F\} = \text{directed} \, \{Q\} + \text{assembled} \, \{f\}. \qquad (15.14.12)$$

Based on linear elasticity, the stiffness matrix [k] of an element is formulated as Equation (15.14.7), while each element might contain several layers of orthotropic laminae with different fiber orientations. {Q} is a vector of forces directly applied. A composite element or super-element could be an economic approach to the calculation and prediction of structural performance, especially for the modal analysis and dynamic transient response. Both composite and super-elements (as super-positioned) are to be addressed.

15.14.2. **3D Coordinate Transformations**

A 3D FEA of composites built with 3D printing based on 8-node or 20-node isoparametric elements needs to acquire proper 3D coordinate transformations between the global coordinates (X, Y, Z), natural coordinates (ξ, η, ζ), and primary material coordinates (1, 2, 3), which are required for the calculations of material stiffness and loading. A method for finding a natural coordinate system for an 8-node linear isoparametric solid element is given in Cook et al. [2001]. The following procedure is employed to formulate the transformation matrix and implement transformation for the 20-node quadratic isoparametric solid element [Chiang and Tang 1995].

1. For each of the six faces, locate the central point by taking the average of all the nodes on the surface.

2. Form three vector components a_1, a_2, and a_3 by connecting the two central points of opposite faces. Compute lengths of all three connected line segments, L_1, L_2, and L_3, and the vector sum $L = L_1 + L_2 + L_3$. Normalize each vector:

$$v_1 = a_1 / L ; \qquad (15.14.13)$$

$$v_2 = a_2 / L ; \qquad (15.14.14)$$

$$v_3 = a_3 / L. \qquad (15.14.15)$$

3. Use the following weighted averaging scheme to compute three new vectors that are closer to being orthogonal than the v_i, i = 1, 2, 3, vectors:

$$u_1 = (L_2 + L_3) (v_2 \times v_3) + L_1 v_1) / L, \qquad (15.14.16)$$

$$u_2 = (L_3 + L_1) (v_3 \times v_1) + L_2 v_2) / L, \qquad (15.14.17)$$

$$u_3 = (L_1 + L_2) (v_1 \times v_2) + L_3 v_3) / L. \qquad (15.14.18)$$

4. Test for convergence. Is $u_1 \approx v_1$, $u_2 \approx v_2$, and $u_3 \approx v_3$? The criterion used here is that each component of vectors u_1, u_2, and u_3 must not differ from its corresponding components in v_1, v_2, and v_3 by more than a tolerance, and 0.001 is a good example. If convergence is met, go to step 7; otherwise, go to step 5.

5. Normalize the u vectors.

6. Replace the v vectors by the normalized u vectors, using the following under-relaxation equation to speed up convergence, for the n_{th} iteration:

$$(a_i)_{n+1} = \frac{1}{2n} (a_i)_n + \frac{2n - 1}{2n} (b_i)_n, \qquad i = 1, 2, 3. \qquad (15.14.19)$$

Return to step 3.

1. Make sure that the unit vectors are orthogonal. Define a 3×3 transformation matrix $[T]$ whose rows are the three vectors of $\mathbf{v_1}$, $\mathbf{v_2}$, and $\mathbf{v_3}$:

$$[T] = \begin{bmatrix} v_{11} & v_{12} & v_{13} \\ v_{21} & v_{22} & v_{23} \\ v_{31} & v_{32} & v_{33} \end{bmatrix}. \tag{15.14.20}$$

2. Calculate the center of an element (x_o, y_o, z_o) by taking the average of overall nodes of this element. Compute the new coordinates for each node with origin at (x_o, y_o, z_o), and then, put them in

$$[X]_g = \begin{bmatrix} x_1 & x_2 & x_3 & \dots & x_{20} \\ y_1 & y_2 & y_3 & \dots & y_{20} \\ z_1 & z_2 & z_3 & \dots & z_{20} \end{bmatrix}. \tag{15.14.21}$$

3. Transform the global coordinate system of each node of this element to the local, natural, coordinate system by

$$[X]_l = [T][X]_g. \tag{15.14.22}$$

4. Transform the force vector, $\{F\}_g$ in the global coordinate system, to that in the local natural coordinate system, $\{F\}_l$,

$$\{F\}_l = [T]\{F\}_g. \tag{15.14.23}$$

5. Form the element stiffness matrix $[k]_l$ and force vector $\{f\}_l$ in the natural coordinate system. Prior to assemblage, transform them back to the global coordinate system, $[k]_g$ and $\{f\}_g$:

$$[k]_g = [T^*]^T [k]_l [T^*], \tag{15.14.24}$$

$$[f]_g = [T^*]^T \{f\}_l, \tag{15.14.25}$$

and

$$[T^*]_{60 \times 60} = \begin{bmatrix} [T] & & & & \\ & [T] & & & \\ & & [T] & & [0] \\ & & & \ddots & \\ & [0] & & & [T] \\ & & & & & [T] \end{bmatrix}. \tag{15.14.26}$$

6. Solve the whole assembled simultaneous equations for the whole global nodal displacements $\{\Phi\}_g$ from which the global nodal displacements $\{\phi\}_g$ of an individual element can be extracted.

7. Strains and stresses are computed from the deformations in the local, natural coordinate system:

$$\{\phi\}_l = [T]\ \{\phi\}_g. \tag{15.14.27}$$

Then, strains and stresses are transformed back to the global coordinate system:

$$\{\varepsilon\}_g = [T_\varepsilon]^T\ \{\varepsilon\}_l\ [T_\varepsilon], \tag{15.14.28}$$

$$\{\sigma\}_g = [T_\sigma]^T\ \{\sigma\}_l\ [T_\sigma]. \tag{15.14.29}$$

15.14.3. Integration for Elemental Matrices

Integrations have to be carried out numerically for elemental matrices and loads induced by finite element formulations [Chiang 2019]. The numerical integrations for various mechanical properties involved in finite element formulations can be implemented based on the Gauss-quadrature formulae in terms of natural coordinates, as

$$\iiint_V f(x, y, z)\, dV = \int_{-1}^{1} \int_{-1}^{1} \int_{-1}^{1} f(\xi, \eta, \zeta)\, J(\xi, \eta, \zeta)\, d\xi\, d\eta\, d\zeta$$

$$= \sum_i \sum_j \sum_k w_i\, w_j\, w_k\, f(\xi_i, \eta_j, \zeta_k)\, J(\xi_i, \eta_j, \zeta_k), \tag{15.14.30}$$

where

$f(\xi_i, \eta_j, \zeta_k)$ and $J(\xi_i, \eta_j, \zeta_k)$ are the values to be computed at sampling point (ξ_i, η_j, ζ_k)

w_i, w_j, and w_k are the weighting functions for the Gauss-quadrature integral

i, j, and k are the subscripts to number the integration order

An integration point is the point within an element, at which integrals are evaluated numerically. These points are chosen in such a way that the results for a particular numerical integration scheme are the most accurate. Subscripts for 20-noded solid elements may range from 1 to 3 for all three directions for full integration, i.e., $3 \times 3 \times 3$ integration and from 1 to 2 for all three directions for reduced integration, i.e., $2 \times 2 \times 2$ integration. Depending on the integration scheme used, the location of these points will vary. The 20-noded solid also has a special type of integration point scheme, called 14-point integration in addition to $3 \times 3 \times 3$ and $2 \times 2 \times 2$ integrations. This scheme places points close to each of the eight corner nodes and close to the centers of the six faces for a total of 14 points [Cook et al. 2001].

References

Abuiedda, D. et al. (2019), "Mechanical Properties of 3D Printed Gyroid Cellular Structures - Experimental and Finite Element Studies," *Materials and Design*, 165, p. 107597.

Agarwala, S. et al. (2018), "Optimizing Aerosol Jet Printing Process of Silver Ink for Printed Electronics," *IEEE Access*, 6, p. 012027.

de Aquino, D. A. et al. (2020), "Investigation of Load Direction on the Compressive Strength of Additively Manufactured Triply Periodic Minimal Surface Scaffolds," *Int'l Journal of Advanced Manufacturing Technology*, 109, pp. 771-779.

Areir, M. et al. (2017), "3D Printing of Highly Flexible Supercapacitor Designed for Wearable Energy Storage," *Materials Science & Engineering, B*, 226, pp. 29-38.

ASTM D3039 (2014-6-4), Standard Test Method for Tensile Properties of Polymer Matrix Composite Materials.

ASTM D638-14 (2022-7-21), Standard Test Method for Tensile Properties of Plastics.

Bajsic, E. G. et al. (2005), "Blends of Thermoplastic Polyurethane and Polypropylene," *Proceeding of the 8th Polymers for Advanced Technologies International Symposium*, Budapest, Hungary, 13-16 September 2005.

Balanis, C. A. (1997), *Antenna Theory: Analysis and Design*, John Wiley & Sons, New York.

Barši Palmić, T. B. et al. (2020), "Process Parameters for FFF 3D-Printed Conductors for Applications in Sensors," *Sensors*, 20(16), p. 4542.

Bathla, P. and Kennedy, J. (2020), "3D Printed Structured Porous Treatments for Flow Control with Applications for Noise and Vibration Control," *Vibroengineering Procedia*, 31, pp. 80-85.

Bijwaard, D. J. et al. (2011), "Industry: Using Dynamic WSNs in Smart Logistics for Fruits and Pharmacy," *Proceedings of the 9th ACM Conference on Embedded Networked Sensor Systems*, Seattle, WA, November 2011, pp. 218-231.

Cantrell, J. T. et al. (2017), "Experimental Characterization of the Mechanical Properties of 3D-Printed ABS and Polycarbonate Parts," *Rapid Prototyping Journal*, 23(4), pp. 811-824.

Casavola, C. et al. (2016), "Orthotropic Mechanical Properties of Fused Deposition Modeling Parts Described by Classical Laminate Theory," *Materials and Design*, 90, pp. 453-458.

Chaturvedi, V. (2009), "Parametric Optimization of Fused Deposition Modeling Using Response Surface Methodology," MS Thesis, Dept. of Mechanical Engineering, National Institute of Technology, Rourkela.

Chen, L. Y. et al. (2014), "Continuous Wireless Pressure Monitoring and Mapping with Ultra-Small Passive Sensors for Health Monitoring and Critical Care," *Nature Communications*, 5(1), p. 5028.

Chiang, Y. J. (2019), *Mechanics and Design for Product Life Prediction*, Chongqing University Press; ISBN: 978-7-5689-1917-6.

Chiang, Y. J. and Tang, C. (1995), "Accuracy Assessment to Applying 20-Node Solid Elements to Pressurized Composite Shells," *Finite Elements in Analysis and Design*, 20, pp. 219-231.

Coogan, T. J. and Kazmer, D. O. (2019), "In-line Rheological Monitoring of Fused Deposition Modeling," *Journal of Rheology*, 63, p. 141.

Cook, R. D., Malkus, D. S., Plesha, M. E., and Witt, R. J. (2001), *Concepts and Applications of Finite Element Analysis*, 4th Edition, John Wiley & Sons, New York.

Duty, C. et al. (2019), "Z-Pinning Approach for Reducing Mechanical Anisotropy of 3D Printed Parts," *Proceedings of the 29th Annual Int'l Solid Freeform Fabrication Symposium*, Austin, TX, 2019, pp. 2405-2412.

Eguren, J., Esnaola, A., and Unzueta, G. (2020), "Modelling of an Additive 3D-Printing Process Based on Design of Experiments Methodology," *Quality Innovation Prosperity*, 24(1), pp. 128-151.

Faes, M., Ferraris, E., and Moens, D. (2016), "Influence of Inter-layer Cooling Time on the Quasi-Static Properties of ABS Components Produced via Fused Deposition Modelling," *Procedia CIRP*, 42, pp. 748-753.

Farah, S. et al. (2016), "Physical and Mechanical Properties of PLA, and Their Functions in Widespread Applications - A Comprehensive Review," *Advanced Drug Delivery Reviews*, 107, pp. 367-392.

Finkenzeller, K. (2003), *RFID Handbook: Fundamentals and Applications in Contactless Smart Cards, Radio Frequency Identification and Near-Field Communication*, 2nd Edition, Wiley, New York; ISBN: 978-0470695067.

Galantucci, L. M., Lavecchia, F., and Percoco, G. (2008), "Study of Compression Properties of Topologically Optimized FDM Made Structured Parts," *CIRP Annals - Manufacturing Technology*, 57(1), pp. 243-246.

Galantucci, L. M. et al. (2015), "Analysis of Dimensional Performance for a 3D Open-Source Printer Based on Fused Deposition Modeling Technique," *Procedia CIRP*, 28, pp. 82-87.

Galusha, J. W. (2008), "Discovery of Diamond-Based Photonic Crystal Structure in Beatle Scales," *Physics Review E: Statistical, Nonlinear, and Soft Matter Physics*, 77, p. 050904.

Gardan, J. et al. (2018), "Improving the Fracture Toughness of 3D Printed Thermoplastic Polymers by Fused Deposition Modeling," *Int'l Journal of Fracture*, 210(7), pp. 1-15.

Gardiner, G. (2018), "3D Printing Composites with Continuous Fiber," Composites World, November 5, 2018.

Goo, B., Hong, C., and Park, A. (2020), "4D Printing Using Anisotropic Thermal Deformation of 3D-Printed Thermoplastic Parts," *Materials and Design*, 188, p. 108485.

Han, D., Lu, Z., Chester, S. A., and Lee, H. (2018), "Micro 3D Printing of a Temperature Responsive Hydrogel Using Projection Micro-Stereolithography," *Scientific Reports*, 8(1), p. 1963.

Hernández, R. P., Pei, E., and Monzon, M. (2017), "Lightweight Parametric Design Optimization for 4D Printed Parts," *Integrated Computer-Aided Engineering*, 34(3), pp. 225-240.

Hibbert, K., Warner, G., Brown, C., Ajide, O., Owolabi, G., and Azimi, A. (2019), "The Effects of Build Parameters and Strain Rate on the Mechanical Properties of FDM 3D-Printed Acrylonitrile Butadiene Styrene," *Open Journal of Organic Polymer Materials*, 9, pp. 1-27.

Hikmat, M. et al. (2021), "Investigation of Tensile Property-Based Taguchi Method of PLA Parts Fabricated by FDM 3D Printing Technology," *Results in Engineering*, 11, p. 100264.

Hippler, M. et al. (2019), "Controlling the Shape of 3D Microstructures by Temperature and Light," *Nature Communications*, 10, p. 232.

Huiskes, R. et al. (1992), "The Relationship between Stress Shielding and Bore Resorption around Total Hip Stems and the Effects of Flexible Materials," *Clinical Orthopedics and Related Research*, 274(1), pp. 124-134.

Hull, C. W. (1986), "Apparatus for Production of Three-dimensional Objects by Stereolithography," U.S. Patent No. 4575330, March 11, 1986.

Jeong, H. Y. et al. (2020), "3D and 4D Printing for Optics and Metaphonics," *Nanophotonics*, 9(5), pp. 1139-1160.

Jiang, C. et al. (2017), "Acoustic Absorption of Porous Materials Produced by Additive Manufacturing with Varying Geometries," *Acoustics 2017*, Perth, Australia, November 19–22, 2017.

Kacergis, L., Mitkus, R., and Sinapius, M. (2019), "Influence of Fused Deposition Modeling Process Parameters on the Transformation of 4D Printed Morphing Structures," *Smart Materials and Structures*, 28(10), p. 105042.

Karimi, M. (2011), "Diffusion in Polymer Solids and Solutions," in *Mass Transfer in Chemical Engineering Processes*, pp. 17-40.

Khan, H. M. et al. (2021), "Weibull Distribution of Selective Laser Melted AlSi10Mg Parts for Compression Testing," *Journal of Advances in Manufacturing Engineering*, 2(1), pp. 14-19.

Ligon, S. M. et al. (2017), "Polymers for 3D Printing and Customized Additive Manufacturing," *Chemical Reviews*, 117(15), pp. 10212-10290.

Liu, G. et al. (2006), "Resource Management with RFID Technology in Automatic Warehouse System," *2006 IEEE/RSJ International Conference on Intelligent Robots and Systems*, Beijing, China, 2006, pp. 3706-3711.

Liu, Z., Zhan, J., Fard, M. and Davy, J. L. (2016), "Acoustic Properties of a Porous Polycarbonate Material Produced by Additive Manufacturing," *Materials Letters*, 181, pp. 296-299.

Lu, X. et al. (2020), "3D Microstructure Design of Lithium-Ion Battery Electrodes Assisted by X-Ray Nano-Computed Tomography and Modelling," *Nature Communications*, 11, p. 2079.

Luthria, K. (2016), "Clearance Analysis of 3D Printed Assemblies Using Fused Clearance Analysis of 3D Printed Assemblies Using Fused Filament Extrusion Filament Extrusion," MS Thesis, Rochester Institute of Technology, NY.

Maskely, I. et al. (2018), "Insights into the Mechanical Properties of Several Triply Periodic Minimal Surface Lattice Structures Made by Polymer Additive Manufacturing," *Polymer*, 152, pp. 62-71.

Materialise Manufacturing (2020), "Datasheets 3D Printing Materials," Updated June 2020, Retrieved July 14, 2020.

Matsuzak, R. et al. (2016), "Three-Dimensional Printing of Continuous-Fiber Composites by In-Nozzle Impregnation," *Scientific Reports*, 6, p. 23058.

McCloskey, B. D. et al. (2012), "Twin Problems of Interfacial Carbonate Formation in Nonaqueous Li-O$_2$ Batteries," *Journal of Physical Chemistry Letters*, 3(8), pp. 997-1001.

MMPDS -08 (2013), Metallic Materials Properties Development and Standardization, Battelle Memorial Institute.

Montero, M., Roundy, S., Odell, D., Ahn, S.-H., and Wright, P. K. (2001), "Material Characterization of Fused Deposition Modeling (FDM) ABS by Designed Experiments," *Proceedings of Rapid Prototyping and Manufacturing Conference*, Cincinnati, OH, May 15, 2001, p. 10.

Moradi, M. et al. (2020), "Post-Processing of FDM 3D-Printed Polylactic Acid Parts by Laser Beam Cutting," *Polymers*, 12(3), p. 550.

Muhsin, S. A. et al. (2019), "Determination of PEEK Mechanical Properties as a Denture Material," *Saudi Dental Journal*, 31, pp. 382-391.

Nazouri, M., Seifzadeh, A., and Masaeli, E. (2020), "Characterization of Polyvinyl Alcohol Hydrogels as Tissue-Engineered Cartilage Scaffolds Using a Coupled Finite Element-Optimization Algorithm," *Journal of Biomechanics*, 99, pp. 109525.

Nguyen, V. H. et al. (2019), "Single and Multi-Objective Optimization of Processing Parameters for Fused Deposition Modelling in 3D Printing Technology," *Int'l Journal of Automotive and Mechanical Engineering*, 17(1), pp. 7542-7551.

Ning, H. et al. (2015), "Holographic Patterning of High-Performance On-Chip 3D Lithium-Ion Microbatteries," *PNAS*, 112(21), pp. 6573-6578.

Normura, A. et al. (2024), "Gravimetric Analysis of Lithium-Air batteries during Discharge/Charge Cycles," *Journal of Power Sources*, 592, pp. 233924.

Novoa, C. and Flores, A. (2019), "Optimizing the Tensile Strength for 3D Printed PLA Parts," *Proceedings of the 30th Annual Int'l Solid Freeform Fabrication Symposium*, Austin, TX, 2019, pp. 745-765.

Ouhsti, M. et al. (2018), "Effect of Printing Parameters on the Mechanical Properties of Parts Fabricated with Open-Source 3D Printers in PLA by Fused Deposition Modeling," *Mechanics and Mechanical Engineering*, 22(4), pp. 895-907.

Pan, J. et al. (1998), "Screen Printing Process Design of Experiments for Fine Line Printing of Thick Film Ceramic Substrates," *Journal of Electronics Manufacturing*, 9(3), pp. 203-213.

Pascual-González, C. et al. (2020), "An Approach to Analyzing the Factors behind the Micromechanical Response of 3D-Printed Composites," *Composites Part B: Engineering*, 186, p. 107820.

Poon, T. C. et al. (2009), "A RFID Case-Based Logistics Resource Management System for Managing Order-Picking Operations in Warehouses," *Expert Systems with Applications*, 36(4), pp. 8277-8301.

Qahtani, M. et al. (2019), "Experimental Design of Sustainable 3D-Printed Poly(Lactic Acid)/Biobased Poly(Butylene Succinate) Blends via Fused Deposition Modeling," *ACS Sustainable Chemistry and Engineering*, 7, pp. 14460-14470.

Qi, H. J. and Boyce, M. C. (2005), "Stress-Strain Behavior of Thermoplastic Polyurethane," *Mechanics of Materials*, 37(8), pp. 817-839.

Qin, Z. et al. (2017), "The Mechanics and Design of a Lightweight Three-Dimensional Assembly," *Science Advances*, 3, p. e1601536.

Rafiee, M. et al. (2020), "Multi-Material 3D and 4D Printing: A Survey," *Advanced Science*, 7(12), p. 1902307.

Rahman, T. et al. (2018), "3D Printed High Performance Strain Sensors for High Temperature Applications," *Journal of Applied Physics*, 123, p. 024501.

Raol, T. S. et al. (2014), "An Experimental Investigation of Effect of Process Parameters on Surface Roughness of Fused Deposition Modeling Built Parts," *Int'l Journal of Engineering Research & Technology*, 3(4), pp. 2270-2274.

Renteria, A. et al. (2019), "Optimization of 3D Printing Parameters for BaTiO$_3$ Piezoelectric Ceramics through Design of Experiments," *Materials Research Express*, 6(8), p. 085706.

Restrepo, S. et al. (2017), "Mechanical Properties of Ceramic Structures Based on Triply Periodic Minimal Surface (TPMS) Processed by 3D Printing," *Journal of Physics Conference Series*, 935(1), p. 012036.

Rosker, E. S. et al. (2018), "Printable Materials for the Realization of High-Performance RF Components: Challenges and Opportunities," *Int'l Journal of Antennas and Propagation*, 2018(3), pp. 1-19.

Schild, H. G. (1992), "Poly(N-Isopropylacrylamide): Experiment, Theory and Application," *Progress in Polymer Science*, 17(2), pp. 163-249.

Schroder-Turk, C. E. et al. (2011), "The Chiral Structure of Porous Chitin with the Wing Scales of Collophrys Rubi," *Journal of Structural Biology*, 174, pp. 290-295.

Sheth, S. (2019), "Material Characterization and Fracture Prediction of FDM Printed Parts," MS Thesis, University of Texas, Arlington, TX, December 2019.

Somireddy, M. et al. (2020), "Mechanical Behavior of 3D Printed Composite Parts with Short Carbon Fiber Reinforcement," *Engineering Failure Analysis*, 107, p. 104232.

Song, Y. et al. (2017), "Measurements of the Mechanical Response of Unidirectional 3D-Printed PLA," *Materials & Design*, 123, pp. 154-164.

Sood, A. K., Ohdar, R. K., and Mahapatra, S. S. (2009), "Improving Dimensional Accuracy of Fused Deposition Modelling Processed Part Using Grey Taguchi Method," *Materials & Design*, 30(1), pp. 4243-4252.

Sun, M. et al. (2020a), "Broadband Acoustic Ventilation Barriers," *Physical Review Applied*, 13(4), p. 044028.

Sun, X. et al. (2020b), "Highly Tunable Bioadhesion and Optics of 3D Printable pNIPAm/Cellulose Nanofibrils Hydrogels," *Carbohydrate Polymers*, 234, p. 115898.

Szykiedans, K. et al. (2017), "Selected Mechanical Properties of PETG 3-D Prints," *Procedia Engineering*, 177, pp. 455-461.

Taufik, M. and Jain, P. K. (2017), "Characterization, Modeling and Simulation of Fused Deposition Modeling Fabricated Part Surfaces," *Surface Topography: Metrology and Properties*, 5(4), p. 045003.

Taylor, G., Anandan, S., Murphy, D., Leu, M., and Chandrashekhara, K. (2019), "Fracture Toughness of Additively Manufactured ULTEM 1010," *Virtual and Physical Prototyping*, 14(3), pp. 277-283.

Tibbits, S. (2014), "4D Printing: Multi-Material Shape Change," *Architectural Design*, 84, pp. 116-121.

Torquato, S. and Donev, A. (2004), "Minimal Surface and Multifunctionality," *Proceedings of the Royal Society of London, A*, 4, pp. 1849-1856.

Torrado, A. R. and Roberson, D. A. (2016), "Failure Analysis and Anisotropy Evaluation of 3D-Printed Tensile Test Specimens of Different Geometries and Print Raster Patterns," *Journal of Failure Analysis and Prevention*, 16, pp. 154-164.

Turner, B. N. and Gold, S. A. (2015), "A Review of Melt Extrusion Additive Manufacturing Processes: II. Materials, Dimensional Accuracy, and Surface Roughness," *Rapid Prototyping Journal*, 21, pp. 250-261.

Velasco, M. A., Narváez-Tovar, C. A., and Garzón-Alvarado, D. A. (2015), "Design, Materials, and Mechanobiology of Biodegradable Scaffolds for Bone Tissue Engineering," *BioMed Research International*, 2015(1), p. 729076.

Vijayavenkataraman, S. et al. (2020), "3D-Printed Ceramic Triply Periodic Minimal Surface Structures for Design of Functionally Graded Bone Implants," *Materials & Design*, 191, p. 108602.

Wang, X. et al. (2019), "Effect of Porosity on Mechanical Properties of 3D Printed Polymers: Experiments and Micromechanical Modeling Based on X-Ray Computed Tomography Analysis," *Polymers*, 11(7), p. 1154.

Wang, S. et al. (2020), "Effects of Fused Deposition Modeling Process Parameters on Tensile, Dynamic Mechanical Properties of 3D Printed Polylactic Acid Materials," *Polymer Testing*, 86, p. 106483.

Wu, D., Wei, Y., and Terpenny, J. (2019), "Predictive Modelling of Surface Roughness in Fused Deposition Modelling Using Data Fusion," *Int'l Journal of Production Research*, 57(12), pp. 3992-4006.

Yao, B. et al. (2019), "Efficient 3D Printed Pseudocapacity Electrodes with Ultrahigh MnO_2 Loading," *Joule*, 3, pp. 1-12.

Yi, X., Wu, T., Wang, Y., Leon, R., Tentzeris, M., and Lantz, G. (2011), "Passive Wireless Smart-Skin Sensor Using RFID-Based Folded Patch Antennas," *Int' Journal of Smart and Nano Materials*, 2(1), pp. 22-38.

Yi, X. et al. (2015), "Sensitivity Modeling of an RFID-Based Strain-Sensing Antenna with Dielectric Constant Change," *IEEE Sensor Journal*, 15(11), pp. 6147-6155.

Yoo, K. et al. (2017), "A Review of Lithium-Air Battery Modeling Studies," *Energies*, 10, p. 1748.

Zaman, U. K. et al. (2019), "Impact of Fused Deposition Modeling (FDM) Process Parameters on Strength of Built Parts Using Taguchi's Design of Experiments," *Int'l Journal of Advanced Manufacturing Technology*, 101, pp. 1215-1226.

Zhang, J. et al. (2020), "3D-Printed Functional Electrodes towards Zn-Air Batteries," *MaterialsToday Energy*, 16(2020), p. 100407.

Zhao, S. et al. (2019a), "3D Printed Actuators - Reversibility, Relaxation, and Ratcheting," *Advanced Functional Materials*, 29, p. 190545.

Zhao, Y., Chen, Y., and Zhou, Y. (2019b), "Novel Mechanical Models of Tensile Strength and Elastic Property of FDM AM PLA Materials: Theoretical and Experimental Analyses," *Materials & Design*, 181, p. 108089.

Problems

P15.1: Data for studying the influence of process parameters on the ultimate tensile strength (MPa) of 3D-printed PLA parts (i.e., listed in column Y), as derived from fused deposition modeling, are listed as follows [Hikmat et al. 2021]:

Run	a	b	c	d	e	f	g	Y (MPa)
1	X (flat)	1	1	1	1	1	1	31.26
2	X (flat)	1	2	2	2	2	2	46.62
3	X (flat)	1	3	3	3	3	3	54.97
4	X (flat)	2	1	1	2	2	3	39.51
5	X (flat)	2	2	2	3	3	1	52.55
6	X (flat)	2	3	3	1	1	2	43.92
7	X (flat)	3	1	2	1	3	2	34.82
8	X (flat)	3	2	3	2	1	3	41.08
9	X (flat)	3	3	1	3	2	1	59.38
10	Y (on edge)	1	1	3	3	2	2	55.18
11	Y (on edge)	1	2	1	1	3	3	56.39
12	Y (on edge)	1	3	2	2	1	1	57.39
13	Y (on edge)	2	1	2	3	1	3	53.91
14	Y (on edge)	2	2	3	1	2	1	58.44
15	Y (on edge)	2	3	1	2	3	2	60.29
16	Y (on edge)	3	1	3	2	3	1	56.12
17	Y (on edge)	3	2	1	3	1	2	58.98
18	Y (on edge)	3	3	2	1	2	3	56.50

How would the individual factors and their interactions affect the ultimate tensile strength? The corresponding design levels (L_1, L_2, and L_3) of individual factors are addressed as follows:

Factor	L_1	L_2	L_3	Coded variable
A: Build orientation	X	Y	—	a = A
B (deg): Raster orientation	0°	30°	45°	b = (B – 30)/15
C (mm): Nozzle diameter	0.3	0.4	0.5	c = (C – 0.4)/0.1
D (°C): Extruder temperature	210	215	220	d = (D – 215)/5
E (%): Infill density (rate)	80	90	100	e = (E – 90)/10
F: Number of shells	2	3	4	f = (F – 3)/1
G (mm/s): Extruding speed	20	40	60	g = (G – 40)/20

P15.2: SLM is an additive manufacturing process to fabricate 3D structures by fusing powder particles using a computer-guided laser source. Fabrication parameters and physical values of AlSi10Mg are given as follows:

(a) Laser power: 370 W

(b) Scanning rate: 1300 mm/s

(c) Hatching: 190 μm

(d) Layer thickness: 30 μm

(e) Spot size diameter: 100 μm

(f) Scanning strategy: Alternate, 670

The compressive strength (MPa) measured from 22 test units is given as follows [Khan et al. 2021]:

585.06, 585.69, 561.83, 571.90, 578.68, 579.87, 586.90, 566.04, 579.67, 588.66, 598.93, 618.11, 594.29, 592.75, 590.89, 574.45, 573.43, 582.35, 592.90, 592.65, 593.38, 592.50

Please identify the applicable statistical distribution of the compressive strength. Why 22 data?

P15.3: Data for studying the influence of three process parameters on the dimensional integrity of 3D-printed ABS parts, as derived from fused deposition modeling, are listed as follows:

Run	Block	a	b	c	A	B	C (%)	δH	δL	δW	R_a
1	1	1	1	−1	500	1.3	20	6.75	0.15	0.56	16.02
2	1	1	−1	1	500	1	100	6.90	0.03	0.76	19.34
3	1	−1	1	1	250	1.3	100	13.65	1.21	5.86	17.24
4	1	−1	−1	−1	250	1	20	8.70	0.59	1.94	2.12
5	2	1	−1	−1	500	1.3	20	8.40	0.11	0.86	22.48
6	2	1	1	1	500	1.3	100	6.90	0.17	0.40	13.62
7	2	−1	−1	1	250	1	100	11.45	0.96	3.84	5.84
8	2	−1	1	−1	250	1	20	4.45	0.69	2.56	1.86
9	3	−1	1	1	250	1	100	14.80	1.13	5.70	13.18
10	3	1	1	−1	500	1.3	20	7.45	0.16	0.32	11.98
11	3	1	−1	1	500	1.3	100	5.85	0.12	0.56	16.82
12	3	−1	−1	−1	250	1	20	5.15	0.39	1.72	2.12
13	4	−1	−1	1	250	1	100	10.40	0.81	4.08	3.72
14	4	−1	1	−1	250	1	20	7.50	0.69	3.02	2.46
15	4	1	1	1	500	1.3	100	8.95	0.78	0.24	13.42
16	4	1	−1	−1	500	1.3	20	5.90	0.16	0.22	17.52

where

A (μm): Layer thickness; coded variable a = (A − 375)/125
B: Speed multiplier; coded variable b = (A − 1.15)/0.15
C (%): Infill density; coded variable c = (C − 60%)/40%
δH (%): Height variation
δL (%): Length variation
δW (%): Width variation
R_a (μm): Surface roughness

Bibliography: Books

Akao, Y. (Editor) (1990), *Quality Function Deployment: Integrating Customer Requirements into Product Design*, Productivity Press, Cambridge, MA, 392 pages; ISBN: 9781563273131.

Allen, P. (2020), *Design of Experiments for 21st Century Engineers*, Lulu.com, 214 pages; ISBN: 978-0244584504.

Altshuller, G., *Creativity as an Exact Science* (New York: Gordon & Breach, 1984).

American Supplier Institute, *Quality Function Deployment* (Dearborn, MI: American Supplier Institute, 1993).

Anderson, M.J. and Whitcomb, P.J. (2016), *RSM Simplified: Optimizing Processes Using Response Surface Methods for Design of Experiments*, Productivity Press, New York, 311 pages; ISBN: 978-1315382326.

Anderson, M.J. and Whitcomb, P.J. (2015), *DOE Simplified: Practical Tools for Effective Experimentation*, 3rd Edition, Productivity Press, London, UK, 268 pages; ISBN: 978-0429258022.

Anderson, V.L. and McLean, R.A., *Design of Experiments: A Realistic Approach* (New York: Marcel Dekker, 1974).

Antony, J., *Design of Experiments for Engineers and Scientists*, 3rd ed. (Amsterdam, the Netherlands: Elsevier, 2023), ISBN:978-0-44-315173-6.

Armold, S.F., *Mathematical Statistics* (Englewood Cliffs, NJ: Prentice-Hall, 1990).

Atkinson, A.C., Donev, A.N., and Tobias, R.D., *Optimum Experimental Designs, with SAS* (Oxford, UK: Oxford University Press, 2007), ISBN:978-0-19-929660-6.

Automotive Industry Action Group, *Measurement System Analysis Manual*, 2nd ed. (Southfield, MI: AIAG, 2010).

Bailey, R.A., *Design of Comparative Experiments* (Cambridge, UK: Cambridge University Press, 2008).

Beckwith, T.G., *Mechanical Measurements* (Reading, MA: Addison-Wesley Publishing Company, 1990).

Bertsche, B. (2008), *Reliability in Automotive and Mechanical Engineering*, Springer, Berlin, 511 pages; ISBN: 978-3540681892.

Bhote, K.R. and Bhote, A.K. (1999), *World Class Quality Using Design of Experiments to Make it Happen*, 2nd Edition, American Management Association, New York, 487 pages; ISBN: 978-0814426425.

Bhote, K. (1996), *Going beyond Customer Satisfaction to Customer Loyalty*, American Management Association, New York, 140 pages; ISBN: 978-0814423622.

Bicheno, J., *New Lean Toolbox: Towards Fast Flexible Flow* (Buckingham, UK: Picsie Books, 2004).

Bjorke, O., *Computer-Aided Tolerancing*, 2nd ed. (New York: ASME Press, 1989).

Blischke, W.R., Karim, M.R., and Prabhakar Murthy, D.N., *Warranty Data Collection and Analysis* (London, UK: Springer, 2011), ISBN:978-0-85729-646-7.

Boothroyd, G., Dewhurst, P., and Knight, W., *Product Design for Manufacture and Assembly*, 3rd ed. (Boca Raton, FL: CRC Press, 2010).

Box, G.E.P. and Draper, N.R. (2007), *Response Surfaces, Mixtures, and Ridge Analyses*, 2nd Edition, John Wiley & Sons, Hoboken, NJ, 857 pages; ISBN: 978-0470053577.

Box, G.E.P. and Draper, N.R. (1998), *Evolutionary Operation: A Statistical Method for Process Improvement*, First Printing Edition, John Wiley & Sons, New York, 237 pages; ISBN: 978-0471255512.

Box, G.E.P., Hunter, J.S., and Hunter, W.G. (2005), *Statistics for Experimenters: Design, Innovation, and Discovery*, 2nd Edition, John Wiley & Sons, Hoboken, NJ, 672 pages; ISBN: 978-0-471-71813-0.

Brook, R.J. and Arnold, G.C. (1985), *Applied Regression Analysis and Experimental Design*, Taylor & Francis Group/CRC Press, 256 pages; ISBN: 0-8247-7252-0.

Brue, G. and Launsby, R.G., *Design for Six Sigma* (New York: McGraw-Hill, 2003).

Bryson, A.E. and Ho, Y.C., *Applied Optimal Control* (New York: John Wiley & Sons, 1975).

Chiang, Y.J., *Mechanics and Design for Product Life Prediction* (Chongqing University Press, 2019), ISBN:978-7-5689-1917-6.

Chiang, Y.J., *Automotive Engineering Materials - Thermomechanical Properties* (Chongqing University Press, 2022), ISBN:978-7-5689-3293-6.

Cochran, W.G. and Cox, G.M., *Experimental Design* (New York: John Wiley & Sons, 1957).

Coles, S., *An Introduction to Statistical Modeling of Extreme Values* (London, UK: Springer-Verlag, 2001).

Condra, L., *Reliability Improvement with Design of Experiment*, 2nd ed. (CRC Press, 2019), ISBN:978-0824705275.

Cogorno, G.R., *Geometric Dimensioning and Tolerancing for Mechanical Design*, 3rd ed. (New York: McGraw-Hill, 2020), ISBN:978-1260453782.

Cooper, W., Lawrence, M., and Joe, Z., *Handbook on Data Envelopment Analysis* (New York: Springer, 2011), ISBN:978-1-4419-6150-1.

Cox, N.D., *How to Perform Statistical Tolerance Analysis* (Milwaukee, WI: ASQC, 2006), ISBN:978-0-87389-010-6.

Czitrom, V. and Spagon, P.D., *Statistical Case Studies for Industrial Process Improvement*, ASA-SIAM Series on Statistics and Applied Probability (Philadelphia, PA: Society for Industrial and Applied Mathematics, 1997).

Das, R.N. (2014), *Robust Response Surfaces, Regression, and Positive Data Analyses*, Chapman and Hall/CRC, Boca Raton, FL, 336 pages; ISBN: 978-1466506770.

Davenport, W.B. and Root, F.S., *An Introduction to the Theory of Random Signals and Noise* (New York: McGraw-Hill, 1980).

Davies, O.L. (Eds), *The Design and Analysis of Industrial Experiments* (New York: McMillan Co., 1971).

Davim, P.J. (Eds), *Design of Experiments in Production Engineering* (Cham: Springer, 2016), ISBN:978-3319238371.

Dean, A., Voss, D., and Draguljic, D., *Design and Analysis of Experiments*, 2nd ed. (Cham: Springer, 2017), ISBN:978-3319522487.

Deming, W.E., *Out of the Crisis* (Cambridge, MA: Massachusetts Institute of Technology, 1986).

Diamond, W.J., *Fractional Experimental Designs* (Belmont, CA: Wadsworth, 1981).

Dodson, B., *The Weibull Analysis Handbook*, 2nd ed. (Milwaukee, WI: American Society for Quality, Quality Press, 2006).

Doebelin, E.O., *Measurement Systems, Applications and Design*, 4th ed. (New York: McGraw-Hill, 1990).

Drake, P. Jr., *Dimensioning and Tolerancing Handbook* (New York: McGraw-Hill, 1999), ISBN:0-07018131-4.

Eriksson, L. et al., *Design of Experiments Principles and Applications*, 3rd ed. (Umeå, Sweden: Umetrics Academy, 2008), ISBN:91-973730-0-1.

Fang, K., Li, R., and Sudjianto, A., *Design and Modeling for Computer Experiments* (London, UK: Chapman & Hall/CRC, 2006), ISBN:978-1584885467.

Fienberg, S.E., *The Analysis of Cross-Classified Categorical Data* (Cambridge, MA: The MIT Press, 1987).

Fisher, R.A. (1966), *The Design of Experiments*, 8th Edition, Hafner Publishing Company, New York, 236 pages.

Fleiss, J.L., *Statistical Methods for Rates and Proportions*, 2nd ed. (New York: John Wiley & Sons, 1981).

Foster, T. (2004), *Managing Quality: An Integrative Approach*, 2nd Edition, Prentice Hall, Upper Saddle River, NJ, 544 pages; ISBN: 978-0138759643.

Fowlkes, W.Y. and Creveling, C.M., *Engineering Methods for Robust Product Design: Using Taguchi Methods in Technology and Product Development* (Reading, MA: Addison-Wesley Publishing Company, 1995), ISBN:978-0133007039.

Funkenbusch, P.D., *Practical Guide to Designed Experiments: A Unified Modular Approach* (CRC Press, 2004), ISBN:978-0824753887.

Good, P.I. and Hardin, J.W., *Common Errors in Statistics (And How to Avoid Them)*, 3rd ed. (Hoboken, NJ: John Wiley & Sons, 2009).

Goos, P. and Jones, B. (2011), *Optimal Design of Experiments: A Case-Study Approach*, Wiley, New York, 459 pages; ISBN: 978-0470744611.

Grant, E.L. and Leavenworth, R.S. (2004), *Statistical Quality Control*, Indian Edition, McGraw-Hill; ISBN 13: 9780070435551.

Grunditz, E.A. (2016), Design and Assessment of Battery Electric Vehicle Powertrain, with Respect to Performance, Energy Consumption and Electric Motor Thermal Capability, PhD thesis, Dept. of Energy and Environment, Chalmers University of Technology, Göteborg, Sweden.

Gunter, B. and Coleman, D. (2014), *A DOE Handbook: A Simple Approach to Basic Statistical Design of Experiments*, Createspace Independent Publishing Platform, 118 pages; ISBN: 978-1497511903.

Gwet, K.L., *Handbook of Inter-Rater Reliability*, 4th ed. (Gaithersburg, MD: Advanced Analytics, LLC, 2014), ISBN:978-0970806284.

Harrington, M. (2011), *The Design of Experiments in Neuroscience*, 3rd Edition, SAGE, Thousand Oaks, CA, 261 pages; ISBN: 978-1412974325.

Hicks, C.R. and Turner, K.V. (1999), *Fundamental Concepts in the Design of Experiments*, 5th Edition, Oxford University Press, New York, 565 pages; ISBN: 978-0195122732.

Hines, W.W. and Montgomery, D.G., *Probability and Statistics in Engineering and Management Science* (New York: John Wiley & Sons, 1990).

Hinkelmann, K. and Kempthorne, O. (2008), *Design and Analysis of Experiments Set*, Wiley, Hoboken, NJ, 1411 pages; ISBN: 978-0-470-38551-7.

Hite, J.A., Schmidt, J.W., and Bennett, G.H., *Analysis of Queuing Systems* (New York: Elsevier, 1975).

Huhn, S. and Drechsler, R., *Design for Testability, Debug and Reliability: Next Generation Measures Using Formal Techniques* (Cham: Springer, 2021), ISBN:978-3030692087.

Huitema, B.E. (2011), *The Analysis of Covariance and Alternatives: Statistical Methods for Experiments, Quasi-Experiments, and Single-Case Studies*, 2nd Edition, Wiley, Chicester, UK, 688 pages.

International Bureau of Weights and Measures, *The International System of Units*, 8th ed. (2006), ISBN:92-822-2213-6.

Joens, B. and Montgomery, D.C. (2019), *Design of Experiments: A Modern Approach*, 1st Edition, Wiley, 272 pages; ISBN: 978-1-119-61119-6.

Juran, J.M. and Gryna, F.M., *Quality Planning and Analysis from Product Development through Use*, 3rd ed. (New York: McGraw-Hill, 1993), 256.

Kalpakjian, S. (Eds), *Manufacturing, Engineering and Technology*, 2nd ed. (Boston, MA: Addison Wesley, 1992).

Kapur, K.C. and Lamberson, L.R. (1991), *Reliability in Engineering Design*, Wiley, 608 pages; ISBN: 978-0-471-51191-5.

Keyte, B. and Locher, D., *The Complete Lean Enterprise: Value Stream Mapping for Administrative and Office Processes* (New York: Productivity Press, 2004).

Khuri, A.I. and Cornell, J.A. (2019), *Response Surfaces Design and Analyses*, Taylor & Francis Group, 536 pages; ISBN: 978-0367401252.

King, J.P. and Jewett, W.S., *Robustness Development and Reliability Growth: Value-Adding Strategies for New Products and Processes* (Upper Saddle River, NJ: Prentice Hall, 2010), ISBN:978-0-13-222551-9.

Kuehl, R.O., *Design of Experiments: Statistical Principles of Research Design and Analysis*, 2nd ed. (Pacific Grove, CA: Duxbury Press, 2000), ISBN:978-0534368340.

Kverneland, K.O., *Metric Standards for Worldwide Manufacturing Engineering* (New York: ASME, 1996).

Lawless, J.F., *Statistical Models and Methods for Lifetime Data*, 3rd ed. (New York: John Wiley & Sons, 2003).

Lawson, J., *Design and Analysis of Experiments with R* (Boca Raton, FL: Chapman & Hall/CRC Press, 2014), ISBN:978-1439868133.

Lorenzen, T.J. and Anderson, V.L., *Design of Experiments: A No-Name Approach* (New York: Marcel Dekker, Inc., 1993).

Liker, J.Z. (2000), *The Toyota Way: 14 Management Principles from the World's Greatest Manufacturer*, 2nd Edition, McGraw-Hill, New York, 448 pages; ISBN: 0071392319.

Liston, C. and Sheth, N.J., *Statistical Design and Analysis of Engineering Experiments* (New York: McGraw-Hill, 1973).

Long, J.S., *Regression Models for Categorical and Limited Dependent Variables* (Thousand Oaks, CA: SAGE Publications, 1997).

Mabie, H.H. and Reinholtz, C.F., *Mechanisms and Dynamics of Machinery*, 4th ed. (New York: John Wiley & Sons, 1987).

Mascitelli, R., *Mastering Lean Product Development: A Practical, Event-Driven Process for Maximizing Speed, Profits and Quality* (Northridge, CA: Technology Perspectives, 2011), ISBN:978-0966269741.

Mantz, H.F. and Waller, R.A., *Bayesian Reliability Analysis* (New York: John Wiley & Sons, 1977).

Mason, R.L., Gunst, R.F., and Hess, J.L. (2003), *Statistical Design & Analysis of Experiments with Applications to Engineering and Science*, 2nd Edition, John Wiley & Sons, Inc., Hoboken, NJ, 760 pages; ISBN: 978-0-471-37216-5.

Mathews, P.G., *Design of Experiments with MINITAB: Homework Problems* (Milwaukee, WI: ASQ Quality Press, 2004).

Meeker, W.Q., Escobar, L.A., and Pascual, F.G. (1998), *Statistical Methods for Reliability Data*, 2nd Edition, John Wiley & Sons, New York, 704 pages; ISBN: 978-1-118-11545-9.

Milliken, G.A. and Johnson, D.E., *Analysis of Messy Data. Vol. 1: Designed Experiments* (New York: Van Nostrand Reinhold, 1984).

MIL-STD-785 (n.d.), Reliability Program for Systems and Equipment Development and Production, DOD, Washington, DC.

Modarres, M., Amiri, M., and Jackson, C. (2017), *Probabilistic Physics of Failure Approach to Reliability: Modeling, Accelerated Testing, Prognosis and Reliability Assessment*, Wiley-Scrivener, Hoboken, NJ, 288 pages; ISBN: 978-1119388630.

Moen, R. et al., *Quality Improvement through Planned Experimentation* (New York: McGraw-Hill, 2012), ISBN:978-0071759663.

Montgomery, D., Rigdon, S., Pan, R., and Freeman, L., *Design of Experiments for Reliability Achievement* (Hoboken, NJ: Wiley, 2022), ISBN:978-11-19237693.

Montgomery, D.C. (2019), *Design and Analysis of Experiments*, 10th Edition, John Wiley & Sons, Inc., Hoboken, NJ, 688 pages; ISBN: 978-1-119-49244-3.

Morris, M.D., *Design of Experiments: An Introduction Based on Linear Models* (Boca Raton, FL: CRC Press, 2017), ISBN:978-1138628021.

Murthy, D.P., Xie, M., and Jiang, R., *Weibull Models* (Hoboken, NJ: John Wiley & Sons, Inc., 2004).

NASA (2010), NASA System Safety Handbook, Volume 1, System Safety Framework and Concepts for Implementation, Version 1, SP-2010-580, Washington, DC.

Nelson, B.L. (2013), *Foundations and Methods of Stochastic Simulation*, Springer, Boston, MA; On-line ISBN: 978-1-4614-6160-9.

Nelson, W.B., *Accelerated Testing: Statistical Models, Test Plans, and Data Analysis* (New York: John Wiley & Sons, 2004), ISBN:978-0471697367.

Nevins, J.L. and Whitney, D.E., *Concurrent Design of Products and Processes* (New York: McGraw-Hill, 1990).

NIST (n.d.), Engineering Statistics Handbook, National Institute of Standards and Technology, Washington, DC.

NIST/SEMATECH (2012), e-Handbook of Statistical Methods, http://www.itl.nist.gov/div898/handbook.

O'Connor, P. and Kleyner, A., *Practical Reliability Engineering* (New York: John Wiley & Sons, 2011).

Oehlert, G.W. (2010), *A First Course in Design and Analysis of Experiments*, 683 pages; ISBN: 9780716735106; Retrieved from the University of Minnesota Digital Conservancy, https://hdl.handle.net/11299/168002.

Ohring, M. and Kasprzak, L. (2014), *Reliability and Failure of Electronic Materials and Devices*, 2nd Edition, Elsevier, London, UK, 734 pages; ISBN: 978-0120885749.

Onyiah, L.C. (2008), *Design and Analysis of Experiments: Classical and Regression Approaches with SAS*, Chapman and Hall/CRC, Boca Raton, FL; ISBN (eBook): 9780429140273.

Peace, G.S., *Taguchi Methods: A Hands-On Approach* (Reading, MA: Addison-Wesley Publishing Company, 1993).

Phadke, M.S., *Quality Engineering Using Robust Design* (Englewood Cliffs, NJ: Prentice Hall, 1989).

Petrov, V. (2019), *Laws of System Evolution: TRIZ*, 57 pages; ISBN: 978-1696068833.

Petroski, H., *Invention by Design How Engineers Get from Thought to Thing* (Cambridge, MA: Harvard University Press, 1996).

Proust, M., *Design of Experiments Guide* (Cary, NC: JMP, A Business Unit of SAS, 2010).

Pyzdek, T. and Keller, P. (2019), *The Six Sigma Handbook*, 5th Edition, McGraw-Hill, New York, 720 pages; ISBN: 9781260121827.

Rhinehart, R.R. (2016), *Nonlinear Regression Modeling for Engineering Applications: Modeling, Model Validation, and Enabling Design of Experiments*, Wiley, Chichester, UK, 400 pages; ISBN: 978-1-118-59796-5.

Rigdon, S.E., Pan, R., Montgomery, D.C., and Borror, C.M. (2022), *Experiments for Reliability Achievement*, Wiley, Hoboken, NJ, 416 pages; ISBN: 978-1119237693.

Rodrigues, M. and Iemma, A. (2014), *Experimental Design and Process Optimization*, CRC Press, Boca Raton, FL, 336 pages; ISBN: 978-0429161865.

Ross, P.J., *Taguchi's Techniques for Quality Engineering* (New York: McGraw-Hill, 1988).

Ross, S.M., *Introduction to Probability and Statistics for Engineers and Scientists* (New York: John Wiley & Sons, 1987).

Rössler, A., *Design of Experiment for Coatings* (Hanover, Germany: Vincentz Network, 2014), ISBN:978-3-86630-885-5.

Rother, M., *Toyota Kata* (New York: McGraw-Hill, 2009).

Rother, M. and Shook, J., *Learning to See* (Lean Enterprise Institute, 2009).

Roy, R.K. (2001), *Design of Experiments Using the Taguchi Approach: 16 Steps to Product and Process Improvement*, Wiley, New York, 560 pages; ISBN: 978-0-471-36101-5.

Russell, K.G. (2021), *Design of Experiments for Generalized Linear Models*, Chapman and Hall/CRC, Boca Raton, FL, 240 pages; ISBN: 978-1032094052.

Sahai, H. and Ageel, M.I., *The Analysis of Variance* (Boston, MA: Birkhauser, 2000).

Samuels, M.L., Witmer, J.A., and Schaffner, A.A., *Statistics for the Life Sciences*, 5th ed. (Boston, MA: Pearson Education, 2016), ISBN:978-1-292-10181-1.

Santner, T.J., Williamns, B.J., and Notz, W.I., *The Design and Analysis of Computer Experiments*, 2nd ed. (New York: Springer, 2018), ISBN:978-1493988457.

Schleich, B., *Skin Model Shapes: A New Paradigm for the Tolerance Analysis and the Geometrical Variations Modelling in Mechanical Engineering* (Düsseldorf, Germany: VDI Verlag, 2017).

Schmidt, S.R. and Launsby, R.G. (1994), *Understanding Industrial Designed Experiments*, 4th Edition, Air Academy Press, Colorado Springs, CO, 768 pages.

Shigley, J.E. and Mischke, C.R., *Fundamentals of Machine Component Design*, 5th ed. (New York: McGraw-Hill, 1989), ISBN:978-1118012895.

Shingo, S., *A Revolution in Manufacturing: The SMED System* (Stamford, CT: Productivity Press, 1985), ISBN:978-0915299034.

Silva, V. (2018), *Statistical Approaches with Emphasis on Design of Experiments Applied to Chemical Processes*, 180 pages; ISBN: 978-953-51-3878-5.

Taguchi, G., Chowdhur, S., and Wu, Y., *Taguchi's Quality Engineering Handbook* (Hoboken, NJ: John Wiley & Sons, Inc., 2005).

Taguchi, G., Chowdhury, S., and Taguchi, S., *Robust Engineering* (New York: McGraw Hill, 2000).

Taguchi, G., Elsayed, E., and Hsiang, T., *Quality Engineering in Production Systems* (New York: McGraw-Hill Book Company, 1989).

Taguchi, G. and Konishi, S., *Taguchi Methods, Orthogonal Arrays and Linear Graphs: Tools for Quality Engineering* (Dearborn, MI: American Supplier Institute, 1987).

Taguchi, G., *System of Experimental Design* (New York: UNIPUB, Kraus International Publications, 1987).

Tamhane, A.C. (2009), *Statistical Analysis of Designed Experiments Theory and Application*, Wiley, Hoboken, NJ, 720 pages; ISBN: 978-0-471-75043-7.

Taylor, W., *Optimization and Variation Reduction in Quality* (New York: McGraw Hill, 1991), ISBN:0-07-063255-3.

Thomke, S.H., *Experimentation Matters: Unlocking the Potential of New Technologies for Innovation* (Boston, MA: Harvard Business School Press, 2003).

Toutenburg, H. and Shalabh (2009), *Statistical Analysis of Designed Experiments*, 3rd Edition, Springer, New York, 633 pages; ISBN: 978-1441911476.

Vuchkov, I.N. and Boyadjieva, N.L. (2001), *Quality Improvement with Design of Experiments: A Response Surface Approach*, Springer, Dordrecht, the Netherlands, 508 pages; ISBN: 978-1402003929.

Watson, G.H., *Strategic Benchmarking: How to Rate Your Company's Performance against the World's Best* (Milwaukee, WI: ASQC Quality Press, 1993).

Weber, D.C. and Skillings, J.H., *A First Course in the Design of Experiments: A Linear Models Approach* (New York: CRC Press, 1999).

Winer, B.J., *Statistical Principles in Experimental Design*, 2nd ed. (New York: McGraw-Hill, 1971).

Wu, C.F.J. and Hamada, M.S., *Experiments: Planning, Analysis and Parameter Design Optimization*, 2nd ed. (New York: Wiley, 2009), ISBN:978-0471699460.

Yates, F., *The Design and Analysis of Factorial Experiments* (Harpenden, England: Imperial Bureau of Social Science, 1937).

Index

About the Authors

Young J. Chiang, PhD The author earned his Bachelor of Science (BS) degree from the Department of Power Mechanical Engineering, Tsing Hua University, Taiwan, in 1976. When studying for his PhD degree (Dept. of Engineering Mechanics, University of Wisconsin, Madison), the author was honored to take two graduate courses from Professor George Box in the early 1980s as part of his minor program (Dept. of Statistics, University of Wisconsin, Madison). He obtained his PhD degree in 1983. Ever since, Dr. Chiang has had opportunities to apply the design of experiments in combination with finite element methods (computer simulation) and experimental mechanics (experimental life tests) to improve product performance and solve complex reliability problems in the automotive industry. He had 30 years of engineering and management experience with product Research and Development (R&D), computer-aided engineering (CAE), design, reliability, quality, manufacturing, and program management in the US automotive industry, including Uniroyal (Michelin), Cummins, Magna, Edscha, Danaher, CTS, Coda, and Chrysler. Dr. Chiang was a member of American Society of Mechanical Engineers (ASME) and SAE International. After retiring from the US automotive industry, he worked as the vice director of the Weichai Power Research Center (Shandong, China) and founded the Weichai Institute of Reliability Research (>100 research engineers). He was also named the vice director of the China Key State Laboratory of Engine Reliability and the executive commissioner of the Reliability Committee, Chinese Mechanical Engineering Society.

Besides technical papers and reports, Dr. Chiang has put forth a great effort into a nonprofit business engaged in the development of potential automotive engineering textbooks for graduate students and practitioners in the automotive industry, ever since the early 1990s. He has published two books on how to improve product performance and grow product reliability based on the physics of failure and material constituents: *Mechanics and Design for Product Life Prediction* (ISBN 978-7-5689-1917-6) and *Automotive Engineering Materials—Thermomechanical Properties* (ISBN 978-7-5689-3293-6) with Chongqing University Press (Chongqing, China). This publication (*Manufacturing Reliability Growth for Automotive Engineering*) is the third in a book series on *Product Design and Reliability Growth* brought forth by the author. The corresponding manuscript was ever used as a textbook-to-be by the author, in the capacity of a full professor, for teaching graduate students in the Department of Automotive Engineering, Chongqing University.

Amy L. Chiang, MS After the author obtained her BS degree in electrical engineering from the University of Michigan, Ann Arbor, in 2012, she worked for Viasat, Inc. (Carlsbad, CA), with much of her focus on product development and testing. While working for the company, she completed the Master of Science (MS) degree of "Master of Advanced Study (MAS) Degree in Wireless Embedded Systems," offered by the Departments of *Electrical and Computer Engineering* and *Computer Science and Engineering* at the University of California, San Diego.

The author is an active participant in businesses engaged in the development of new technologies and multidisciplinary products. With insights into viable paradigms and advancements in practice, she assists product teams in developing wireless embedded software, program strategy, and methodology of statistical thinking.